Pollination Biology of Cultivated Oilseeds and Pulse Crops

The book provides in detail information on pollination biology of oilseed and pulse crops. The book presents information on improving productivity of oilseed crops and pulses through planned pollination and safety of pollinators from pesticides. Covering the latest information on various major world oil crops and pulses, this book brings the latest advances together in a single volume for researchers and advanced-level students. The book will enlighten the readers with the latest technological developments in pollination of oilseed crops and pulses and shall be useful to agricultural and applied scientists, extension workers, policy planners, and policymakers to improve rural economy and conservation of global biodiversity.

SALIENT FEATURES

1. Covers the latest information on various aspects of pollination biology of oilseed and pulses crops and brings the latest advances together in a single volume for researchers and advanced-level students.
2. An excellent source of advanced study material for academics, researchers, and students and program planners.
3. Provides an excellent source of livelihood through enhanced productivity of oilseed and pulse crops.
4. Aims to promote a large, diverse, sustainable, and dependable bee pollinator workforce that can meet the challenge for optimizing production of oil and pulse crops well into the twenty-first century.
5. The pollination requirements of most of the pulse crops have been reported to benefit production of pulse crops both qualitatively and quantitatively.

This book will be useful for pollination biologists; honeybee biologists; scientists working in agriculture, animal behavior, conservation, biology, and ecology; entomologists; environmental biologists; etc.

Pollination Biology of Cultivated Oilseeds and Pulse Crops

Dharam P. Abrol

CRC Press
Taylor & Francis Group
Boca Raton London New York

CRC Press is an imprint of the
Taylor & Francis Group, an **informa** business

First edition published 2024
by CRC Press
2385 NW Executive Center Drive, Suite 320, Boca Raton FL 33431

and by CRC Press
4 Park Square, Milton Park, Abingdon, Oxon, OX14 4RN

CRC Press is an imprint of Taylor & Francis Group, LLC

ISBN: 978-1-032-65670-0 (hbk)
ISBN: 978-1-032-65671-7 (pbk)
ISBN: 978-1-032-65672-4 (ebk)

DOI: 10.1201/9781032656724

Typeset in Times
by Apex CoVantage, LLC

Dedicated to the everlasting memory of my parents late Shri Bishan Dass Abrol and Shrimati Ram Pyari Abrol. Your unwavering love, endless encouragement, and unspoken sacrifices have been the foundation upon which my dreams have flourished. Your guidance and support have shaped me into the person I am today. This book stands as a tribute to your enduring spirit and the profound impact you've had on my life.

Contents

SECTION 2 Pulses

Preface

Pollination Biology of Cultivated Oilseeds and Pulse Crops presents key insights into pollinator–plant interaction that plays an important role in enhancing the quality and quantity of oil and pulse seed production. Oilseeds are the major source of edible oils. From a nutritional perspective, oils offer a concentrated source of calories, vitamins, and essential fatty acids that are easily absorbed by the body. They are employed in culinary practices to improve the taste and consistency of dishes. Additionally, the leftover solid residue, known as oil cake, contains significant amounts of protein and serves as a valuable ingredient in animal feed production.

A growing and increasingly wealthy human population is expected to double the demands for food between 2005 and 2050. The production of food is a significant contributor to global environmental change, resulting in adverse impacts on climate, water resources, soils, and the diverse biodiversity found worldwide.

Humanity stands in front of a great challenge to tackle these issues and needs to develop the food production system as part of the solution for a sustainable world.

It has been estimated that while 30% of malnourished children in developing countries suffer from a lack of both energy-producing and protein foods, the other 70% suffer from a lack of calories, which could be obtained largely from oils and fats. In tropical Africa, the average individual's fat consumption is approximately one-fourth of that in North America. Few developing nations possess excess vegetable oil quantities available for export.

Malaysia and Indonesia export palm oil, the Philippines export coconut oil, and Brazil and Argentina export soybean oil. In many developing nations, the scarcity of vegetable oils has become a pressing issue due to increasing demand primarily driven by population growth. This situation results in the depletion of limited foreign exchange reserves through the necessity of imports. India holds a significant position in the global oilseeds landscape, accounting for approximately 12–15% of the total land area dedicated to oilseed cultivation, producing around 6–7% of the world's vegetable oil, and consuming approximately 9–10% of the global edible oil supply. Furthermore, India's vegetable oil imports contribute to approximately 13.6% of the worldwide total.

The concept of ecological intensification in agriculture proposes a method to achieve higher crop yields without exacerbating environmental degradation by increasing inputs. While there is a suggestion to enhance yields through increased insect pollination in oilseed crops, the integration of this approach into crop management practices remains uncommon. To accurately assess the value of augmented crop pollination for ecological intensification, it is crucial to obtain reliable estimates on how insect pollination affects yield. Furthermore, it is necessary to evaluate the impact on native flower-visiting insect populations when introducing honeybees to enhance crop pollination. Approximately one-third of our overall diet relies on insect-pollinated plants, directly or indirectly. Oilseeds, such as coconuts, cotton, oil palm, olives, peanuts, rapeseed, soybeans, and sunflower, contribute to half of our fats and oils requirements, many of which greatly benefit from or depend on insect pollination. Insufficient pollination can lead not only to reduced yields but also to delayed yield and a significant proportion of substandard or defective fruits.

Despite good agronomic practices, the level of productivity of oilseeds is far below than the expected. The low productivity of oilseeds/oil crops can be attributed to various factors, such as heavy infestation of pests and diseases. Out of the various factors, one of the most important factors for low production of oilseed crop is due to the failure of seed setting attributable to the lack of proper pollination.

Insect pollination is important for global food nutrition. One often-highlighted ecosystem service is insect pollination. The loss of biodiversity among pollinators has raised questions about whether the pollination services they provide are at risk.

The services provided by pollinators, such as pollination and its crucial role in food production, face various threats, including changes in land use, intensified agricultural practices, climate change, pesticide usage, the presence of pathogens, genetically modified organisms (GMOs), and invasive species.

Insect pollination improves seed germination, oil content, and chlorophyll content of oilseed rape. The economic market value of pollination in crop production globally is estimated to be $235 billion to $577 billion annually. Honeybees have a significant role in pollinating oilseed crops among the various pollinating agents. The honeybee pollination not only results in higher yields but also gives a better quality of produce, and the efficient pollination of flowers also serves to protect the crops against pests.

Though a number of books on oilseed crops are available, they have a limited approach covering one aspect or another. The role of pollinators in oilseed production has not been documented in a comprehensive manner in a single place. The proposed book provides complete information and fulfils the gap in literature, putting all the information about oilseed crops in one book.

Besides oilseeds, pulses occupy a unique place in India. Due to the presence of millions of impoverished and malnourished individuals in India, there is an urgent need for the country's pulses sector to make a significant recovery.

Pulses are an important and relatively inexpensive source of protein for human and animal nutrition in India. Their importance as a builder and restorer of soil fertility in arid areas is well recognized. Pulses occupy an important position in food and nutritional security in India. Besides other factors, the pollination requirements of most of the pulse crops are unknown, whether they are self-fertile or self-sterile. Pollination has been reported to benefit production of pulse crops both qualitatively and quantitatively. Pulse crops rely on varying degrees of open pollination, which necessitates the transfer of pollen between flowers for successful seed production. While mechanisms like wind can facilitate pollen transfer and cross-pollination, the most efficient method involves utilizing insects to carry pollen from one flower to another.

The objective of this book is to consolidate the most recent scientific literature into principles and practices applicable to individuals involved in crop pollination in tropical and temperate regions. The book strives to foster a robust, diverse, sustainable, and reliable workforce of bee pollinators that can effectively contribute to optimizing food production well into the twenty-first century.

The book presents in detail the importance of oilseed crops in Indian vis-à-vis world scenario. The book explores in detail the constraints in production of oilseed crops, such as brassicas, sunflower, sesamum, safflower, groundnut, soybean, cotton, castor, oil palm, coconut, and olive, and pulses, such as pigeon pea, cowpea, chickpea, and various other legumes. The book presents information on improving productivity of oilseed crops through planned pollination. In response to the increasing demand, the scientific community has embraced the challenge by focusing a greater amount of research on achieving sustainable production for both edible and industrial purposes. Covering the latest information on various major world oil crops, this book brings the latest advances together in a single volume for researchers and advanced-level students.

I am extremely thankful to Renu Upadhyay, Commissioning Editor and Jyotsna Jangra Editorial assistant, CRC Press, who took great pains and keen interest in publication of this book in a very impressive way. Last but not the least, my sincere thanks are due to my family members for their endurance and help while writing this book.

I am sure that the book will enlighten the readers with the latest technological developments in pollination of oilseed crops and shall be useful to agricultural and applied scientists, extension workers, policy planners, and policymakers to improve rural economy and also in the conservation of global biodiversity.

Jammu **Dharam P. Abrol**

Foreword

We dwell in the midst of yet another world food shortage that exacerbated by escalating prices the world over. With the observed volatility of climate cycles in recent times, a significant portion of our expanding human population is rapidly nearing the brink of famine. Concomitantly, this situation includes rampant declines in honeybee populations across three continents, for as yet incompletely resolved reasons, and with no remedy or end in sight. These problems have been further aided and abetted by a lack of prophylactic progress in the conservation of biodiversity and increased agricultural production. Pollinators and pollination play a vital role in the operation of nearly all land-based ecosystems, including agricultural systems, as they are essential for sustainable plant reproduction and productivity. In today's global trade, the products derived from major oil crops hold significant value as valuable commodities. Over the past couple of decades, oilseed production has increased to become the most important world source of vegetable oils, in response to the rising world population and living standard.

Ecological intensification of agriculture is proposed as a method to achieve higher crop yields without increasing inputs that may harm the environment. While increased insect pollination in oil crops has been suggested to boost yields, it is rarely incorporated into crop management practices. To evaluate the significance of enhanced crop pollination as a means of ecological intensification, it is essential to obtain reliable estimates of how insect pollination affects crop yield.

Oilseeds such as coconuts, cotton, oil palm, olives, peanuts, rape, soybeans, and sunflower contribute to over half of the world's fats and oils in our diet. Many of these plants rely on or benefit from insect pollination. Considering both animal and plant products, it appears that roughly one-third of our overall diet depends, either directly or indirectly, on insect-pollinated plants. Another advantage of pollination lies in its impact on crop quality and efficiency of production. Insufficient pollination can lead not only to reduced yields but also to delayed yield and a high percentage of subpar or defective fruits. Even self-fertile flowers can produce some seeds and fruit in the absence of animal visitation.

However, animal visitation increases either (a) the proportion of fruits set or (b) the quality of fruits set, because fruit size depends on the number of seeds set or the size of seeds. Seed size is often greater after cross-pollination, which can only be achieved by insect visits. This group contains a number of very important crop plants, including oilseed rape, flax- and linseed, sunflowers, cotton, soybean, and olives.

Despite good agronomic practices, the level of productivity of oilseeds is far below than the expected. The low productivity of oilseeds/oil crops can be attributed to various factors, such as heavy infestation of pests and diseases. Out of the various factors, one of the most important factors for low production of oilseed crop is the failure of seed setting, attributable to the lack of proper pollination. Among the various pollinating agents, insect pollinators play a predominant role in increasing the yield of oilseed crops. Besides oil crops, the pollination requirements of most of the pulse crops are unknown, whether they are self-fertile or self-sterile. Pollination has been reported to benefit production of pulse crops both qualitatively and quantitatively. The book *Pollination Biology of Cultivated Oilseeds and Pulse Crops* by Dharam P. Abrol is first of its kind, dealing exclusively with pollination of oil crops. The proposed book intends to provide a description of basic and applied aspects of the major oil and pulse crops and their pollination requirements, pollinators, and improvement in quality and quantity through pollination. For each crop, information is provided on importance of pollination, important pollinators, and their role in crop production. The objective of the book is to encourage the development of a robust and reliable workforce of bee pollinators that is both diverse and sustainable. This workforce will play a crucial role in optimizing the production of oil crops throughout the twenty-first century.

I congratulate the author for his efforts to provide the timely contribution.

Michal Woyciechowski

Author Biography

Dr. Dharam P. Abrol, Ex-Dean, Faculty of Agriculture, Sher-e-Kashmir University of Agricultural Sciences, has served university in various capacities. He has served as Head of the Division of Entomology from August 2008 to February 2015 and Controller of Examinations and Director of Education during 2012. Prof. Abrol has served university for 35 years in various capacities. He has specialized in pest management, honeybee management, and pollination biology. Prof. Abrol has an impressive publication record, having authored 22 books, 12 manuals, and contributed to 60 book chapters. Furthermore, he has published over 250 original research papers in national and international journals. He has successfully completed over nine externally funded projects and engaged in two collaborative research projects with international organizations based in Poland and Switzerland. Prof. Abrol's expertise has led him to visit countries such as Switzerland, Poland, South Korea, Malaysia, Saudi Arabia, and others as a special invitee. His contributions have been acknowledged with letters of appreciation from various organizations. Prof. Abrol is first from Jammu and Kashmir elected to be Fellow of the National Academy of Agricultural Sciences, India, and Fellow of the Royal Entomological Society London, UK.

He is a recipient of the Young Scientist Award (1992), conferred by the Jammu and Kashmir State Council for Science and Technology, a prestigious state award, for his outstanding contributions in the field of agricultural sciences.

He is also the recipient of Pran Vohra Award 1993, a prestigious Young Scientist Award conferred by the Indian Science Congress Association, Calcutta, for his outstanding and innovative research in the field of agricultural sciences.

He was also conferred the Prof. T. N. Ananthakrishnan Award 1997–1998, a prestigious national award, for his outstanding contributions in the field of entomology, by the T. N. Ananthakrishnan Foundation, G. S. Gill Research Institute, Chennai. Prof. Abrol won the Dr. Rajendra Prasad Puruskar 1999–2000 award, a prestigious national award from the Indian Council of Agricultural Research, New Delhi, for his Hindi book on beekeeping titled *Madhmakhi Palan-Sidhant Evam Vidhian*. He received the 11th Apicultural Association Award (2010) for outstanding contributions in apiculture. King Saud University conferred on him a gold medal for development of apiculture in Asia. At present, Prof. Abrol is Member, PRT team for accreditation, of several universities; Member, QRT team, for evaluation of AICRP projects; Expert Member, Neonicotinoids Group Asian Countries. He serves as an expert for various selection committees

Section 1

Oilseeds

1 Oil Crops: An Introduction

1.1 INTRODUCTION

Since green revolution, the production of oilseeds has witnessed very little growth as compared to the exponential growth of cereal crops like paddy, wheat, maize, etc. Continuous increase in human population coupled with rise in income has increased the consumption of oilseeds. Due to this, India has become the largest importer of oilseeds, a crop which India produced in surplus during the 1960s. One of the major causes for low production of oilseeds in India is low yield, which exists due to multiple factors, such as non-availability of varieties having climate resilience, low oil content, lack of biotic resistance, unsuitability for mechanical harvesting, and lack of post-harvest process development. The existing types of oilseeds have specific drawbacks, such as low oil yield, poor quality of the extracted oil, shattering, synchronized maturing, and similar issues. Evidently, there is a need for genetic improvement, which involves evaluation of germ plasm, selection of lines resistant to dieaseses and pest, early maturing and resistance to shattering, quality of oil, suitability for mechanization, etc.

Oilseeds have become an essential part of our diet in today's world. They consist of seven types of edible oil, namely, rapeseed-mustard (*Brassica* spp.), soybean (*Glycine max*), groundnut (*Arachis hypogaea*), sunflower (*Helianthus annuus*), sesame (*Sesamum indicum*), safflower (*Carthamus tinctoris*), and niger (*Guizotia abyssinica*). Additionally, there are two non-edible crops, castor (*Ricinus communis*) and linseed (*Linum usitatissimum*), which are cultivated in the country. Furthermore, oil palm, coconut, cottonseed, rice bran, tree-borne oilseeds, and solvent-extracted oil serve as secondary sources. Due to the onset of green revolution, the oilseeds have been largely neglected and thus are grown in rainfed conditions and unproductive land. This has resulted in the stagnation of total production. In the last 50 years, the production of oilseeds has shown very little growth, whereas the production of cereal crops such as paddy, wheat, maize, etc. has risen exponentially in the same span of time.

India has made much progress in the production of oilseeds. India's total oilseed production is expected to rise 4% to 39.9 million metric tons (MMT), based on expectations of near normal oilseed crop yields. Total oilseeds production in the country during 2019–2020 was at 33.42 million tons, which is higher by 1.90 million tons than the production of 31.52 million tons during 2018–2019. Although domestic oilseeds performed well, they were unable to meet the growing per capita demand rate of approximately 6% because of increased per capita consumption. This heightened consumption, driven by population growth and higher per capita income, amounts to 18 kg of oil per year.

Considering the growing domestic demand for edible oils, it has now been planned to achieve a production of 45.64 Mt (million tons) from nine (9) annual oilseed crops by 2022–2023, expecting an additional production of about 15.58 Mt over and above the 30.06 Mt production.

Rapeseed-mustard group of oilseed crops is second to soybean in area and production. Due to high oil recovery and wider adaptability, it is considered as the most important commodity of Indian oilseed sector. Indian mustard (*Brassica juncea;* AABB; 2n = 36), commonly known as mustard, is a predominant species among rapeseed-mustard group of crops in India and accounts to more than 90% in its acreage.

Oilseed crops are essential component of human life processes and play a prominent role in agricultural economies. Oil crops are high-value agricultural commodities which occupy a prominent place in agriculture-dominated economics wherein the demand for high-quality seed oils continues to grow as the global population increases (Wittkop et al. 2009). They belong to numerous plant

DOI: 10.1201/9781032656724-2

families, and their seeds are used not only as a source of oil but also as raw materials, which act as a renewable source of energy in power generation (Jankowski and Budzynski 2003) for various oil-based chemical industries. Oilseed crop production has influenced the economic development of many nations throughout the world and plays a great role in the diversification of major cropping systems especially in developing countries. The crop is grown primarily for oil, but the value addition in oilseed crops provides an important source of food for livestock, employment generation, and livelihood security by fetching more remuneration for the human being. Apart from this, the remaining after oil extraction can also be utilized for the preparation of different types of cakes and soil amendments for organic cultivation for sustainable agriculture. India has the largest oil economies in the Southeast Asian region, wherein diverse types of annual oilseed crops are grown.

1.2 OILSEED CROPS

Oilseeds crop plants include more than 40 different oilseeds whose oil can be consumed, but only a few are significant in the total world trade (Lennerts 1983). The increase in production has occurred mainly due to rising demand for oilseed products, increase in area under the crop, supplemented with advanced scientific production technologies, which have resulted in high levels of per unit productivity. Soyabean is a major contributor of oilseed, followed by rapeseed-mustard, cotton, peanut, and sunflower. The most important tropical oil is the coconut, palm, and groundnut. America and Europe together account for more than 60% of the world production of oilseeds, whereas substantially small production (<5%) comes from tropical areas, such as Africa, Malaysia, and Indonesia. Oilseed crop contains a number of very important crop plants, including rapeseed and mustard, flax- and linseed, sunflowers, cottonseed, castor seed, soybean, sesamum, safflower, groundnut, olives, oil palm, and coconut.

In spite of good agronomic practices and other technological interventions, the level of productivity of oilseeds is far below than the expected. Crop production is simultaneously confronted by two problems which are antagonistic to each other. One is protection of crop against pests, and another safety of pollinators for pollination of crops. Evidently, there is a need to explore the integrated approach protecting crop on the one hand and safety of pollinators on the other. Apart from these, the low productivity of oilseed crops can be attributed to various biotic (such as heavy infestation of pests and diseases) and abiotic factors (such as seasonal temperature, relative humidity, rainfall, sunshine hours, etc.).

India is a vegetarian country where more than 80% of the people depend upon vegetarian food exclusively, wherein oilseed crops constitute a major component of their diet. However, the crop is attacked by an array of insects/pests at the time when a large number of pollinating insects responsible to enhance their productivity are also active. Out of the various factors, one of the most important factors for low production of oilseed crop is the failure of seed setting, attributable to the lack of proper pollination. Our country is endowed with the varied habitats and climatic conditions that have encouraged the cultivation of a number of oilseed crops on commercial scales. The cultivation of oilseed crop production stretches from subtropics to intermediate and up to temperate areas. They are the prime source of income generation among the farming communities. The commercial cultivation of these oilseed crops is more remunerative than the cereals and fetches higher income, employment, and livelihood. However, the production of these oilseed crops in India compared to the other countries is meager.

India is heavily dependent on imports to meet its edible oil requirements and is the largest importer of vegetable oils in the world, followed by China and USA. Of all the imported edible oils, the share of palm oil is about 60%, followed by soybean oil with a share of 25%, and sunflower at 12%. Import growth in respect of edible oils during the last decades is about 174%. The import figure of edible oils during 2019–2020 reveals that India imported a total of 13.35 million tons of vegetable oils, costing Rs. 61,559 crore. The per capita consumption is around 19 to 19.80 kg per person per annum over the last five years. Domestic edible oil production has not been able to keep pace with the growth in consumption. During 2019–2020, domestic production of edible oils was 10.65 million tons from both primary (oilseeds) and secondary sources (coconut, oil palm, rice bran oil, cottonseed oil, and TBOs).

1.3 GLOBAL SCENARIO OF OIL CROPS

According to the OECD/FAO 2018 report, the global production of oilseeds is projected to expand at approximately 1.5% per annum, which is lower than the growth rates observed in the previous decade. In 2017, the production of vegetable oil showed an increase compared to 2016, although the growth rate was smaller than in previous years. This was primarily due to the slow recovery in palm oil production following the El Niño event in 2015. The consumption of vegetable oils per capita continued to rise in both developed and developing countries, with developing countries experiencing a much faster growth rate. Global vegetable oil production relies on both the crush of oilseeds and on the production of perennial tropical oil palm. Production of palm oil is concentrated in Indonesia and Malaysia, which together account for more than a third of world vegetable oil production.

By 2027, Brazil and the United States are anticipated to emerge as the top producers, with their production of canola (*Brassica napus*) reaching 129 Mt and 131 Mt, respectively. Canola ranks as the second leading oilseed crop worldwide (Maheshwari et al. 2011). However, the extensive cultivation of *Brassica napus* on large land areas poses challenges for the ecosystem as it disrupts the balance between pollinators and the crop. This imbalance arises from a decline in bumblebee population and an increase in nectar robbers (Diekötter et al. 2010).

China is a major producer of rapeseed and groundnuts, while the European Union produces rapeseed and sunflower seed. These regions are projected to yield approximately 32 Mt and 30 Mt of other oilseed crops, respectively, by 2027. Additionally, Brazil and the United States will be the largest producers of soybeans. The demand for vegetable oil is expected to grow at a slower pace due to the modest increase in per capita food consumption in developing nations and the anticipated stagnation in demand for biodiesel feedstock.

Indonesia and Malaysia are major participants in the exportation of vegetable oil, while the Americas dominate the exports of soybean, protein meal, and other oilseeds. Vegetable oil encompasses various types, such as soybean and other oilseeds (which account for approximately 55% of global vegetable oil production), palm oil (35%), as well as palm kernel, coconut, and cottonseed oils. Despite a slowdown in the expansion of established oil palm plantations, Indonesia (with a projected annual growth rate of 1.8% compared to 6.9% in the previous decade) and Malaysia (with a projected annual growth rate of 1.4% compared to 1.3% in the previous decade) are expected to experience significant production growth. However, the demand for vegetable oil is anticipated to grow at a slower pace in the next decade due to two factors: (1) reduced growth in per capita food consumption in developing countries (1.2% annually compared to 3.2% annually in the previous decade) as consumption levels approach saturation, and (2) the expected stagnation in demand for vegetable oils utilized in biodiesel production.

Among all agricultural commodities, vegetable oil holds a significant trade share, with a production share of 41%. This proportion is projected to remain constant during the outlook period (2017–2026), and it is anticipated that global exports of vegetable oil will reach 96 million metric tons by 2027. The dominant players in vegetable oil exports will continue to be Indonesia and Malaysia, accounting for approximately 70% and over 80%, respectively.

1.4 GAPS IN PRODUCTIVITY

Technology Mission on Oilseeds and Palm (TMOP), started in 1985–86, has played a vital role in increasing area and production of rapeseed-mustard. Afterward, productivity has increased significantly by continuous efforts made by different research institutions and state agricultural universities. There is a huge diversity in soil types and climatic conditions in different states of India. Keeping this fact in view, mustard-growing regions of the country have been divided into five zones. It is concluded from research experiments that average yield of rapeseed-mustard can be increased significantly by adopting suitable varieties and their matching production and protection technologies. Climatic conditions and level of adoption of package and practices determine the productivity levels in different agro-climatic zones. Improved varieties of mustard have been developed specific to soil types and climatic conditions of different zones, but optimum yield per unit area has not been achieved due to the following reasons: cultivation of old varieties despite availability of new improved varieties, cultivation of varieties not specified for a particular zone, sowing of improved varieties either before or after the recommended time, susceptibility of varieties to pests and diseases, using broadcasting method, non-availability of improved seeds, and good quantity of irrigation water.

1.5 FACTORS AFFECTING PRODUCTIVITY

Despite availability of high-yielding modern-day varieties with a potential yield of about 4t/ha, the average yield has been hovering around 1–1.5t/ha since more than a decade. Production and productivity of mustard is highly variable across crop-growing states of India due to variable agro-climatic conditions, cropping systems, production technologies, and biotic and abiotic stresses. More than 24% of the area under this crop does not have irrigation facilities. Abiotic stresses alone cause significant reduction in crop productivity (due to drought [17%], high temperature [40%], and salinity [50–90%]). In the regions of Bharatpur and Jobner, Chauhan et al. (2007) documented a range of 4.3% to 60% decrease in seed yield across various genotypes when grown under rainfed conditions.

Prevalence of high temperature at the seedling stage also prevents the early sowing of mustard, which is a recommended practice because of many advantages it offers. Higher temperatures coupled with limited water are expected to reduce crop yields, allow damaging weeds and insects to spread, and shift precipitation patterns worldwide.

Salinity is also an important abiotic stress. Arid and semi-arid areas in different states also have saline underground water, which is being used for irrigation purposes. Average yield in salt-affected soils gets reduced to 460 kg/ha, thus resulting in huge economic loss. Out of total mustard-growing area in the country, nearly 2 mha areas is salt-affected. Frost is another abiotic stress that affects the crop every five to seven years and causes huge losses (up to 100% in frost-prone areas). Therefore, majority of mustard-growing farmers don't want much to invest on production and protection inputs.

On the other hand, the mustard crop is attacked by many disease-causing organisms, including bacteria, fungi, viruses, etc. These pathogens cause huge crop losses and always challenged the productivity and sustainability of Indian oilseed sector. More than 22 diseases have been reported to affect rapeseed-mustard group of crops in India (Ghosh et al. 2018). Rapeseed-mustard harbors nearly 38 insect pests from field to storagein India. Among these 38 insect pests, *mustard* aphid, *Lipaphiserysimi* (Kaltenbach) and painted bug, *Bagradahilaris* (Burmeister) are the most destructive insect pests in major mustard-growing regions of India, and due to increasing global temperature, these insect pests are expected to increase their kitty in yield losses caused by these pests in the future. Aphids are reported to cause damage from 10–90%, depending upon the climatic conditions (Dhillon et al. 2018). In addition to the direct losses caused by this insect, it has also been reported to transmit 13 different viruses.

Method adopted for seeding and soil moisture conditions at the time of sowing determines the productivity of the mustard crop. Improper germination and poor seedling establishment lead to poor plant stand, thus are major impediments in realizing higher yields. Significant differences were observed among different methods adopted for seeding this crop. Broadcasting of seed in lowland and upland conditions without tillage, broadcasting of seed followed by tillage, and mechanical sowing using calibrated seed drill largely determine the percent germination. Sowing of mustard is undertaken under pre-irrigated, conserved moisture, no moisture (rainfed), dry sowing, followed by sprinkler irrigation, and post-rain moisture conditions also determine the rate of growth and plant establishment in mustard. Sowing the crop at proper moisture using seed drills has been advocated at all platforms; however, the socio-economic and agro-ecological pressures never allowed this practice to be largely accepted by the farmers, especially in fragile environments. *Orabanche*, a parasitic weed, is also challenging productivity of Indian mustard. Many farmers have left cultivating mustard in heavily infested fields and opting for alternative crops.

1.6 PRESENT STATUS OF OIL SCENARIO IN INDIA

India holds a significant position in the global oilseeds landscape, both in terms of land coverage and in production. It ranks as the fourth largest edible oil economy worldwide, following the United States, China, and Brazil. India contributes approximately 10% of the global oilseeds production, 6–7% of the global vegetable oil production, and nearly 7% of protein meal production. This sector holds a crucial position within the Indian agricultural industry, encompassing an area of around 26.5 million hectares and yielding over 29 million tons of oilseeds in the three-year period ending in 2011–12. This accounts for roughly 14.8% of the country's total cropped area. In terms of agricultural output value, oilseeds represent about 9.8% (at 2004–05 prices) of the overall agricultural production during the same three-year period. Additionally, India maintains a noteworthy presence in the global oilseeds scenario, with 12–15% of the total land area dedicated to oilseed cultivation, contributing 6–7% to vegetable oil production, accounting for 9–10% of the world's edible oil consumption, and representing 13.6% of vegetable oil imports. The country cultivates a diverse range of oilseed crops across various agro-climatic regions.

Three main oilseeds, namely, groundnut, soybean, and rapeseed-mustard, accounted for over 88% of total oilseeds output during the TE 2011–12. Soybean is the most important crop, with an estimated production of 11.6 million tons in TE 2011–12, grown mainly in Madhya Pradesh, Maharashtra, and Rajasthan, accounting for more than 95% of total production. The second most important oilseed crop is rapeseed-mustard (7.1 million tons), mainly grown in Rajasthan, Madhya Pradesh, Haryana, Uttar Pradesh, West Bengal, and Gujarat, with an estimated share of about 93% in total production in the country.

In the 1990s, groundnut, the largest oilseed crop at that time, experienced a decline in its market share and is currently the third most significant oilseed. It has an average production of 6.9 million tons in TE 2011–12 and is predominantly cultivated in Gujarat, Andhra Pradesh, Tamil Nadu, Rajasthan, Karnataka, and Maharashtra. These states collectively contribute approximately 91% to the global production in the country.

Other notable edible oilseeds include sesamum, sunflower, and safflower. In addition to West Bengal (21.3%) and Rajasthan (21.2%), Madhya Pradesh (16.8%) and Gujarat (14.1%) are major producers of sesamum in the country. Sunflower production is primarily concentrated in Karnataka (37.3%), Andhra Pradesh (27.2%), and Maharashtra (14.6%), accounting for around 80% of the total production. However, sunflower production has remained relatively stable with considerable variability over the past decade and a half. Safflower production has consistently declined, with Maharashtra (54.7%), Karnataka (27.9%), and Gujarat (12.7%) being the primary producers, collectively contributing over 95% to the total production.

The oilseeds sector occupies a distinct position in the Indian agriculture sector after cereals, sharing for 13% of the country's gross cropped area, accounting for 3% of gross national product and 10% of the value of agriculture product. India, with its diverse agro-economical conditions, is suitable for the production of nine important oilseed crops: groundnut, sesame, castor, linseed, safflower, rapeseed-mustard, sunflower, soybean, and niger. India also produces two perennial oilseeds crop—coconut and oil palm—besides the secondary oilseed crops, such as maize and cotton. India ranks first in the world in the production of groundnut, sesamum, castor, and linseed and safflower. It is second in rapeseed production and fourth in the production of sunflower and soybean. The productivity, however, has been less than two-thirds (2/3) of the world's average productivity.

India, the world's second-largest consumer nation, following closely behind China, and the leading global importer of vegetable oil, is projected to maintain a significant annual per capita consumption growth rate of 3.1% and reach 24 kg per capita by 2027. The consumption of vegetable oil in India is expected to rise to 37 million metric tons by 2027, a notable increase from 24 million metric tons recorded during the period of 2015–2017. This substantial growth will be met by a combination of expanding domestic production through intensified oilseed cultivation and a further surge in imports, primarily of palm oil sourced from Indonesia and Malaysia (OECD 2018).

India has a great diversity of climates and occupies a distinct position in Southeast Asia in the production of oilseeds. The prevailing agro-ecological conditions have been favorable for growing several important annual oilseeds, including edible (namely, groundnut, rapeseed-mustard, soybean, sunflower, safflower, and sesame) and non-edible oilseeds (namely, castor and linseed). The other minor crops which substantiate significantly include oil-bearing tree species. India occupies a prominent position in acreage and production of oilseed crops in the world. It constitutes the second-largest agricultural commodity next to cereals in India, sharing 14% of the country's gross cropped area, accounting for nearly 5% of the gross national product and 10% of the value of all agricultural products. Oilseed crops are grown in an area of 16.5 Mha, with a production of 10 MMt, which accounts for one-tenth of the total cultivated area in India. The major oilseed crop–growing states are Madhya Pradesh (24.1%), Gujarat (11.6%), Andhra Pradesh (11.5%), Rajasthan (11.4%), Maharashtra (10.9%), Karnataka (9.5%), Uttar Pradesh (6%), and Tamil Nadu (4.5%). More than 15 million people are engaged in production and processing of oilseeds (Hegde 2000). In India, oilseed crop importance is ranked thus: groundnut, rapeseed-mustard, sesame, linseed, safflower, castor, sunflower, and niger. Oilseeds contribute more than half of the dietary requirements in terms of fats and oils (Guidry 1964).

India is also the largest producer of substantial sesame (31.2%) and groundnut (25.1%). India is the major producer of castor and sesame and the second-largest producer of groundnut and mustard, next to China. Castor productivity in India is highest than in any other country of the world. Castor is grown in almost all states of the country; however, Gujarat is the major producer, accounting for more than 83% of total production in the country.

1.7 NATIONAL MISSION ON OILSEED

The role played by agricultural development in oilseeds cultivation over an area of 25.3 million hectares is of paramount importance to the Indian economy. This group of crops not only occupies a significant portion of agricultural land next to food grains but also is an essential food item. In spite of their important position in national agricultural economy and the multiplicity of crops and crop-growing situations, the country's domestic output of oilseeds is lagging far behind the requirement. Nevertheless, the Technology Mission on Oilseeds (TMO), started in 1986 and comprised of technologies related to genetic enhancement, plant protection, and crop and resource management, has substantially increased the production of oilseeds.

The major aim was to reduce dependence on imports through increase in domestic oilseeds production. There are many factors responsible for this growth in area and productivity (extended cultivation to non-traditional area and of non-conventional crop and value addition), which could be related with inputs, infrastructural development, and price incentives besides the technological interventions with time. We are still lagging behind in food shortage and food and nutrition insecurity, exacerbated by ever-increasing prices throughout the world. Every year, the growing menace of climatic variability results into severe famine and floods which affect one-fourth of the growing human population in different parts of the world. These frequent climatic vagaries lead to crop failures and thereby rampant declines in pollinator populations across the continents. The problems have been further aggravated due to a lack of prophylactic progress in the conservation of biodiversity and increased agricultural production.

Among various oilseed crops, soybean, sesame, safflower, sunflower, groundnut, and castor are the preferred ones (Weiss 2000; Ranganatha et al. 2012a, 2012b, 2012c), while sunflower, soybean, and canola offer good management options for irrigation reduction and thereby enhancing the benefits of reduced input costs of these oilseed crops (Aiken and Lamm 2006).

More than half of the world's diet of oils and fats comes from oilseeds, such as coconuts, cotton, oil palm, olives, peanuts, rape, soybeans, and sunflower. Many of these plants are dependent upon insect-mediated pollination, which improves the qualitative and quantitative characteristics of crop production. Inadequate pollination can result not only in reduced yields but also delayed yield, with high percentage of inferior-quality seeds/fruits. Even self-fertile flowers can set some seed and fruit in the absence of animal visitation. However, animal visitation increases either (a) the proportion of fruits set or (b) the quality of fruits set. Seed size is sometimes greater after cross-pollination, which can only be achieved by mediation by foraging of insect visitors. Pollinators and pollination are vital in the functioning of the agricultural ecosystem as they are instrumental in enhancing productivity through mediating the cross-pollination and improving the gene flow of crops. Major world oil crops and their products are among the most valuable commodity to meet the demands of people as well as to earn livelihood security. Over the past couple of decades, oilseed production has increased to become an important source of vegetable oils, in response to the rising world population and ameliorating their socio-economic conditions.

1.8 AREA PRODUCTION AND PRODUCTIVITY

Oilseed cultivation is undertaken across the country in about 27.04 million hectares, largely under rainfed areas covering 72% and producing around 33.42 million tons of oilseeds during 2019–20. Nine oilseeds are the major source of vegetable oil in the country. Among the nine major oilseeds, soybean (33.5%), groundnut (30%), and rapeseed and mustard (27%) contribute to more than 90% of total oilseeds production in the country. However, in terms of vegetable oil production, mustard, soybean, and groundnut contribute 27%, 34%, and 30%, respectively. Due to the concerted efforts of the government of India (GOI), the production of oilseeds has been maintained at around 31 million tons, with increased productivity. During 2019–20, the highest production of 33.42 million tons was recorded from an area of 27.04 million hectares, yielding 1,236 kg per hectare, which was largely driven by productivity increase. Madhya Pradesh, Rajasthan, Maharashtra, and Gujarat are the major oilseeds-producing states, contributing more than 78% of oilseeds production in the country.

1.9 ADVANTAGES OF OILSEED CROPS

The oilseed plants are the major sources of so many useful products, such as vegetable oil, fatty acids, feedstock, raw materials for various industries, etc. and may be promising in manufacturing biodiesel, an alternative fuel in the petroleum industry (Harwood et al. 2013). Some of the non-edible oilseed crops have been explored as the source of biofuel (Balat 2011; Yadav et al. 2012). To overcome ecological contamination and economic constraint using petrochemicals, use of biofuels from agronomic biomass is recommended (Nigam and Singh 2011). The supply of fossil fuels is becoming limited, which requires the use of biofuels, which consist of biohydrogen, bioethanol, biodiesel, and biomethanol. Biodiesel can be produced from either non-edible or edible oils (Yusuf et al. 2011). Oilseed crops provide a major source for biodiesel production. Brazil holds the fourth rank in production of biodiesel around the world and is the second-largest producer of soybean. There are some oilseed crops not engaged in the production of food sources which can function as the preferred feedstock for biofuel manufacture.

Applications of oilseed crops are shown in Figure 1.1.

Oilseed crops play an important role in the food industry due to their rich nutritional content. *Guizotia abyssinica*, niger, has rich nutritional and antioxidative properties (Ramadan 2012).

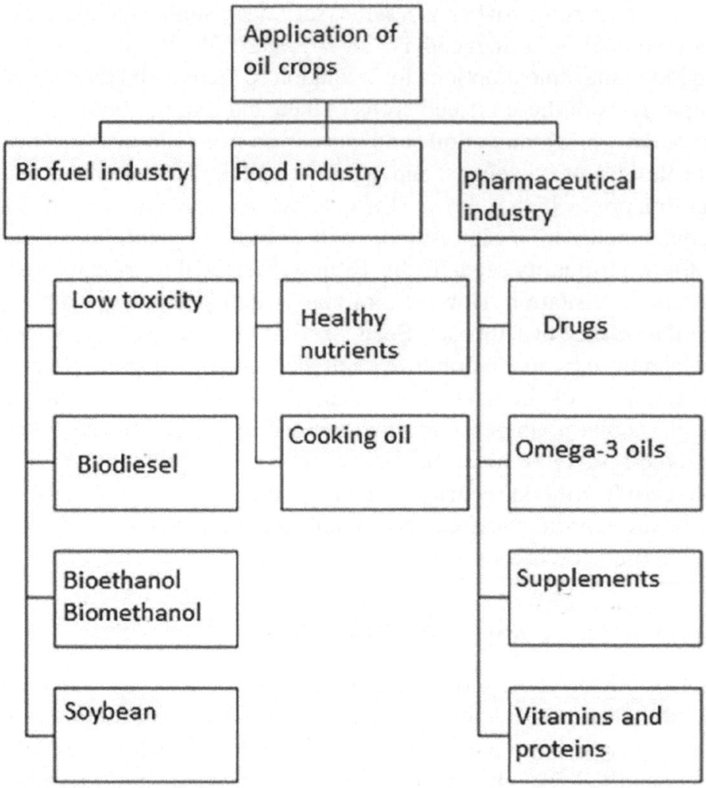

FIGURE 1.1 Applications of oilseed crops in different industries.

Source: Waseem et al. (2017).

TABLE 1.1
Oilseed Crops and Their Uses

Name of Plants	Scientific Name	Family	Type of Oil	Uses
Rapeseed	*Brassica napus*	Brassicaceae	Vegetable oil	Vegetable oil, animal feeds as cakes, protein supplement, biodiesel
Soybeans	*Glycine max*	Fabaceae	Vegetable oil	Protein supplement, cooking oil, flour, infant formula, pharmaceutical industry
Sunflower	*Helianthus annuus*	Asteraceae	Seed oil	Cooking oil
Safflower	*Carthamus tinctorius*	Asteraceae	Vegetable oil	Edible oil used to reduce blood sugar, cholesterol, and skin inflammation
Sesamum	*Sesamum indicum*	Pedaliaceae	Edible oil	Seeds that contain protein, fiber, and healthy fats; sesame seeds also provide calcium, B vitamins, vitamin E, and antioxidants
Linseed	*Linum usitatissimum*	Linaceae	Edible oil as a source of alpha-linolenic and omega-3 fatty acid	Food and fiber plant

TABLE 1.1 (*Continued*)
Oilseed Crops and Their Uses

Name of Plants	Scientific Name	Family	Type of Oil	Uses
Niger	*Guizotia abyssinica*	Asteraceae	Edible oil	Rich in nutrition; good source of protein, carbohydrate, and fiber
Castor	*Ricinus communis*	Euphorbiaceae	Vegetable oil	Used in manufacturing of soaps, lubricants, brake fluids, paint dyes, coatings
Cotton	*Gossypium* spp.	Malvaceae	Refined seed oil extracted from kernels used in cooking	Cake or meal after oil extraction used in poultry and livestock feed
Coconut	*Cocos nucifera*	Arecaceae	Vegetable oil and biofuel	Culinary uses, decoration, food industry, chocolates, vinegar, cooking, nectar
Palm oil	*Elaeis guineensis*	Arecaceae	Edible vegetable oil	Used in cooking, manufacturing food products
Olive	*Olea europaea*	Oleaceae	Used in cooking	Used in salad dressing, cosmetics, pharmaceuticals and soaps, etc.
Groundnut/ peanut	*Arachis hypogea*	Fabaceae	Cooking oil	Peanut butter, cooking, cosmetics, plastics, dyes, textile material, flour

Source: Modified afterWaseem et al. (2017).

Oilseed crops are a rich source of protein content for humans as well as animals. Among oilseed crops, *Jatropha* is rich in amino acids, with the exception of lysine.

Oilseed crops find tremendous use in pharmaceutical industries. Among oilseed crops, niger has extreme medicinal and antioxidant properties (Ramadan 2011). *Jatropha* spp. are rich in oil content and play an important role in drug preparations (Sosa-Segura et al. 2014). Production of biodiesel can be increased by producing it from soybean, as large amounts of diesel can be prepared with just a low concentration of crop. Moreover, the leftover part of this crop can be used as feedstock for animals (Padula et al. 2012).

1.10 ROLE IN FOOD AND NUTRITION

Oilseed crops constitute one of the major crops in the world with high nutritional content. In the industrial sector, oilseed crops offer distinct benefits compared to other crops, owing to the wide range of products they can yield, such as food, biodiesel fuel, fertilizer, fiber, and more. Such products include, among others, food for humans and by-products of feed formulation for animals that will eventually form a good source of protein for humans, biodiesel fuel, fertilizer for land rejuvenation, fibers, artificial fibers, plastics produced in the textile and plastic industries, etc. Oilseed contains about 20–40% protein and about 20–50% fats, which enable them to be a superb candidate as a significant source of human and animal nutrition.

According to FAO (2005), in the Census of Agriculture, oilseed crops were classified as group 4, after cereals, vegetables and melon, and fruits and nut. The world sources of edible seed oils include soybean, sunflower, rapeseed, cotton, olive, coconut, oil palm, and peanut. Oilseeds are important crops for low-income families in the semi-arid tropics, as they contribute 40% of the total calorie intake in their diet (Gunstone 2002; Graham and Vance 2003). These crops are cheap, are readily available, and grow in various soils. Seed oils from flax (linseed) and castor bean are used for industrial purposes (Sarwar et al. 2013). Globally, oilseeds are being modified for high nutrition, improved oil quality, and composition. Oilseed-defatted meal serves as a good source of protein in human nutrition. The oil contains polyunsaturated fatty acid, which is beneficial to human health and industrial application.

1.11 APPROACHES TO IMPROVE PRODUCTIVITY

1.11.1 Enhancing Genetic Potential

The genetic architecture of the plant imposes inherent ceiling on the genetic gains. The phenomenon yield is determined by the corollary between harvest index (HI) and biomass. When breeders try to improve HI beyond present level (≈ 0.25), there is reduction in biomass, and if biomass is improved beyond a limit, the HI reduces. Therefore, to improve genetic potential beyond present level, new plant ideotype needs to be defined and created. Effort is required to optimize these two factors to realize higher yields. The defined ideotype may also require different plant geometry to produce more. Information from distant relatives and cultivated types shall be helpful in understanding such relationship and defining ideotype in the future.

Heterosis in oilseed *Brassica* can be exploited commercially using highly efficient hybrid seed production mechanisms, such as cytoplasmic male sterility–fertility restoration (CMS-FR) system. In India, *Raphanus sativus (ogura)* and *Moricandria arvensis (mori)* have been widely used to develop commercial Indian mustard hybrids. If concerted efforts are made, hybrids have the potential to improve production due to their higher yielding capacity and better tolerance to abiotic stresses.

1.12 YIELD STABILIZATION

Genetic potential can be achieved by cultivating notified varieties with recommended production and plant protection technologies. Indian mustard possesses better level of tolerance against most of the abiotic stresses, such as drought, heat, salinity, etc., in comparison to most other crops cultivated in India. Indian mustard is also known to be a high yielder and possess better tolerance against drought as compared to *B. napus* and *B. rapa*, thus has enormous cultivation potential in semi-arid areas. Due to this fact, proportion of area under *B. juncea* has significantly increased, whereas the area under other species like *B. rapa*c vs. yellow sarson, brown sarson, and *B. napus* has declined. Though the level of drought tolerance is better in Indian mustard than other field crops, the crop-growing environments still demand better level of tolerance. Mustard being a crop of marginal lands under rainfed conditions, genotypes with inbuilt mechanism to tolerate scanty moisture conditions is the need of the hour. We have very limited genetic variation in the cultivated germplasm which can tolerate water stress at different stages, thus results in fluctuation in national production and productivity. Genotypes with high water use efficiency can be exploited for the improvement of mustard for such conditions. There have been very little efforts to understand genetic and molecular mechanisms, use of wild and related species for trait introgress, and breed genotypes possessing high water use efficiency.

Poor plant stand is another important reason for non-realization of actual yield potential in early and timely sown crops, mainly due to high temperature at seedling stage. The late-sown crop faces high temperature at the reproductive stage, thus results in forced maturity and reduction in yield and oil content. Hence, genotypes having inbuilt tolerance to high temperature at seedling stage as well as terminal heat tolerance are needed. These genotypes mature in about 100–120 days and are highly suitable for early (September) sown conditions. Five genotypes, viz EJ-22, NPJ-113, NPJ-124, Pusa Bahar, and 5011 (Pusa Agrani x Laxmi), identified using the aforementioned protocol, have also been characterized using various morphological and biochemical parameters in order to establish effective heat tolerance selection criteria (Azharudheen et al. 2013). Further, the related species, like *B. taurnifortii* and *B. carinata*, possessing drought and heat tolerance, are valuable genetic resources and can be used for improving the level of heat tolerance in *B. juncea*.

Screening techniques have been standardized for salinity tolerance at germination stage in hydroponics, pots, and microplots and in target environments. Salinity is becoming one of the limiting factors in *Brassica* production, and genotypes tolerant to heavy metals and enhanced CO2 utilization also need attention in the times to come.

Genetic variability in *Brassicas* and their related species has been studied and also identified the sources of resistance or partial resistance against all the aforementioned pathogens. Despite that, efforts directed toward the development of resistant varieties have not been much successful. Resistant sources and varieties succumb to regional pathotypes/races due to variable resistant or susceptible reaction; as a result, the resistance in the developed variety is often challenged by farmers and the scientific community. Pathogen biology and epidemiology have been well understood.

The pathogen variability has been reported by many workers despite that host differential sets have not been developed for most of these pathogens; thus, genes could not be identified and designated.

Furthermore, the reliability of identified molecular markers is always challenged during genotyping of populations using different isolates every time. Thus, *Brassica* breeders rely mainly on selecting resistant segregants under disease epiphytotic conditions without having much knowledge about race spectrum and genes. Due to limited knowledge about the underlying genes, their precise location on chromosomes, reaction to different pathotypes/races, and availability of reliable linked molecular markers, the pyramiding of genes is not possible.

Delineating pathotypes/races, on the basis of host differentials, is pre-requisite for identification of *R*-genes and establishing gene-for-gene hypothesis. This shall further help in separating the effect of genes and revealing their host- and stage-specific differential expression.

Combining race-specific and race-nonspecific genes together in one cultivar can significantly improve durability of *R*-genes deployed within *Brassica* species. In general, *B. carinata* resist white rust, *Alternaria* leaf blight, and sclerotinia stem rot diseases; tolerate drought and high temperature; and can be easily hybridized with *B juncea*. Efforts are being made to enrich the genome of *B. juncea* with *B. carinata* for productivity and adaptability traits. A good amount of genetic variability has been generated and is being used for further improvement of *B. juncea*. In addition to this, efforts have also been made to improve *B. juncea* using *B. elongate*, *B. fruticulosa*, *B. rapa*, *Sinapis alba*, *Sinapis arvensis*, *Eruca sativa*, *Eruca vesicaria*, *Diplotaxis erucoides*, *Camelina sativa*, *Raphanus sativus*, etc. Recently, genome of *Brassica juncea*, *B. napus*, *B. rapa*, and *B. oleracea* has been sequenced, which can help in the discovery of novel *R*-genes and in the enhancement of our understanding about genetic relationships among different *R*-genes within and between members of *Brassica* species. Availability of DNA sequences of *Brassicas* and progress in functional genomics will further lead to identification of many novel genes involved in pathways of different stress tolerance mechanism. Hybridization between divergent groups of Brassicaceae harboring numerous genes is already demonstrated in India and elsewhere, which can further complement the incorporation of identified genes/QTLs from one species to another. Once the identified genes/QTLs are tagged and mapped with molecular markers, their pyramiding in single genetic background would be possible. This shall also open new opportunities for development of multiple disease-resistant Indian mustard genotypes.

1.12.1 Infrastructure Development for Achieving Higher Productivity

Regional breeding programs in the eastern and northeastern zones of India need to be strengthened. Institutional and financial support for establishing oil mills in the non-conventional area shall be helpful in increasing area under this crop. Creating irrigation facilities in areas with limited water availability shall boost the yield levels. Sowing with seed drill following soil pulverization in rice fallow conditions of eastern India shall help in improving area and productivity under mustard cultivation. The seed hubs need to be created and strengthened in regions having low productivity.

1.13 SUSTAINABLE CULTIVATION OF OILSEED CROPS TO MEET THE FUTURE DEMANDS

The vegetable oil complex is today at a crossroads, and there is a need to make concerted and coordinated efforts to withstand the pressures built up by the globalization and WTO regime.

1.14 INTERNATIONAL OIL PRICES AND CONSEQUENT IMPACT ON DOMESTIC AVAILABILITY AND DEMAND

The international prices of edible oil depend on their demand and supply. Continuous rise in human population and increase in income would demand more edible oils. Unless domestic production is increased, the dependence on imports would increase, resulting in rise in prices.

1.15 IMPORTING OILSEEDS INSTEAD OF OIL

China, besides being the largest oilseeds producer in the world, has taken a different route to tackle the problems of growing demand for oils, by importing oilseeds rather than edible oil. Heavy duty on import of seed rather than edible oil by China, in comparison to India, logically favors the edible oil, crushing industries of the former, and provides meal to meet the booming animal feed demand.

1.16 INCENTIVES FOR MOTIVATING FARMERS TO SHIFT TO OILSEED CULTIVATION

In the early 1990s, there was a controlled approach to maintaining the minimum support price (MSP) for food grains compared to oilseeds. This strategy significantly boosted domestic oilseed prices in relation to competing crops, leading to a 70% increase in oilseed production between 1987–88 and 1994–95. Nevertheless, starting from the mid-1990s, oilseed prices began to decline compared to other crops, primarily due to increased domestic oilseed supplies and the liberalization of edible oil imports that commenced in 1994. Since the mid-1990s, the MSP for oilseeds has been raised, aiming to incentivize growers to focus on oilseed crops. However, the increasingly favorable returns from rice and wheat have attracted agricultural land away from oilseeds, resulting in a decrease in oilseed production. From 1998 to 1999 onward, though, the MSP of major edible oilseeds has been consistently increasing in an effort to encourage growers to shift their focus back to oilseed cultivation.

1.17 RESEARCH INITIATIVES FOR ENHANCING YIELD RATES OF OILSEEDS

The ICAR and State Agricultural Universities have developed a number of high-yielding varieties (HYVs) along with the production management technologies suited to various agro-climatic conditions of the country. This has helped increase productivity of oilseeds. Availability of enough breeder seed production and its further multiplication through foundation and certified seed production are the key constrains for the availability of quality seed at farmers' level. Further, the industries/private houses should support goal-oriented basic and strategic research to enhance research and development (R&D) activities, especially in frontier research areas like biotechnology for diseases and pest management of oilseeds crops.

1.18 AGRICULTURAL REFORMS POLICY

The future of this industry will not be bright unless it upgrades and modernizes its technology to attain world standards. The large number of intermediaries tended to reduce the farm gate price of the farmer. Appropriate domestic policy measures such as contract farming, marketing-warehousing-storage facilities, etc. are needed to increase the farm gate price realization of the farmer. A novel policy framework for the processing industry will have to balance the interests of four constituent factors: an incentive price for farmers, an affordable price for consumers, reasonable profit margins for industry with the incentives to modernize, and satisfactory levels of employment, income, exports, and public revenue. Srinivasan (2005) has suggested that a multi-commodity, partial-equilibrium model shall be instrumental in solving prices and quantities in markets simultaneously for five oils, four oilseeds, and four oil meals. Of several alternatives to protect farmers' interests, he found, through his sophisticated econometric model, that "between the (other) two alternatives, import tariff on edible oil and government subsidy in cash, the latter turns out to be more attractive."

1.19 STRATEGIES OF OILSEEDS PRODUCTION

Strategies for enhancing the productivity (and profitability) of oilseed-based production systems are prepared annually for oilseeds and oil palm in the country. The interventions/strategies proposed in oilseeds are time-tested with scale neutrality that can be grounded for enhancing the productivity of the oilseed-based production system with necessary institutional support/hand-holding. The proposed strategies are categorized as follows:

1. Horizontal (area expansion) and vertical (productivity increase) expansion of oilseeds crop.
2. Increasing seed production and distribution of newly released varieties.
3. Low-cost technologies with high impact on productivity, resulting in higher income.
4. Technologies with high impact that involve reasonable investment with high return on investment (ROI), with emphasis on eco-friendliness and high input use efficiency.
5. Strategies with emphasis on quality improvement and value addition, leveraging technologies with a bearing on employment through skill/entrepreneurship development.
6. Strategies to increase additional area and production of oilseeds through rice fallow, intercropping, and crop diversification and in non-traditional areas, with focused district/cluster approach.

1.20 CONCLUSION

The oilseed sector in India *vis-à-vis* different states is at a crossroads and facing several challenges. Most of the state is facing low yield rates and high cost of oilseeds production per unit area. Yield improvements and increases in the oil and protein content of the seeds are particularly crucial for oilseeds to compete more effectively with other crops for the use of scarce resources, such as land and labor, and other critical inputs, such a fertilizer, electric power, irrigation water, whose prices are likely to go up, to better reflect their true economic costs. So there is a need for policy reforms in the country both at the central level and at the state level to promote production of oilseeds in the country, providing warehouse facilities, ensuring proper marketing channels for bulk transportation facilities, setting up food safety standards to combat frauds, restricting imports, mostly of palm oil and many others which impact the small and marginalized farmers in a positive way, to increase the production of oilseeds in the country.

REFERENCES

Aiken, R. M. and Lamm, F. R. 2006. Irrigation of oilseed crops. Proceedings of 18th Annual Central Plains Irrigation Conference and Exposition. February 21–22, 2006, Colby, K.S., pp. 162–172.

Azharudheen, M., Gowda, M. V. C., Sattigarahalli, L., and Bhat, A. R. S. 2013. Evaluation of sister homozygous lines for oil content and oil quality in groundnut. *Karnataka Journal of Agricultural Sciences* 26(1), 1–5.

Balat, M. 2011. Production of bioethanol from lignocellulosic materials via the biochemical pathway: A review. *Energy Conversion and Management* 52(2), 858–875.

Chauhan, J. S., Bhadauria, V. P. S., Singh, M., Singh, K. H., and Kumar, A. 2007. Quality characteristics and their interrelationships in Indian rapeseedmustard (Brassica sp) varieties. *Indian J. Agril. Sci.* 77, 616–620.

Dhillon, M. K., Singh, N., Tanwar, A., Yadava, D. K., and Sujata, V. 2018. Standardization of screening techniques for resistance to *Lipaphiserysimi* (Kalt.) in rapeseed-mustard under field conditions. *Indian Journal of Experimental Biology* 56(9), 674–685.

Diekötter, T., Kadoya, T., Peter, F., Wolters, V., and Jauker, F. 2010. Oilseed rape crops distort plant-pollinator interactions. *J. Appl. Ecol.* 47, 209–214.

Ghosh, T., Biswas, M. K., and Aikat, K. 2018. A review on seed borne mycoflora associated with different oilseed crops and their management. *Int. J. Pure Appl. Biosci.* 6, 1526–1538.

Graham, P. H. and Vance, C. P. 2003. Legumes: Importance and constraints to greater use. *Plant Physiology* 131(3), 872–877, ISSN: 0032–0889.

Guidry, N. P. 1964. Graphic summary of world agriculture. *U. S. Dept. Agr. Misc. Pub.* 705, 64.

Gunstone, F. D. 2002. *Vegetable Oils in Food Technology, Composition, Properties, and Uses.* John Wiley & Sons, West Sussex, UK, p. 337.

Harwood, J. L., Ramli, U. S., Tang, M., Quant, P. A., Weselake, R. J., Fawcett, T., and Guschina, I. A. 2013. Regulation and enhancement of lipid accumulation in oil crops: The use of 1150 metabolic control analysis for informed genetic manipulation. *European Journal of Lipid Science and Technology* 115, 1239–1246.

Hegde, D. M. 2000. Technology for higher yields. Directorate of Oilseeds Research. Survey of Indian Agriculture, pp. 65–69.

Jankowski, K. J. and Budzynski, W. S. 2003. Energy potential of oilseed crops. *Electronic Journal of Polish Agricultural Universities* 6(2), #03. Available at www.ejpau.media.pl/volume6/issue2/agronomy/art-03.html

Lennerts, D. R. L. 1983. *Oelschrote, Oelkuchen, pflanzliche Oele und Fette, Herkunft, Gewinnung, Verwendung.* VerlagAlfred Strothe, Hannover, Germany.

Maheshwari, P., Selvaraj, G., and Kovalchuk, I. 2011. Optimization of Brassica napus (canola) explant regeneration for genetic transformation. *New Biotechnology* 29(1), 144–155.

Nigam, P. S. and Singh, A. 2011. Production of liquid biofuels from renewable resources. *Prog. Energ. Combust. Sci.* 37(1), 52–68.

OECD/Food and Agriculture Organization of the United Nations. 2018. Oilseeds and oilseed products. In: *OECD-FAO Agricultural Outlook 2018–2027.* OECD Publishing, Paris/Food and Agriculture Organization of the United Nations, Rome. DOI: https://doi.org/10.1787/agr_outlook-2018-7-en

Padula, A. D., Santos, M. S., Ferreira, L., and Borenstein, D. 2012. The emergence of the biodiesel industry in Brazil: Current figures and future prospects. *Energy Policy* 44, 395–405.

Ramadan, M. F. 2011. Artificial neural networks: A novel tool for detecting GMO. *Journal of Consumer Protection and Food Safety* 6, 13–23.

Ramadan, M. F. 2012. Functional properties, nutritional value, and industrial applications of Niger oilseeds (*Guizotia abyssinica* Cass). *Crit. Rev. Food Sci. Nutr.* 52(1), 1–8.

Ranganatha, A. R. G., Lokesha, R., Tripathi, A., Tabassum, A., Paroha, S., and Shrivastava, M. K. 2012b. Sesame improvement—present status and future strategies. *Journal of Oilseeds Research* 29, 1–26.

Ranganatha, A. R. G., Rajdeep, M., and Aasfa, T. 2012c. Seed production of sesame and niger. Training Manual on Oilseeds Seed Production, Directorate of Oilseeds Research, Hyderabad.

Ranganatha, A. R. G., Tripathi, A., Paroha, S., Tabassum, A., and Nema, A. 2012a. Evaluation of genetic stocks for development of superior export types in Sesame. Proceedings of National Seminar on Plant Genetic Research for Eastern and North Eastern India, IARI, New Delhi, p. 57.

Sarwar, M. F., Sarwar, M. H. Sarwar, M., Qadri, N. A., and Moghal, S. 2013. The role of oilseeds nutrition in human health: A critical review. *J. Cereals Oilseeds* 4(8), 97–100.

Sosa-Segura, M. P., Oomah, B. D., Drover, J. C. G. et al. 2014. Physical and chemical characterization of three non-toxic oilseeds from the Jatropha genus. *J. Food Nutr. Res.* 2(1), 56–61.

Srinivasan, P. V. 2005. Impact of trade liberalisation on India's oilseed and edible oil sector, Research Report, Indira Gandhi Institute for Developmental Research, Mumbai.

Waseem, S., Imadi, S. R., Gul, A., and Ahmad, P. 2017. Chapter 1: Oilseed crops present scenario and future prospects. In: Ahmad, P. (ed.) *Oilseed Crops*. John Wiley and Sons Ltd, West Sussex, UK, p. 312.

Weiss, E. A. 2000. *Oilseed Crops*, 2nd edn. Blackwell Science, Malden, pp. 259–273.

Wittkop, B., Snowdon, R. J., and Friedt, W. 2009. Status and perspectives of breeding for enhanced yield and quality of oilseed crops for Europe. *Euphytica* 170(1–2), 131–140.

Yadav, S., Kumar, S., Husain, Z., and Dutta, M. 2012. *Guizotia abyssinica* (L.f.) Cass.: An untapped oilseed resource for the future. *Biomass Energy* 43, 72–78.

Yusuf, N., Kamarudin, S. K., and Yaakub, Z. 2011. Overview on the current trends in biodiesel production. *Energ. Conv. Manag.* 52(7), 2741–2751.

2 Production Constraints

2.1 INTRODUCTION

The introduction of high-yielding varieties resulted in increased rapeseed production since the 1980s in the main growing areas. However, since the 1990s, production declined steadily, which has been attributed to different biotic stresses. A global survey has revealed that a total of 16 diseases, 37 insect pests, several species of nematodes, and slugs have been affecting rapeseed production (Zheng et al. 2020). Oilseed rape is attacked by several diseases that attack oilseed crops, which include soil-borne pathogens (*Plasmodiophora brassicae*, *Verticillium longisporum*), or by pathogens that attack the stem (*Sclerotinia sclerotiorum*, *Leptosphaeria maculans*, *L. biglobosa*, *Alternaria* spp., *Pseudocercosporella capsellae*, *Pyrenopeziza brassicae*). Sclerotinia stem rot, or white mold, is present in all major oilseed rape–growing areas. These diseases result in significant yield losses (Kirkegaard et al. 2006; Del Río et al. 2007). The causal agent, *Sclerotinia sclerotiorum*, is reported to infect more than 400 plants (Mizubuti 2019), and its sclerotia can survive in the soil for more than four years. Its ascospores can be produced and spread during flowering by wind, insects, and rain (Link and Johnson 2007).

There is growing evidence that pollinator-dependent crops with insect pollination can give higher yields. However, the interacting agricultural inputs need to reconcile with food production and conservation of biodiversity. Catarino et al. (2019) found that greater yields could be achieved by increasing bee abundance and reducing use of pesticides. Pollinator abundance has been rapidly declining worldwide due to indiscriminate use of pesticides, which results in the mortality of pollinators and reduction in their efficiency. Insect pollination increases yield of entomophilous crops (Free 1970; McGregor 1976; Sihag 1983a, 1983b, 1983c, 1986). However, the indiscriminate use of pesticides in modern agriculture results in the reduction of natural and managed pollinators (Kevan 1975; Johansen 1977; Miller and Kevan 1979; Logan and Schofield 1984), reducing crop production (Kevan 1977; Kevan and Opperman 1980). Sihag (1986) reported that either factor alone gave small yield increases, but the two combined interacted significantly to give major increases. Therefore, both protective application of pesticides and use of bee pollination are essential for maximum crop yields.

Since integrated pest management (IPM) tools such as crop rotation, soil management, resistant cultivars, or biocontrol are ineffective or not available, insect control largely relies on insecticides. There is need for an efficient and chemical low-management strategy for sustainable rapeseed production in the future. Oilseed crops play a dominant role in the economy of the world. They constitute an essential part of human diet and their role in industrial application. The production of oilseeds has increased tremendously through improvement in technological interventions. Several factors are responsible for low production of oilseed crops, which include attack of different insect pests and diseases (Prakash and Hinata 1980; Lamb 1989). The two most important oilseed crops are *Brassica campestris* var. *oleifera* and *B. napus* var. *oleifera* (Weiss 1983). The other oilseeds crops are groundnut, sunflower, soybean, safflower, and sesame, yielding high-grade edible oil, and the non-edible-oil-producing species include castor and linseed.

2.2 SCENARIO OF OILSEED CROPS IN THE WORLD

There are more than 40 different oilseed crops which are consumed as oil in the world (Lennerts 1983). Oilseed crops belong to three categories, which include annual—soybeans, sunflower, groundnuts, and rapeseed—and perennial—coconuts and oil palms. The other oilseed crops are of minor importance but play a significant role as raw materials for several products (Sharma et al.

DOI: 10.1201/9781032656724-3

2012). *Brassica* species constitute the second-largest category of oilseed crops globally, exceeding peanut (*Arachis hypogaea* L.), sunflower (*Helianthus annuus* L.), and cottonseed (*Gossypium hirsutum* L.). This information is supported by various sources, such as OECD and FAO (2010), Raymer (2002) and Sharma et al. (2014). Among the 37 species in the Brassica genus, *Brassica rapa* L., *Brassica juncea* (L.) Czern and Coss, *Brassica napus* L., and *Brassica carinata* A. Braun (Raymer 2002; Rakow 2004; Sovero 1993) are the four most commonly cultivated species for oilseed and vegetable production. Rapeseed is the most favored vegetable oil. The leading producers are Canada, China, India, Germany, and Ukraine, along with Poland, UK, and Australia, which contribute about 77% of the total rapeseed-mustard production of the world (OECD and FAO 2010).

Oil palm contributes about 36% of the total world oil production. The major palm oil–producing countries are Indonesia, Malaysia, Thailand, Nigeria, and Colombia, which together contributed about 93% of total world palm oil production in the year 2008–09. Cotton is cultivated in 70 countries worldwide (OECD and FAO 2010). Over a quarter of the world cotton area is in India, followed by USA (16%), China (14%), and Pakistan (8%). Countries like Turkey, Australia, Greece, Brazil, and Egypt account for the rest. Groundnut (*A. hypogaea* L.) is the most important oilseed crop grown in about 84 counties on an area of 24.52 mha, with a production of 34.15 mmt and average productivity of 13.92 q/ha. Among the groundnut-producing nations, India sows the highest area (around 28% of total world area) under this crop and is the second-largest producer, next only to China. China, India, Nigeria, USA, and Myanmar contributed about 85% of the total world production of groundnut in 2008.

Among the oilseed crops, sunflower (*H. annus* L.) occupies the fourth position in area and production in the world. The top five countries with respect to production (21.49 mmt) and area (15.93 mha) in the world are Russian, Ukraine, Argentina, India, and China. Europe alone accounts for about 50% of world area and total production. Asia and Africa share about 21 and 6% of the total sunflower area in the world, respectively. The olive tree, *Olea europaea* L., is a subtropical evergreen tree native to the Mediterranean region. It is the sixth most important oil crop in the world. World olive oil production is around 2.9 mmt, the Mediterranean countries being the major contributors (International Olive Council 2009, www.internationaloliveoil.org/downloads/ production1_ang. PDF). Spain has the largest area (2.60 mha) and highest production (6.22 mmt), followed in production by Italy, Greece, Turkey, Tunisia, and Greece. These five countries accounted for 80.84% of the total world production (18.32 mmt) in 2008. Sesame (*Sesamum indicum* Lin.) is known as the "queen of oilseeds" and one of oldest oilseed crops grown in India. India represents 30% of the area of the world under production of sesame.

2.3 SCENARIO OF OILSEED CROPS IN INDIA

India presents a great diversity in the production of oilseed crops. The prevailing agro-ecological conditions have been favorable for growing several important annual oilseeds, including edible (namely, groundnut, rapeseed-mustard, soybean, sunflower, safflower, and sesame) and non-edible oilseeds (namely, castor and linseed). India occupies a significant position in terms of acreage and production of oilseed crops. It constitutes the second-largest agricultural commodity next to cereals in India, sharing 14% of the country's gross cropped area and accounting for nearly 5% of the gross national product and 10% of the value of all agricultural products. Today, oilseed crops are cultivated on about 16.5 million hectares, with total production of 10 million tons, which constitutes approximately one-tenth of the total cultivated area in India. The major oilseed crop–growing states are Madhya Pradesh (24.1%), Gujarat (11.6%), Andhra Pradesh (11.5%), Rajasthan (11.4%), Maharashtra (10.9%), Karnataka (9.5%), Uttar Pradesh (6%), and Tamil Nadu (4.5%) (Hegde 2000). Important oilseed crops grown in the country are groundnut, rapeseed and mustard, sesame, linseed, safflower, castor, sunflower, and niger. According to Guidry (1964), more than half of the world's diet of fats and oils is derived from coconuts, oil palm, olives, cotton, peanuts, rape, soybeans, and sunflower.

India holds a significant proportion of global castor production, accounting for 76.9%. Additionally, it contributes substantially to the production of sesame, with a share of 31.2%, and groundnut, with a share of 25.1%. Castor is grown in Gujarat, Andhra Pradesh, Tamil Nadu, Rajasthan, and Karnataka, accounting for about 51, 29, 6, 5, and 3% of the total area under the crop, respectively. The Gujarat state of India alone contributes to 83% of the country's total production.

2.4 CONSTRAINTS IN OILSEED CROP PRODUCTION

Oilseed crops are attacked by a large number of insects, which cause serious damage. The most serious pests include leaf miner, thrips, and tobacco caterpillar; head borer in sunflower; aphids and *H. armigera* in safflower; leaf webber and capsule borer in sesame; aphids, sawfly, cabbage butterfly, and painted bug in rapeseed-mustard; semilooper and Bihar hairy caterpillar in castor; and bud fly and *H. armigera* in linseed. Polyphagous pests such as red hairy caterpillars, Bihar hairy caterpillars, root grubs, termites, cutworms, and serpentine leaf miners frequently target these crops. The yield losses due to these pests could range from 50 to 90% across oilseed-growing areas. Soybean crop is severely attacked by about a dozen of insect pests, causing damage ranging from 20 to 100%. However, with implementation of IPM schedule, 30–35% additional yield can be obtained (Sharma et al. 2014). Knowledge on insect pests and their interaction with plants is essential to evolve high-yielding varieties with resistance to insect pests. Most of the available oilseed crop cultivars are susceptible to major insect pests; however, some have shown escape to some insect pests. Although the use of insecticides is quite effective, their excessive and indiscriminate use has resulted into imbalance in agro-ecosystem and health hazards. Therefore, the answer lies in the development of resistant/tolerant varieties, exploitation of biological control agents, and use of bio-rational approaches against the major insect pests of oilseed crops. IPM offers multiple strategies that can be considered, such as utilizing *Trichoderma* sp. for seed treatment, releasing Trichogramma, employing pheromone traps, and implementing targeted pesticide application. The integration of new technologies has greatly influenced the progress of IPM practices. Rapidly growing human population has increased the demand for production of increased oilseeds. As land resources became limited, efforts were made to increase productivity by fighting the losses inflicted by insects, weeds, and plant pathogens through the interventions of pesticides. The use of pesticides has resulted in serious problems, such as contamination of food, pest resistance, and outbreak of pests, requiring the use of safe and eco-friendly pest control methods (Abrol and Shankar 2014). The IPM principle does not preclude chemical pesticide use but rather uses it as one of the tools in the management package, to be used prudently and integrated with other tools (Goodell 1984). IPM strategies incorporate various approaches, such as (1) enhancing monitoring techniques to achieve more precise assessments of pest populations; (2) employing trap crops to draw and concentrate pests, thereby diminishing pest density and minimizing harm to the primary crop; and (3) encouraging conservation biological control, with specific focus on minimizing tillage and abstaining from insecticide applications during flowering stages (Cook and Denholm 2008).

The crop is attacked by an array of insect pests at the time when a large number of pollinating insects responsible to enhance their productivity are also active. Thus, crop production is simultaneously confronted by two problems which are antagonist to each other. There are two key aspects to consider: safeguarding crops from insect pests and ensuring the safety of pollinators for effective crop pollination. It is clear that an integrated approach is required to address both the protection of crops and the well-being of pollinators.

2.5 ENHANCING OILSEED PRODUCTION AND PRODUCTIVITY

Some of the areas of country after cultivation of paddy in eastern India are left uncultivated due to irrigation unavailability. By raising early growing rice varieties, these areas, with proper management and integrated pest management, can be utilized for oilseed cultivation with availability of ample natural resources and fertile land (Singh et al. 2017). Intercropping of oilseeds can help increase area under oilseeds; the adoption of improved varieties and the latest technologies to overcome deficit under-rainfed areas, such as promoting microirrigation, can help enhance oilseed production. Efforts are needed to promote varieties with short durations, short stature, and high oil content. Application of primary nutrients such as N, P, and K and secondary and micronutrients such as zinc, sulfur, and boron can enhance oil productivity. Good agronomic practices and pest management practices should be given due consideration. Government incentives like crop insurance and minimum support prices can attract growers for more production of oilseeds.

2.6 CONCLUSIONS AND FUTURE STRATEGIES

The multitude of insect pests occurring in different locations in different oilseed crops causes heavy losses in the production of oilseeds in an already oilseed-deficient world. Pests occurring at different stages of crop growth are tackled by chemical control methods, which drastically affect the population of beneficial insects, such as natural enemies, predators, parasitoids, and pollinators. It is therefore recommended to adopt the most widely acceptable tools of integrated pest management for sustainable production of oilseeds, by regular monitoring of the pest situation along with the weather condition of a particular area so that timely protection from pest and prophylactic measures, if any, may be initiated, followed by conservation and augmentation of beneficial species of parasites (Powell 1986; Blande et al. 2008), predators, and microorganisms, by selective use of chemical pesticides and training of field functionaries/farmers on awareness about biological control agents, coupled with IPM demonstrations in farmers' field.

REFERENCES

Abrol, D. P. and Shankar, U. 2014. Pesticides, food safety and integrated pest management. In: Pimental, D. and Peshin, R. (eds.) *Integrated Pest Management-Pesticide Problems*. Springer, Dordrecht, The Netherlands, pp. 167–199.

Blande, J. D., Pickett, J. A., and Poppy, G. M. 2008. Host foraging for differentially adapted brassica-feeding aphids by the braconid parasitoid Diaeretiella rapae. *Plant Signal Behav.* 3(8), 580–582.

Catarino, R., Bretagnolle, V., Perrot, T., Vialloux, F., and Gaba, S. 2019. Bee pollination outperforms pesticides for oilseed crop production and profitability. *Proc. R. Soc. B* 286, 20191550. http://dx.doi.org/10.1098/rspb.2019.1550

Cook, S. M. and Denholm, I. 2008. Ecological approaches to the control of pollen beetles in oilseed rape. *IOBC/Wprs Bulletin* 38, 110–113.

Del Río, L. E., Bradley, C. A., Henson, R. A., Endres, G. J., Hanson, B. K., McKay, K. et al. 2007. Impact of sclerotinia stem rot on yield of canola. *Plant Dis.* 91, 191–194. doi: 10.1094/PDIS-91-2-0191

Free, J. B. 1970. *Insect Pollination of Crops*. Academic Press, London.

Goodell, G. 1984. Challenges to international pest management research and extension in the third world: Do we really want IPM to work? *Bulletin of the Entomological Society of America* 30, 18–26.

Guidry, N. P. 1964. Graphic summary of world agriculture. U. S. Dept. Agr. Misc. Pub. 705, 64 pp.

Hegde, D. M. 2000. Oilseeds situation. Directorate of Oilseeds Research, ICAR, Hyderabad.

Johansen, C. A. 1977. Pesticides and pollinators. *A. Rev. Entomol.* 22, 177–191.

Kevan, P. G. 1975. Forest application of the insecticide Fenitrothion and its effect on wild bee pollinators (Hymenoptera: Apoidea) of low bush blueberries *(Vaccinium* spp.) in southern New Brunswick, Canada. *Bioi. Conserv.* 7, 301–309.

Kevan, P. G. 1977. Blueberry crops in Nova Scotia and New Brunswick—pesticides and crop reductions. *Can. J. Agric. Econ.* 25, 61–64.

Kevan, P. G. and Opperman, E. B. 1980. Blueberry production in New Brunswick, Nova Scotia and Maine: A reply to wood et a!. *Can. J. Agric. Econ.* 28, 81–84.

Kirkegaard, J. A., Robertson, M. J., Hamblin, P., and Sprague, S. J. 2006. Effect of blackleg and Sclerotinia stem rot on canola yield in the high rainfall zone of southern New South Wales, Australia. *Aust. J. Agric. Res.* 57, 201–212. doi: 10.1071/AR05179

Lamb, R. J. 1989. Entomology of oilseed *Brassica* crops. *Ann. Rev. Entomol.* 34, 211–229.

Lennerts, L. 1983. Oelschrote, oelkuchen, pfl anzliche Oele und Fette, Herkunft, Gewinning, Verwendung, Bonn 1983. Alfred Strothe, Hannover.

Link, H. V. and Johnson, K. B. 2007. White mold (Sclerotinia). Available at www.apsnet.org/edcenter/disandpath/fungalasco/pdlessons/Pages/WhiteMold.aspx (accessed 15 September 2020).

Logan, J. W. M. and Schofield, P. 1984. *The Environmental Effects of Insecticides on Non-Target Organisms in Tropical and Sub-Tropical Developing Countries.* Tropical Development and Research Institute, London.

McGregor, S. E. 1976. Insect pollination of cultivated crop plants. Agric. Handb. U.S. Dep. Agric. No. 496.

Miller, D. and Kevan, P. G. 1979. Insecticides and non-target arthropods on *Artemisiajilifolia* (Compositae). *South West Nat.* 24, 439–446.

Mizubuti, E. S. G. 2019. Special issue on white mold—*Sclerotinia sclerotiorum. Trop. Plant Pathol.* 44, 1–2. doi: 10.1007/s40858-019-00276-1

Organization for Economic Co-operation and Development (OECD) and the Food and Agriculture Organization (FAO). 2010. FAO Agricultural Outlook 2010–19. Available at https://www.fao.org › docs › ENGLISH_outlook

Powell, W. 1986. Enhancing parasitoid activity in crops. In: Waage, J. and Greathead, D. (eds.) *Insect Parasitoids. 13th Symp. Royal Ent. Soc. London, 18–19 Sept. 1985.* Acad. Press, London, pp. 19–340.

Prakash, S. and Hinata, K. 1980. Taxonomy, cytogenetics and origin of crop brassicas, a review. *Opera Bot.* 55, 1–57.

Rakow, G. 2004. Species origin and economic importance of Brassica. In: Pua, E. C. and Douglas, C. J. (eds.) *Biotechnology in Agriculture and Forestry, vol. 54, Brassica.* Springer, Berlin, pp. 3–11.

Raymer, P. L. 2002. Canola: An emerging oilseed crop. In: Janick, J. and Whipkey, A. (eds.) *Trends in New Crops and New Uses.* ASHS, Alexandria, VA, pp. 122–126.

Sharma, A. N., Gupta, G. K., Verma, R. K., Sharma, O. P., Bhagat, S., Amaresan, N., Saini, M. R., Chattopadhyay, C., Sushil, S. N., Asre, R., Kapoor, K. S., Satyagopal, K., and Jeyakumar, P. 2014. Integrated pest management for Soyabean. Directorate of Plant Protection, Quarantine & Storage, CGO Complex, NH IV, Faridabad-121001, p.41.

Sharma, P., Saini, M. K., Deep, S., and Kumar, V. 2012. Biological control of groundnut root rot in farmer's field. *Journal of Agricultural Science* 4(8), 48–59.

Sihag, R. C. 1983a. Foraging behaviour and pollination ecology of different bee pollinators in relation to cultivated cruciferous crops. *29 Illl. Beekeep. Congr. Budapest*, 334–335.

Sihag, R. C. 1983b. Foraging behaviour and pollination ecology of different bee pollinators in relation to cultivated umbelliferous crops. *29 Ill. Beekeep. Congr. Budapest*, 335.

Sihag, R. C. 1983c. Foraging behaviour and pollination ecology of different bee pollinators in relation to onion crop. *29 Illl. Beekeep. Congr. Budapest*, 336.

Sihag, R. C. 1986. Insect pollination increases seed production in cruciferous and umbelliferous crops. *J. Apic. Res.* 25, 121–126.

Singh, A. K., Singh, A., Choudhary, A. K., and Kumari, A. 2017. Towards oilseeds sufficiency in India: Present status and way forward. *J. Agrisearch.* 4, 80–84.

Sovero, M. 1993. Rapeseed, a new oilseed crop for the United States. In: Janick, J. and Simon, J. E. (eds.) *New Crops.* Wiley, New York, pp. 302–307.

Weiss, E. A. 1983. Rapeseed. In: Wrigley, G. (ed.), *Oilseed Crops.* Longman Group, New York, pp. 161–215.

Zheng, X., Koopmann, B., Ulber, B., and von Tiedemann, A. 2020. A global survey on diseases and pests in oilseed rape—Current challenges and innovative strategies of control. *Front. Agron.* 2, 590908. doi: 10.3389/fagro.2020.590908

3 Pollination in Oil Crops

3.1 INTRODUCTION

Oilseed crops have been the backbone of the agricultural economy of India since time immemorial and play a vital role in agricultural industries and trade. India occupies a prominent position in the world, both in regard to acreage and production. Oilseeds form the second-largest agricultural commodity after cereals in India, sharing 14% of the country's gross cropped area and accounting for nearly 5% of the gross national product and 10% of the value of all agricultural products. Today, these crops are cultivated on about 16.5 million hectares, with total production of 10 million tons, which constitutes approximately one-tenth of the total cultivated area in India. According to Hegde (2000), approximately 14 million people are involved in the production of oilseeds, while around 1 million individuals are engaged in the processing of oilseeds.

The important oilseed crops grown in the country are groundnut, rapeseed and mustard, sesame, linseed, safflower, castor, sunflower, and niger. Oilseeds, such as coconuts, cotton, oil palm, olives, peanuts, rape, soybeans, and sunflower, account for over 50% of the global consumption of fats and oils (Guidry 1964). Most of these crops are either fully dependent on or benefit from insect pollination. Approximately one-third of our food consumption relies on plants that are directly or indirectly pollinated by insects. Insufficient pollination not only leads to decreased crop productivity but also produces lower-quality fruits.

Even self-fertile flowers can set some seed and fruit in the absence of animal visitation. Animal visitation helps increase the proportion of fruits set and/or the quality of fruits set, because fruit size depends on the number of seeds set or the size of seeds. Seed size is sometimes greater after cross-pollination, which can only be achieved by insect visits (Charlesworth and Charlesworth 1987; Richards 1997). Important crops include oilseed rape (Free and Ferguson 1983; Williams et al. 1987), flax- and linseed, sunflowers (Free and Williams 1977), cotton, soybean, and olives.

Pollinators play an important role in most terrestrial ecosystems that are vital to the maintenance of both wild plant communities (Ashman et al. 2004; Aguilar et al. 2006) and agricultural productivity (Klein et al. 2007; Ricketts et al. 2008). The planting of mass-flowering crops, such as oilseed rape and sunflowers, could provide valuable nectar and pollen resources for pollinators (Westphal et al. 2003). However, such super-abundant resources are only available for brief periods of time, and as a consequence, they might have little effect in sustaining viable pollinator populations (Kremen et al. 2007). The honeybee (*Apis mellifera*) is the primary pollinator that is commonly managed worldwide to increase agricultural yields. However, in specialized scenarios, other bee species, like the leaf-cutting bee, *Megachile rotundata*, are also employed. The honeybee, which has been well studied compared to other bee species, has been documented to be capable of increasing yield in 96% of animal-pollinated crops (Klein et al. 2007). Managed honeybee colonies have severely declined in the USA (59% loss of colonies between 1947 and 2005) (van Engelsdorp et al. 2008) and Europe (25% loss of colonies in central Europe between 1985 and 2005) (Potts et al. (2010), threatening the dependence of agricultural crops and wild plants.

Atmowidi et al. (2007) studied the diversity pollinator insects and their effect on seed set of mustard (*Brassica rapa*) crop in West Java. They found more than 19 species of insects visiting mustard flowers. Honeybees *Apis cerana*, *Ceratina* sp., and *Apis dorsata* were more abundant. Insect pollination was found to increase the number of pods, seeds per pod, seed weights per plant, and seed germination. Pollinators provide key ecosystem services to both natural and agro-ecosystems, and a positive relationship has been found between floral resources and pollinator diversity (Steffan-Dewenter et al. 2002; Klein et al. 2003). On sunflowers, wild bees may contribute more pollination services through enhancement than that of direct pollination, as they concentrate on male flowers

DOI: 10.1201/9781032656724-4

rather than female ones. Bees visiting only male flowers make no direct contribution to its pollination but do contribute through enhanced pollination (Greenleaf and Kremen 2006). Animal pollinators are thought to contribute in 15 up to 30% of global food production (Roubik 1995), and bees are recognized to be the most important pollinating taxon (Delaplane and Mayer 2000). Farmers obtain adequate pollination services by bringing large numbers of honeybees to crop fields (Greenleaf and Kremen 2006). Destruction and fragmentation of habitats is one of the important factors responsible for the decline of solitary wild bees and bumblebees (Daily 1997; Kremen et al. 2002; Steffan-Dewenter et al. 2002).

Brassica rapa L., a member of the Cruciferae family, holds significant agricultural value in Asia (Rubatzkky and Yamaguchi 2000). The majority of Crucifer species, around 95%, rely on cross-pollination, although certain cauliflower varieties can self-pollinate. However, self-pollination without cross-pollination generally leads to reduced seed yield, smaller seed size, and decreased yields in subsequent generations (Delaplane and Mayer 2000). Plants that are self-incompatible require the transfer of pollen from one plant to another (Wallace et al. 2002). Honeybee pollination has been found to increase yields in male sterile oilseed rape (*B. napus*) (Westcott and Nelson 2001). Similarly, insect pollination in sarson (*B. campestris*) has been observed to enhance seed yield, promote the formation of well-shaped and larger grains, and result in more viable seeds (Abrol 2009). Oilseed rape is very attractive to honeybees, bumblebees, and solitary bees due to mass-flowering crops and abundance of nectar and pollen. However, bees foraging on the crop are adversely affected by insecticides when these are applied during flowering to control inflorescence pests. Effects may be lethal or sub-lethal; the latter has been little studied, but there is growing evidence that insecticides affect many aspects of bee behavior and physiology and other activities such as division of labor, foraging and orientation, reproduction and respiration.

Despite good agronomic practices, the level of productivity of oilseeds in India is far below than the average productivity in the world. The low productivity of oilseeds/oil crops can be attributed to various factors, such as heavy infestation of pests and diseases. Out of the various factors, one of the most important factors for low production of oilseed crop is the failure of seed setting, attributable to the lack of proper pollination (Rao et al. 1980; Free 1993; Abrol 2007, 2008, 2009).

Therefore, cross-pollination of entomophilous crops by honeybees is considered as one of the most effective and cheapest methods for increasing crop yields. Applied pollination, pollinator management, and managed pollination are the common efforts recently practiced for maximization of production in cross-pollinated crops and to bring pollinators to target crops. Important oil crops that need attention are grouped under two headings, as given here:

Oilseed Crops

1. Rapeseed, mustard, and canola
2. Sunflower
3. Safflower
4. Sesame
5. Groundnut
6. Soybean
7. Castor bean flower
8. Cottonseed
9. Niger
10. Taramira
11. Linseed

Oil from Nuts and Fruits

1. Coconut
2. Oil palm
3. Olive

TABLE 3.1
Impact of Animal Pollination on Production of Oil Crops*

Crop Species	Commodity	World Production (Mt)	Animal Pollination	References
Glycine max, G. soja	Soybean	204,429,429	Increase	Koelling et al. 1981; Free 1993; Moreti et al. 1998; Nogueira-Couto et al. 1998; Chiari et al. 2005a, 2005b
Elaeis guineensis	Oil palm	163,141,697	Increase	Free 1993; Dhileepan 1994; Westerkamp and Gottsberger 2000; Tandon et al. 2001; Krantz and Poinar 2004; Mayfield 2005
Gossypium hirsutum, G. barbadense, G. arboreum, G. herbaceum	Seed cotton	69,849,042	Increase	Free 1993; Rhodes 2002
Cocos nucifera	Coconut	54,708,169	Increase	Free 1993; Da Conceicao et al. 2004; Meléndez-Ramírez et al. 2004
Brassica napus	Rapeseed, oilseed rape	46,770,903	Increase	Free 1993; Adegas and Nogueira Couto 1992; Abel and Wilson 1999; Bürger 2004; Manning and Boland 2000; Abel et al. 2003; Morandin and Winston 2005
Arachis hypogaea	Peanut, groundnut	35,894,864	Increase	Crane 1991; Free 1993
Helianthus annuus	Sunflower	26,460,824	Increase	Bichee and Sharma 1988; Crane 1991; Free 1993; DeGrandi-Hoffman and Martin 1993; Moreti et al. 1996; Heard 1999; DeGrandi-Hoffman and Watkins 2000; Dag et al. 2002; Greenleaf and Kremen 2006b
Olea europaea	Olive	17,176,231	No increase	Free 1993; Singh 1997
Phoenix dactylifera	Date palm	6,907,093	No increase	Crossa-Raynaud 1984

* Data compiled from different sources.

The breeding system of different oil crops varies from one crop to another. Some of the crops that are highly cross-pollinated depend exclusively on insect pollination for any significant seed/fruit yield, whereas other compatible ones also improve qualitatively and quantitatively when pollinated by insects. The impact of insect pollination on production of different oil crop is presented in Table 3.1. The data reveal that there is significant increase in the production of each crop when cross-pollinated.

The pollination requirements, the type of pollinators, their behavior and efficiency in the pollination, and the production of different crops are briefly described next.

3.2 RAPESEED, MUSTARD, AND CANOLA (*BRASSICA* SPP.)

Many species of *Brassica*, such as rape (*Brassica napus*), sarson (*Brassica campestris* var. sarson), toria (*Brassica campestris* var. toria), Indian mustard or rai (*Brassica juncea*), white mustard (*Brassica alba*), and black mustard (*Brassica nigra*), are cultivated widely as oilseed crops throughout the world. Most of these crops bloom during February to March for over a month. The bright-yellow fragrant flowers are produced in long terminal racemes.

These crops are usually self-pollinated. however, some degree of cross-pollination occurs in *Brassica campestris*. The flower produces a very good amount of nectar and pollen, which are highly attractive to honeybees and other insect visitors. Cross-pollination by honeybee increases yield, quality, and oil content of seed. The average sugar concentration in the afternoon ranges between 40 and 45%, whereas in the morning it is 16% only (Sharma 1958; Murrell and Nash 1981). Mohammed (1935) recorded 117 visiting species of insects belonging to seven orders in West Punjab at Lyallpur. Among them, *Andrena ilerda*, *Apis florea*, and *Halictus* spp. comprised 82% of insect visitors. He also observed that stigma of the crops remained receptive for two days after opening, and in five days, their fertility was lost, whereas pollen remained viable for seven days.

Auto-fertilization of these crops was studied by Mohammed (1935) for six years by enclosing individual plants in muslin bags; in toria, seed set was 12.3%; in brown sarson, 20.3%; and in yellow-seeded sarson, 91.0%. Then he compared self-pollination to cross-pollination in toria. He bagged flowers for cross-pollination and hand-pollinated them; there was 100% pod formation. Sihag (1986) studied seed set in toria plots caged and not caged and left open for natural cross-pollination; there developed 25 and 1,060 pods per plant, 6.3 and 18.2 seeds per pod, and 4 and 385 seeds per plant. Similar observations on sarson gave 59 and 95% pods containing 3.5 and 12.7 seeds per pod (Mishra et al. 1988). Observations made in field plots of both sarson and toria showed that when crop was caged to exclude insects, yield was 68 g; when caged with *Apis cerana*, it gave 219 g; and in open field, where other insect pollinators were also present, yield was 244 g (Latif et al. 1960). In general, there is quite a high degree of self-incompatibility in these crops, and good yields are obtained by cross-pollination with the aid of natural fauna in most places, where good weather prevails, and probably yield can be further increased with the help of honeybee colonies to the extent of 10–30% (Rahman 1940; Singh 1954), covering for inclement weather. Singh and Singh (1992) reported that bee-pollinated plants in field cages as compared with self-pollinated plants in individual bags produced 3 times heavier pods, 4 times more seed per pod, 50 times more seeds, and 84 times more seeds yield per plant.

Mishra et al. (1988) reported that percentage of pod setting was significantly higher in insect-pollinated (95.21%) compared to open-pollinated (70.27%) or isolated from insects (58.81%). In comparison to nylon-caged (5.55 seeds/pod) and muslin-bagged (3.05 seeds/pod) flowers, open-pollinated flowers exhibited a significantly higher number of seeds per pod (912.62 seeds/pod). Pods obtained from open-pollinated flowers contained a significantly greater proportion of healthy seeds (84.78%) compared to those from net-caged (41.45%) or muslin-bagged (48.37%) flowers. However, the latter two did not show a significant difference between each other. The average weight of 100 healthy dry seeds from muslin-bagged flowers (542.20 mg) was significantly higher than that of seeds from open-pollinated (478.50 mg) and net-caged flowers (459.40 mg). The percentage of oil content in healthy seeds from muslin-bagged flowers (41.64%) was significantly higher than that from open-pollinated (39.02%) flowers.

However, oil content of seeds from muslin-bagged flowers was equal to that from net-caged flowers. When all the yield parameters were taken into consideration for oil content (milligram per pod) calculation, it was observed that the total oil yield increased by 9.76 times for open-pollinated plants and 1.55 times for net-caged plants compared to muslin-bagged plants.

Varieties of *Brassica campestris* L. are generally considered to require insect pollination for seed production (Free 1970; McGregor 1976). Mohammed (1935) and Rahman (1940) concluded that *Andrena ilerda* Cam., *Halictus* spp., and *Apis florea* were the most important pollinators, whereas Latif et al. (1960), Kapil et al. (1971), and Bhalla et al. (1983) reported *Apis cerana indica* to be the principal pollinator in *B. campestris* L. *var. sarson*. In a similar study, Singh et al. (2000) found that there was significant improvement in qualitative and quantitative parameters of rape when pollinated by *Apis cerana Himalaya*. In bee-pollinated plots, siliqua set was 239.77% over no pollination and 2.65% over open pollination. Similarly, in other parameters, siliqua length was

14.72% over no pollination and 8.65% over open pollination. Other parameters such as number of seeds per siliqua, weight of 1,000 seeds per gram, seed yield (in q/ha), seed germination (%), and oil content (%) were significantly higher in bee-pollinated plots compared to open pollination and those deprived of bee pollination.

Pollination by bees in rapeseed results in increased productivity, uniformity in seed setting, and early establishment of the pod (Abrol 2007). According to Sabbahi et al. (2005), this results from the abundant production of flowers during the flowering period, which ensures proper pollination and thus a high and relatively early production of siliques. Therefore, the absence of cross-pollination, which occurs when free plants in the presence of pollinators interchange pollen, is responsible for the low production of well-developed siliques.

Chhuneja et al. (2007) studied the role of *Apis mellifera* in the seed production of *Brassica campestris* var. toria. They found that the ratio of bees engaged in pollen foraging was greater within a distance of 50–75 m from the bee colony. They also observed higher presence of pollen foragers in the morning and noon hours compared to the evening. Likewise, the mean number of siliquae per plant, number of seeds per 20 siliquae, thousand seed weight, seed yield per plant, and oil content was significantly higher, up to 50 and 75 m distances. One colony of *Apis mellifera* with ten-frame bee strength was sufficient for effective pollination of one hectare of crop area. Araneda et al. (2010) evaluated the yield achieved by *B. napus* cv. Artus pollinated by *A. mellifera* in Chile. He showed that the parameter least affected by bee intervention was the grains per silique variable. In contrast, siliques per plant and 1,000 grain weight parameters presented significant differences, contributing to a yield greater than 5 t ha^{-1}, which represented a figure 50.34% higher than in the treatment without bees. He further concluded that the inclusion of bees in crops is fully justified as a production tool.

Deaodikar and Suryanarayana (1972) found enhanced yield in oilseed crops when pollinated by bees. They reported that in *Brassica napus* L. (rape), bee-pollinated plots produced 12.8–139.3% more seeds than self-pollinated ones. In *Brassica campestris* L. var. toria (toria), this increase was 66.0–220.9%, and in *Brassica campestris* L. var. dichotoma (sarson), 222% over self-pollinated plots.

Langridge and Goodman (1975) also found significant differences in production of rapeseed in open and caged plots. Plots receiving unrestrained visits had significantly higher plant density, seed yield, seed/plant, weight of 1,000 seeds, weight of seeds per plant, germination, and percentage of oil content as compared to those isolated from insect visits. Abrol (2009) studied the pollination requirement of different oilseed crops and expected increase in yield due to bee pollination, which have been summarized in Table 3.2.

The yields of many field, fruit, and seed crops are greatly enhanced by bumblebee visitation (Corbet et al. 1991; Osborne and Williams 1996). Field beans are largely dependent on pollination by longer-tongued species, such as *B. pascuorum* and *B. hortorum*, and without them, yields are poor (Free and Williams 1976). Bumblebees are acknowledged to be more reliable pollinators than honeybees, particularly because they will continue foraging even when it is cold and wet.

Rapeseed canola, renowned for its high-quality oil suitable for human consumption (Porter and Crompton 2008; Amjad 2010), is an oleaginous crop. While wind plays a significant role in pollen dispersal over long distances for rapeseed (Hoyle et al. 2007), bees are the primary pollinators, as they are abundant in rapeseed cultivars and varieties used for seed production (Westcott and Nelson 2001; Pordel et al. 2007). Introducing honeybee colonies, such as A. *mellifera*, for controlled pollination has proven to enhance crop production, particularly for rapeseed (*Brassica napus*) (Sabbahi et al. 2006). This plant offers a rich source of pollen and nectar, attracting various insects, notably bees (Smith 2002; Sabbahi et al. 2006).

Successful pollination is crucial for significant seed production in most oilseed crops, which often rely on cross-pollination. Studies have shown that insect pollination can increase seed yield in mustard (Mohammed 1935; Latif et al. 1960). Latif et al. (1960) reported a 60% increase in seed yield when *Apis cerana* colonies were present near sarson and toria fields. Similar improvements have been observed in oilseed rape (Langridge and Goodman 1975; Kisselhagen 1977) and

TABLE 3.2
Expected Increase in Crop Yield Due to Bee Pollination

Crop	Pollination Requirements	Expected Percent Increase in Yield due to Cross-Pollination
Coconut	Pollen transfer is essential	5
Mustard rape	Mostly cross-pollinated	20
Sesamum	Often cross-pollinated	20
Sunflower	Cross-pollinated	20
Cotton	Often cross-pollinated	10
Brassica napus	Often cross-pollinated	50–72
Brassica napus	Often cross-pollinated	35–51
Brassica rapa	Often cross-pollinated	78–89
Brassica rapa	Often cross-pollinated	39–69
Sinapis alba	Often cross-pollinated	51–75
Helianthus annuus	Often cross-pollinated	70–86
Mustard	Often cross-pollinated	13–222
Safflower	Often cross-pollinated	4–114
Sunflower	Often cross-pollinated	21–3,400
Sesame	Often cross-pollinated	24–40
Niger	Often cross-pollinated	17–45
Linseed	Mostly cross-pollinated	2–49

Source: Abrol (2009).

Brassica campestris var. Jambuck (Kubisova et al. 1980). Bisht et al. (1980) discovered that rapeseed flowers visited by *Apis* species experienced higher pod set, an increased number of seeds per pod, and greater seed weight compared to flowers without pollinator visits.

Mishra et al. (1988) found that in *Brassica campestris*, percent pod setting, number of pods per plant, and proportion of healthy seeds were significantly higher in open-pollinated flowers than in net-caged and muslin-bagged ones. Similarly, average weight of seeds and oil content were higher in open-pollinated flowers. *Apis cerana* was the most common pollinating species. The other pollinators observed included *A. mellifera*, syrphid flies, etc. In a similar study, Prasad et al. (1989) found that pollination *of Brassica juncea* by *A. cerana* resulted in more siliqua setting, increased length of siliqua, increased seed weight yield, and a pronounced effect on oil contents and germination. Evidently, pollination is one of the most important natural factors enhancing crop production. Among the various pollinating agents, insect pollinators hold a dominant role in increasing the yield of oilseed crop.

From the plant breeder's point of view, the methods employed for the improvement of a crop depend, to a large extent, on the mode of pollination of the crop concerned. Pollination methods need to be critically studied to select lines of effecting improvement in a crop. The embryo, which gives rise to an adult plant, results from the fusion of two male and female gametes or reproductive cells. The constitution of which may be identical or different. In case these two gametes have identical constitution, the plant is self-fertilized. On the other hand, if two gametes fusing together are different in constitution, a large amount of cross-fertilization occurs in the plant, the one generation being different from the other.

Cresswell et al. (2004) observed that the flower of *Brassica napus* L. demonstrates characteristics typically associated with animal pollination (zoophily). This is attributed to its visually appealing petals, sturdy stigma, and nectaries. While wind pollination is possible, its effectiveness is uncertain due to the intricate interactions between the flower and windborne particles. The petals of the flower hinder wind pollination by reducing the efficiency of pollen deposition on the stigma when facing the direction of the wind. In the case of flowers facing away from the wind, pollen reception was minimal. Several features of the plant's structure (floral architecture, pollen cohesiveness, inflorescence structure) clearly indicate its reliance on animal pollination (Cresswell and Osborne 2004). Based on estimates, wind pollination is unlikely to play a significant role in the long-distance dispersal of *B. napus* genes, including those from genetically modified varieties.

This research demonstrates the potential of computational fluid dynamics (CFD) as a valuable tool for future investigations on wind pollination. However, considering the importance of these crops, insufficient data is known about their pollination requirements. Bee pollination has been observed to significantly benefit oilseed rape plants. Noticeable variations in seed yield characteristics were observed when comparing plots with bee pollination to enclosed plots.

3.3 SUNFLOWER (*HELIANTHUS ANNUS* L., FAMILY COMPOSITAE)

Sunflower (*Helianthus annuus* L.) is a globally cultivated major oil crop appreciated for its high-quality, low-cholesterol oil. Its productivity is low, which may be due to pests/diseases damage, poor soil fertility, or water stress, but there is strong evidence that insufficient pollination can significantly minimize the yields (Free 1999). Sunflower benefits from insect pollination. Honeybee (*Apis mellifera* L.) pollination increases the seed yield by 30% and oil content by more than 6% in hybrid varieties (Langridge and Goodman 1974; Furgala et al. 1979; Jyoti and Brewer 1999; Hoffmann 1994). Other bees, especially the non-*Apis* bees, are also known to frequent sunflower but are not considered as reliable pollinators due to their low activity (Radford et al. 1979). DeGrandi-Hoffman and Watkins (2000) highlighted a potential secondary function of non-*Apis* bees, as evidenced by their capacity to improve pollination through the presence of *Apis mellifera*.

Sunflower crop are visited by a number of insects, including Hymenoptera, Lepidoptera, Coleoptera, and Diptera. Honeybees, wild honeybees, bumblebees, and solitary bees are the most important groups. Sunflowers are considered by beekeepers to be a fair source of pleasant-flavored, yellow-colored honey (Burmistrov 1965).

Nderitu et al. (2008) observed that individuals belonging to 14 insect species visited (*Helianthus annuus* L.) floral heads in Eastern Kenya. Insect-pollinated plots produced on average 53% more seed yield compared with those where insect visitors were excluded.

Even though there is mechanical possibility for the achievement of self-pollination, intra-floral pollination does not occur in nature. However, some of the varieties may be partially self-fertile. The role of honeybees and other insects is essential for intra-floral cross-pollination. It was demonstrated that bumblebees achieved 35% seed set, whereas there was only 1% seed set in flower heads covered with cloth bags. Cross-pollination takes place by the transfer of pollen by insects from another floret in the same flower head or from florets on other flower heads (Free and Simpson 1964). There are important advantages of cross-pollination, high percentage of seed set and crop yield; as a result of hybrid vigor, seeds are more and raise healthy plants; greater percentage of oil is obtained from hybrid seed (Furgala 1954a, 1954b).

All-day observations on sunflowers showed that *Apis mellifera* visited a flower 261 times, *A. dorsata* visited 11 times, and a moth, a butterfly, a housefly, and a beetle each visited once only (Goyal and Atwal 1973a, 1973b). Honeybees, among all the insects, acquire a pre-eminent position because the colonies can be moved to the desired fields for pollination, and at the same time, surplus honey can be gathered. Furgala (1954b) conducted thorough research in Europe and North America that delved into different facets regarding cross-pollination and its impact on crop yield. As part of this endeavor, Furgala sought to investigate the influence of honeybee visits on sunflower seed production in Manitoba. To accomplish this, 2.5 colonies per hectare were positioned by Furgala (1954b) along the eastern perimeter of the sunflower fields.

The seed production per hectare in these two fields was much higher compared with four fields located 4.8 km away from the honeybee colonies. Furthermore, he noted that seed yield in the

of peanut nectar by bees (Graham 1941; Pellett 1947), there is no functional nectar present within the flower. The key question revolves around whether cross-pollination leads to an increase in seed production. Srinivasalu and Chandrasekaran (1958) observed that different varieties of peanuts exhibit variations in cross-pollination, primarily determined by the extent to which the stigma protrudes from the keel. Leuck and Hammons (1969) discovered that two cultivars possess flower structures that hinder self-pollination but encourage cross-pollination facilitated by bees. In their research, Leuck and Hammons (1965a) found no hybridization when plants were kept in cages, but a significant number of hybrids were obtained from uncaged plants. Subsequently, they (1965b) reported that approximately 80% of peanut flowers in open settings were effectively tripped for pollen by bees.

There is little uncertainty regarding the potential for cross-pollination of peanut flowers, primarily facilitated by bees. According to Hammons (1963), Lasioglossum, Megachile, Bombus, and Apis species were observed to interact with peanut flowers. Additional species such as Anthidium and Melissodes were identified by Hammons et al. (1963) and Leuck and Hammons (1969), but the main credit for cross-pollination was given to halictids and megachilids. In the Malawi region of Africa, Gibbons and Tattersfield (1969) documented visits to the flowers by *A. mellifera adansonii, Nomia* spp., and *Megachile* spp. Hammons et al. (1963) pointed out that halictids and megachilids were most abundant during the cool morning hours, when optimal pollination of peanuts occurs, while honeybee activity was distributed throughout the day. The possibility of enhancing honeybee visitation by concentrating their population in the area was not taken into consideration. Given that Girardeau and Leuck (1967) attributed a 6– 11% increase in yields to bee pollination, it is important to consider bolstering the bee population in large peanut plantings. One potential approach could involve "saturation pollination" with honeybees, provided that its practicality can be demonstrated.

3.7 SOYBEAN (*GLYCINE MAX* L. MERR, FAMILY LEGUMINOSAE)

The soybean (*Glycine max*) is a species of legume native to East Asia. Soybean products are now a fundamental component of the diets of over 16% of the global population (PIBA 2001). Although soybean is generally considered a self-pollinating species, yield increases of about 10–40% have been demonstrated when comparing honeybee-pollinated vs. self-pollinated plants (Erickson et al. 1978). Further, honeybees can produce substantial honey crops when they visit soybeans flowers (Erickson et al. 1978). Flowering usually continues for four to six weeks, and there may be half a million florets per acre (McGregor 1976). Yield increases of between 10% and 40% have been recorded in soybean when comparing honeybee-pollinated plants against self-pollinated plants (Robacker et al. 1983), whilst cage exclusion trails have shown up to 15% increase in production (Erickson et al. 1978).

A more recent study by Chiari et al. (2005) also found that seed production was higher (P = 0.0001) in covered areas with honeybee colonies (50.64%) and uncovered areas (57.73%) than in covered areas without honeybee colonies; they concluded that honeybees were responsible for 95.5% of the pollination accomplished by insects (3.12). The pod number in covered treatment with honeybees was also 61.38% higher (P = 0.0002) than in the covered treatment without honeybees (Chiari et al. 2005).

3.8 CASTOR BEAN FLOWER *(RICINUS COMMUNIS* L., FAMILY)

Inflorescence of castor bean is available throughout the year in dense terminal clusters, with female flowers just above the male flowers. This species is clearly monoecious, with separate male and female flowers on the same individual. Each male flower has a cluster of many stamens which literally smoke as they shed pollen in a gust of wind. Pollination takes place with the help of wind. According to Roetheli et al. (1991), castor has the capacity to emerge as the leading vegetable oil crop for industrial oil production worldwide. Castor is considered an excellent choice for generating high-value industrial oil feedstocks due to its seeds' remarkably high oil content (48–60%), its potential for producing large quantities of oil (500–1,000 L per acre), and its distinctive capability to produce oils with exceptionally high levels (80–90%) of ricinoleic acid (Brigham 1993). Castor bean is both self- and cross-pollinated by wind, varying from 5 to 36% depending on the weather conditions. Pollen sheds readily between 26 and 29°C, with a relative humidity of 60%. For single cross-hybrid seed production, strains giving a 1:1 ratio or pistillate and heterozygous monoecious plants are used, the latter being rogued one to five days before flowering begins. Three-way cross-hybrids can also be used. For open-pollinated types, roguing of all off-types is done after the last cultivation, and for pure seed production, isolation necessity depends on the wind velocity. Stands of hybrid and open-pollinated varieties are typically separated by distances ranging from 300 to 720 m. However, in areas with lower wind velocity, shorter distances might be adequate. The specific pollination requirements and agents remain unidentified. Purseglove (1968) thought that although castor is mostly wind-pollinated, insects may play some part, as extrafloral glands on leaves below the flowers produce much nectar. However, Alex (1957a) reported that, although honeybees visit staminate flowers of castor and obtain large amounts of pollen, tests showed they are of no value as pollinators. Stafford (1976) reported that castor is largely cross-pollinated.

3.9 COTTONSEED (*GOSSYPIUM* SPP., FAMILY MALVACEAE)

During the middle of the season, honeybees tend to visit flowers of cotton at a relatively low rate, while at the beginning and end of the season, their visitation rate is considerably higher, as indicated by multiple observations. In cotton, nectar is profuse, secreted by one floral and four extrafloral sets of nectaries. The floral nectar inside the base of petals is difficult to reach by small insects, but the extrafloral nectar attracts many insects. Nectar has sugar concentration of 18–24% in various cultivars and, owing to the evaporation concentration, may reach 20–54% in floral and 60–82% in extrafloral nectars (Vansell 1944). Mound (1962) observed that leaf nectaries secrete at different times of the day, viz, in the morning hours and in the afternoon. Sugar concentration reaches the maximum between 1400 and 1600 h.

A total of 41 species from 23 families and 7 orders were documented in Ludhiana as visitors to *G. hirsutum*. Many of these species either did not visit the floral nectaries or avoided contact with the sexual parts of the flowers. It was observed that each flower received visits from an average of 47 insects, with *A. dorsata*, *A. cerana*, and *A. florea* being the key pollinators. These bees visited flowers at rates of 6.7, 6.6, and 5.8 per minute, respectively, and made 141, 128, and 120 flower visits per colony. Notably, only experimental colonies of *A. cerana* were accessible during the study conducted by Sidhu and Singh (1961).

On *G. arboretum*, *A. cerana*, *A. dorsata*, and *A. mellifera* formed 32% of insect visitors, and solitary bees were 27%; the three honeybees visited 104–124, 94–102, and 87–99 flowers per trip, respectively. The plants caged with honeybees (*A. cerana* or *A. mellifera*) gave 10% more bolls and 15% more yield and 17% more seed as compared with those caged without bees. It was observed that *A. cerana* did not touch the anthers but brushed part of the stigma in 60% of the visits, whereas *A. mellifera* touched anthers and brushed past stigma in 70% of the visits (Tanda 1983, 1984; Tanda and Goyal 1979a, 1979b), indicating the possibilities of cross-pollination. *A. mellifera* was present with beekeepers, and *A. cerana* came only from experimental colonies.

3.10 NIGER (*GUIZOTIA ABYSSINICA* CASS, FAMILY COMPOSITAE)

Niger blooms from September to October for over a month and produces deep-yellow flowering heads. A flowering head consists of two types of florets: ray florets and disc florets. Ray florets are a conspicuous yellow and consist of an inferior ovary without stamens or pistils. Disc florets are hermaphrodite (bisexual) and consist of five united petals, five stamens with united anthers, and a pistil having a one-chambered ovary with one ovule, a style, and a bifid stigma. Disc florets produce plenty of nectar and pollen (Pandey and Dhakal 2004). Pollination is accomplished by insects, particularly honeybees. Honeybee pollination increases both yield and quality of seed. To produce high yields with high oil content, place three to five colonies of *Apis mellifera* or six to eight colonies of *Apis cerana* evenly per hectare. It is a comparatively new crop in South India and provides forage to honeybees during the dearth period. It covers an area of 4.8 lac ha, and seed production is 1.2 lac tons. It is grown in Madhya Pradesh, Maharashtra, Bihar, Orissa, and Tamil Nadu.

The composite head has 40–60 tubular hermaphrodite flowers and a row of marginal florets. Spread over seven to eight days, the flowers in the outer rows open and dehisce in the morning, and the style lobes of stigma grow during the day and curl backward but do not touch their own pollen; therefore, cross-pollination is common through external agencies like insects. In an experiment in Maharashtra, 40 flower heads in two plots were caged, one with *Apis cerana* and the other without bees; in the former, on average, 40 seeds were produced, but in the latter, there were only 15 (Bhambure 1958). In another experiment, yield was compared under different situations; in measured plots without *A. cerana* bees, it was 0.9 kg, those caged with bees 1.2 kg, and in the open with natural pollinators was 1.1 kg (Panda et al. 1988). In the open, bees were observed at 1.8 to 8.1/m^2 per minute. It indicates that there is some self-fertilization, but the crop benefits from pollinators that bring about cross-pollination.

3.11 TARAMIRA (*ERUCA SATIVA* LAM, FAMILY CRUCIFERAE)

This oilseed crop is grown under unirrigated conditions by itself or mixed with chickpea. Its flower opens in the morning and remains so for three days. Anthers dehisce as soon as the flower opens, and nectar is available sometimes later on; therefore, pollen gatherers visits first and nectar gatherers follow them. Cobley (1956) reported that taramira has been identified as self-sterile, relying on external agents for seed production. Sihag (1986) conducted a study to determine its pollination requirements. In his research, Sihag enclosed some plants in cages to prevent insect pollination and compared them with plants exposed to natural pollinators. The results showed that the enclosed plants produced 79 pods with an average of 2.2 seeds per pod, while the exposed plants yielded 1,075 pods, with an average of 15.5 seeds per pod. Additionally, the enclosed plants produced 4 g of seeds per plant, whereas the exposed plants produced 321 g of seeds per plant.

Insect visitors were surveyed (58 samples of 30 sweeps each) in Punjab (including Haryana and Himachal Pradesh) for three seasons during 1965–68. Of all the visitors, *Apis florae* (50.8%) was the most numerous, followed by *Apis dorsata* (21.2%). *Andrena ilerda* and allied species (18.5%) and *Colletes* spp. (2.2%) and all the rest of bees (6.9%) and Halictidae were also recorded, but the presence of Colletes on this crop was unique. On the whole, population density was less than half of that recorded on other oilseed Cruciferae. In the drier southern district, *Apis florea* held first position, and *Andrena* spp. The second position. However, in central districts, *Apis dorsata* was the most abundant, followed by *A. florea* and *Andrena* (Atwal 1970; Kapil et al. 1971). It was reported that *Andrena* in West Punjab had comparatively higher population, perhaps owing to the drier climate there. The overall density of population on taramira was 0.9 insect per net sweep in East Punjab.

3.12 LINSEED/FLAX (*LINUM USITATISSIMUM* L., FAMILY LINACEAE)

Linseed is cultivated globally for the purpose of extracting linseed oil and linseed meal, which are utilized in livestock feed. Apart from being known as flaxseed, this plant is cultivated in various countries for its fibrous stalk, which is used in the production of linen. Honeybees derive a limited amount of pollen and nectar from flax, and the extent of their visitation and collection depends on the specific region and the presence of alternative floral sources. According to Scullen and Vansell (1942), flax is considered a relatively poor source of both pollen and nectar. Alex (1957) concluded that bees primarily collect pollen, while Smirnov (1954) suggested that their main focus is nectar.

Pellett (1947) mentioned that numerous bees are observed on flax only in areas where there is a high concentration of bee colonies nearby.

Flax is typically regarded as a predominantly self-pollinating plant, although there is some degree of cross-pollination (ranging from 1 to 6%) that occurs, especially among the larger-flowered varieties (Dillman 1938; Dillman and Goar 1937; Masuo 1958). Rubis (1970) conducted research using a male sterile line with disk-shaped flowers and noted minimal cross-pollination between the male sterile lines and lines with tubular flowers. However, he observed significant cross-pollination, resulting in good seed production, when other lines with large disk-shaped flowers were involved. Multiple studies have demonstrated that bee pollination enhances seed yields in fiber flax (Alles 1977). Bezdenezhnykh (1956) conducted experiments in Russia and reported a 22.5% increase in seed production when honeybees were introduced in cages compared to plants in bee-less cages. Gubin (1945) also investigated the impact of bee pollination on fiber flax in Russia and found that bees increased seed production by 22.5 to 38.5%. Luttso (1957), another researcher from Russia, reported a 29% increase in seed production, an 18% increase in the number of seeds per capsule, and an 11% increase in seed weight when bee pollination was present, compared to fields without bee pollination

Hassanein (1955) concluded that honeybee pollination improved both the quantity and quality of seeds. However, Pritsch (1965) and Alex (1957) did not observe any significant benefits in terms of increased seed production, and insect pollination is generally considered unnecessary for flax grown in the United States. Alles (1961) determined that the number of honeybee visits per flax flower significantly influenced seed set and weight, but no specific quantity of bees required was specified.

3.13 COCONUT (*COCOS NUCIFERA* L., FAMILY PALMACEAE)

Coconut is one of the most important crops throughout the tropics (Ohler 1999), and in some areas, it is cultivated far inland. In palms, pollination is typically carried out by insects, specifically bees (mellitophily), flies (myophily), and beetles (cantharophily) (Henderson 1986), according to Silberbauer-Gottsberger (1990). Coconut palms are monoecious, producing large inflorescences consisting of numerous unisexual flowers that offer pollen and nectar or only nectar to insect pollinators, particularly bees (pistillate flowers). Like other mellitophilous palms, coconut inflorescences are protandrous, releasing pollen before the pistillate flowers become receptive (McGregor 1976; Free 1993; Ashburner 1995; Roubik 1995; León 2000). However, in most tall ecotypes, there is an overlap between the staminate and pistillate phases (Sholdt and Mitchell 1967; Rognon 1976). Throughout tropical Asia and other regions, various insect species have been observed visiting *C. nucifera* flowers, with bees generally considered the primary pollinators (Sholdt and Mitchell 1967; Jay 1974; McGregor 1976; Free 1993; Macfarlane 1995; Heard 1999). Bees from the *Apis* spp. and Meliponinae families have been documented in numerous countries (McGregor 1976; Engel and Dingemans-Bakels 1980; Pardede et al. 1986; Kevan 1989; Free 1993). While other insects such as wasps, ants, flies, butterflies, black earwigs, and beetles are commonly found on *C. nucifera* flowers, they are not considered significant in pollination (Sholdt and Mitchell 1967).

Whitehead (1965) asserted that insect pollination was predominant in coconuts. However, Huggins (1928) regarded honeybees and various other Hymenoptera as important pollinators, excluding ants. Ochse et al. (1961) recognized both insect and wind pollination. In a study conducted in Hawaii, Sholdt (1966) identified 51 insect species on coconut inflorescences, with ants, bees, earwigs, flies, and wasps being the most frequently encountered. Sholdt and Mitchell (1967) emphasized the importance of both wind and insects in pollination, highlighting honeybees as the most significant insects due to their frequent visits to flowering inflorescences, which resulted in high nut yields. They even suggested introducing honeybee colonies into coconut plantations.

The male flowers are in thousands, and only a few or even one female flower may be present at the base of spadix. The male flowers on opening shed pollen in a day or two, and all of them dehisce about a month earlier than the female flowers open at the base of spadix. The female flowers open three to six days after the male phase. There is nectar both in male and female flowers, and insects visit them. In the tall cultivars, male flowers dehisce at various intervals on different trees so that pollen is available when female flowers open. With the aid of insects, both self-pollination and cross-pollination occurs, but wind plays a very little role. Since pollen remains viable for two to nine days, and with proper technique, it can be stored for several months; hand-pollination is successful particularly in dwarf varieties. In different parts of the world, honeybees (*A. cerana* and *A. mellifera*), *Trigona* sp., wasp (*Polistes* spp.), and ants forage primarily on pollen throughout the year. It has been observed that wasps tend to drive away solitary bees and honeybees (Free et al. 1975). If female flowers are bagged, the fruit set is 3–4%, as against the expected 22% of flowers in the open. Sugar concentration in nectar was recorded 17–23%, and Jay (1974) observed a honeybee visited 20 female flowers in 3 min; 67–99% of the honeybees carried pollen on their bodies. They collected pollen all day, and there were two peak periods, at 1000–1100 h and again in the afternoon. Coconut pollen plays an important role in successful beekeeping near rubber plantations in Malaysia and South India (Mardan and Osman 1982). Meléndez-Ramírez et al. (2004) found that honeybees (*Apis mellifera*) visited both pistillate and staminate flowers and were the most efficient pollinators, as they carry pollen on their ventral surface. Ants were present on flowers day and night but had no effect on pollination. Self-pollination by geitonogamy produced almost 19% of the fruit set, but cross-pollination (xenogamy) was the most important contribution (c. 30%). Anemophilous cross-pollination only accounted for 10% of fruit set.

3.14 OIL PALM (*ELAEIS GUINEENSIS* JACQ, FAMILY PALMAE)

The male and female inflorescences are separate but on the same palm; trees flower at different times, spread over months. Male flower on a spikelet start opening at the base and complete the process in two to four days, producing 25–50 g pollen per inflorescence. In Malaysia, the female inflorescence of the oil palm consists of approximately 200 spiral-arranged spikelets, each bearing 10–30 flowers. These flowers gradually open over the course of one week, and their stigma remains receptive for 36–48 h. The recent introduction of the weevil species *Elaeidobius kamerunicus* Faust (Curculionidae) has effectively facilitated the pollination of oil palm in Malaysia. This weevil, originally from West Africa (the native region of oil palm), was brought to Malaysia specifically for the purpose of pollination, as it is the natural pollinator of the oil palm species (Syed et al. 1982; Kevan et al. 1986).

Syed (1979, 1982) in Malaysia has proven that profuse light pollen is not carried by wind for pollination, but insects carry it and deposit it on stigma. It was observed that 88% of the insects on palm belonged to the groups *Apis cerana*, *Apis dorsata*, *Apis florea*, and *Trigona* sp., and only 5% of them were recorded on female flowers. He also observed that thrips played an important role in cross-pollination, but the greatest success was achieved by the introduction from Cameroon (Syed 1982). The kernel and oil yield has now increased from 20 to 53%. The beetles are perhaps also present in Kerala, India. Oil palm is an important source of pollen for honeybee colonies. In oil palm, much research has focused on the uses of focal species for pollination (e.g., the pollinating weevil *Elaeidobius kamerunicus*) (Greathead 1983).

3.15 OLIVE (*OLEA EUROPAEA* L., FAMILY OLEACEAE)

Olive trees flower profusely, and adequate commercial crops are obtained when only 1–5% of the flowers set fruit. The anthers of most hermaphrodite flowers are sufficiently close to the stigmas so that on their dehiscence, pollen falls or are thrown onto the stigma and self-pollination is accomplished. However, other cultivars have spreading anthers, and in these flowers autopollination is not ensured. Some cultivars are self-sterile, others partially self-sterile, and others self-fertile. Even some self-fertile cultivars have been reported to benefit from cross-pollination

(Bradley et al. 1961; Hartmann and Opitz 1966; Griggs et al. 1975), so it is usually necessary or advisable to interplant different cultivars in the same plantation. Lavee and Datt (1978) estimated that in good weather conditions, pollen is adequately distributed up to 40 m from its source in the direction of the prevailing wind, but little to windward.

Honeybees collect pollen rather sparingly from the olive even though it is present in great abundance at flowering time. Sometimes, an olive honey flow is reported by beekeepers. They found that if pollen tube growth was too slow, the embryo sac began to degenerate before the tube reached it; therefore, no fertilization would result. They concluded that "the chances of fertilization were greater in cross than in self-pollinations, as indicated by the higher percentages of pistils in which a pollen tube reached the embryo sac."

It is generally supposed that olive flowers are wind-pollinated, and although honeybees often visit the flowers to collect pollen, they are not considered essential for pollination (Morettini 1950; Griggs et al. 1975). However, insect pollination may supplement wind pollination. In one experiment, Griggs et al. (1975) found that olive tree branches bagged to exclude bees had a high fruit set. But in another experiment, adjacent trees of cultivars "Mission" and "Oblonga" exposed to honeybee visits had a greater set (7.4 and 18.8%, respectively) than the same two cultivars caged to exclude bee visits (1.8 and 8.6%) or caged separately in the open and so subject to wind pollination (4.0 and 9.4%) or in a greenhouse (2.3 and 3.7%). More experiments of this type, but with better controls, are needed. Possibly, in the absence of adequate wind pollination, bee pollination may contribute to yield.

Even self-compatible varieties give enhanced yields and improve qualitatively when cross-pollinated (Table 3.3). Therefore, pollination in oil crops assumes utmost significance for boosting production of oil crops in a world deficient of oil/oilseeds.

TABLE 3.3
Seed Yield and Yield Parameters in Different Oilseed Crops as Influenced by Bee Pollination*

Crops (Variety)	Yield Attributes	Pollination: With Bees	Pollination: Without Bees
Mustard (var. M-27)	Pod set (%)	71.90	48.60
	Seeds/pod	10.80	5.90
	1,000 seeds weight (g)	5.00	2.00
	Seed yield (q/ha)	13.90	1.20
	Oil content (%)	36.40	32.60
Niger (var. M-15)	Seed set (%)	45.80	28.40
	Seed yield (q/ha)	3.79	3.11
	Total florets	45.75	33.42
	Oil content (%)	35.20	32.00
	Average seed set %	23.80	14.18
	Seed weight (g/head)	0.80	0.07
Sesame (var. Kalika)	Average yield (q/ha)	8.47	5.64
	Oil content (%)	36.80	36.40
Sunflower (var. modern)	Seed yield (g/head)	23.80	11.60
	No. of seeds/head	540.80	418.25
	Filled seed/head (%)	62.40	58.65
	Seed yield (q/ha)	27.54	9.78

* Compiled from different sources.

Abel, C. A., Wilson, R. L., and Luhmann, R. L. 2003. Pollinating efficacy of *Osmia cornifrons* and *Osmia lignaria* subsp. *lignaria* (Hymenoptera: Megachilidae) on three Brassicaceae species grown under field cages. *J. Kans. Entomol. Soc.* 38, 545–552.

Abrol, D. P. 1991. Path analysis of environmental factors influencing daily flight activity of Apis dorsata F. *Acta Ecol. Oecol. Gen.* 12(6), 619–826.

Abrol, D. P. 2007. Foraging behaviour of Apis mellifera and Apis cerana as determined by the energetics of nectar production in different cultivars of Brassica campestris var. toria. *Journal of Apicultural Science*, 51(2), 5–10.

Abrol, D. P. 2008. Plant insect interaction in crucifers. In: Dr. Gupta, S. K. (ed.) *Biology and Breeding of Crucifers*. CRC Press, Taylor and Francis, Boca Raton, FL, pp. 129–149.

Abrol, D. P. 2009. *Bees and Beekeeping in India*, 2nd edn. Kalyani Publishers, New Delhi, p. 720.

Adegas, J. E. B. and Nogueira Couto, R. H. 1992. Entomophilous pollination in rape (*Brassica napus* L. var. *oleifera*) in Brazil. *Apidologie* 23, 203–209.

Aguilar, R., Ashworth, L., Galetto, L., and Aizen, M. A. 2006. Plant reproductive susceptibility to habitat fragmentation: Review and synthesis through a meta-analysis. *Ecol Lett.* 9(8), 968–980.

Alex, A. H. 1957. Pollination of some oilseed crops by honeybees. Progr. Rep. Texas Agric. Exp. Stn., No. 1960.

Alles, P. T. 1961. Influence of the mixtures of flower pollen, of microelements and of vitamin b in the process of formation of the seeds of different varieties of sunflower and flax. 18th Internatl. Apic. Cong. Proc., Summaries of the Congress, Madrid, pp. 5–6.

Alles, P. T. 1977. Effect of intra- and inter-verietal pollination on the quality and yield of fiber flax seed. In: Mel'nichenko, A. N. (ed.) *Pollination of Agricultural Crops by Bees*, Vol. 3. Amarind Pub. Co., New Delhi, India, pp. 223–229.

Amjad, M. 2010. Canola variety guide in WA 2010. Canola Research Officer, Department of Agriculture and Food, Western Australia. Available at http://agric.wa.gov.au/objtwr/imported_assets/content/fcp/gmcrops/fn406_canolavariety.pdf

Araneda, X. D., Ulloa, R. B., Carrillo, J. A., Contreras, J. L., and Bastidas, M. T. 2010. Evaluation of yield component traits of honeybee pollinated (*Apis mellifera* L.) rapeseed canola (*Brassica napus* L.). *Chilean Journal of Agricultural Research* 70(2), 309–314.

Ashburner, G. R. 1995. Reproductive biology of coconut palms. In: Oropeza C. S., Howard, F. W., and Ashburner, G. R. (eds.) *Lethal Yellowing: Research and Practical Aspects*. Kluwer Academic Publishers, Dordrecht, The Netherlands, pp. 111–121.

Ashman, T.-L., Knight, T. M., Steets, J. A., Amarasekare, P., Burd, M., Campbell, D. R., and Dudash, M. R. 2004. Pollen limitation of plant reproduction: Ecological and evolutionary causes and consequences. *Ecology* 85, 2408–2421.

Atmowidi, T., Buchori, D., Manuwoto, S., Suryobroto, B., and Hidayat, P. 2007. Diversity of pollinator insects in relation to seed set of mustard (Brassica rapa L.: Cruciferae). *HAYATI Journal of Biosciences* 14(4), 155–161.

Atwal, A. S. 1970. *Insect Pollinators of Crops*. Punjab Agricultural University Press, Ludhiana.

Bai, A. R. K. and Reddy, C. C. 1977. Laboratory toxicity of some insecticides to *Apis cerana indica*. *Journal of Apicultural Research* 16(3), 142, 161–162.

Bezdenezhnykh, S. M. 1956. Pollination of fiber flax and training of bees. In: Krishchunas, I. V. and Gubin, A. F. (eds.) *Pollination of Agricultural Plants*. Moskva, Gos. Izd-vo. Selkhoz Litry, pp. 21–24.

Bhalla, O. P., Verma, A. K., and Dhaliuwal, H. S. 1983. Insect visitors of mustard bloom (*Brassica campestris var* sarson), their number and foraging behaviour under mid-hill conditions. *J. Ento. Res.* 7, 15–17.

Bhambure, C. S. 1958. Effect of honey bee activity on niger seed production. *Indian Bee J.* 20, 189–191, 195.

Bichee, S. L. and Sharma, M. 1988. Effect of different modes of pollination in sunflower *Helianthus annuus* L. (Compositae) sown on different dates by *Trigona iridipennis* Smith. *Apiacta* 23, 65–68.

Bisht, D. D., Naim, M., and Mehrotra, K. N. 1980. Studies on the role of honeybees in rapeseed production. In: Proc. 2nd Int. Conf. Apiculture in Tropical Climate, Indian Agricultural Research Institute, New Delhi, pp. 491–496.

Bradley, M. V., Griggs, W. H., and Hartmann, H. T. 1961. Studies on self- and cross-pollination of olives under varying temperature conditions. *Californian Agriculture* 15, 4–5.

Brigham, R. D. 1993. Castor: Return of an old crop. In: Janick, J. and Simon, J. E. (eds.) *Progress in New Crops*. John Wiley & Sons, New York, pp. 380–383.

Bürger, C. 2004. Effects of landscape structure on bee diversity and pollination at different spatial scales. PhD thesis. Institute of Agroecology, University of Göttingen, Germany.
Burmistrov, A. N. 1965. The melliferous value of some sunflower varieties. 20th International Apicultural Congress of Apimondia Bucharest, Apimondia Publishing House, Bucharest, Romania, pp. 320–323.
Charlesworth, D. and Charlesworth, B. 1987. Inbreeding depression and its evolutionary consequences. *Annual Review of Ecology and Systematics* 18, 237–268.
Chhuneja, P. K., Singh, J., Gatoria, G. S., and Blossom, K. 2007. Assessing the role of honey bee (*Apis mellifera* L.) in the seed production of *Brassica campestris* var. toria in the Punjab. *Indian Journal of Crop Science* 2(2), 34–40.
Chiari, W. C., de Alencar Arnaut de Toledo, V., Ruvolo-Takasusuki, M. C. C., Attencia, V. M., Costa, F. M., Kotaka, C. S., Sakaguti, E. S., and Magalhaes, H. R. 2005a. Floral biology and behavior of Africanized honeybees *Apis* mellifera in soybean (*Glycine max* L. Merril). *Braz. Arch. Biol. Technol.* 48, 367–378.
Chiari, W. C., de Alencar Arnaut de Toledo, V., Ruvolo-Takasusuki, M. C. C., de Oliveira, A. J. B., Sakaguti, E. S., Attencia, V. M., Costa, F. M., and Mitsui, M. H. 2005b. Pollination of soybean (*Glycine max* L. Merril) by honey bees (*Apis mellifera* L.). *Braz. Arch. Biol. Technol.* 48, 31–36.
Chiari, W. C., Toledo, V. D. A. A. D., Ruvolo-Takasusuki, M. C. C., Oliveira, A. J. B. D., Sakaguti, E. S., Attencia, V. M., Costa, F. M., and Mitsui, M. H. 2005. Pollination of soybean (*Glycine max* L. Merril) by honeybees (*Apis mellifera* L.). *Brazilian Archives of Biology and Technology* 48, 31–36.
Cobley, L. S. 1956. *An Introduction to the Botany of Tropical Crops*. Longmans, London.
Cockerell, T. D. A. 1914. Bees visiting *Helianthus*. *Canad. Ent.* 46, 409–415.
Corbet, S. A., Williams, I. H., and Osborne, J. L. 1991. Bees and the pollination of crops and wild flowers in the European Community. *Bee World* 72, 47–59.
Crane, E. 1991. *Apis* species of tropical Asia as pollinators and some rearing methods for them. *Acta Hort.* 288, 29–48.
Crane, E. and Walker, P. 1983. Pollination of tropical and subtropical crops by bees. In: Crane, E. and Walker, P. (eds.) *The Impact of Pest Management on Bees and Pollination*. IBRA, Cardiff, pp. 5–21.
Cresswell, J. E., Davies, T. W., Patrick, M. A., Russell, F., Pennel, C., Vicot, M., and Lahoubi, M. 2004. Aerodynamics of wind pollination in a zoophilous flower, *Brassica napus*. *Functional Ecology* 18, 861–866.
Cresswell, J. E. and Osborne, J. L. 2004. The effect of patch size and separation on bumblebee foraging in oilseed rape: Implications for gene flow. *Journal of Applied Ecology* 41, 539–546.
Crossa-Raynaud, P. 1984. Quelques productions fruitières dépendant d'une pollinisation anémogame: Noyer, noisetier, olivier, palmier-dattier, pistachier. In: Pesson, P. and J. Louveaux (eds.) *Pollinisation et productions végétales*. INRA, Paris, France.
Da Conceicao, E. S., Delabie, J. H. C., and De Costa Neto, A. 2004. The entomophily of the coconut tree in question: The evaluation of pollen transportation by ants (Hymenoptera: Formicidae) and bees (Hymenoptera: Apoidea) in inflorescence. *Neotropical Entomol.* 33, 679–683.
Dag, A., Lior, E., and Afik, O. 2002. Pollination of confection sunflowers (*Helianthus annuus* L.) by honey bees (*Apis mellifera*). *Am. Bee J.* 142, 443–445.
Daily, G. C. 1997. *Nature's Service: Social Dependence on Naturalecosystem*. Island Press, Washington, DC. ISBN 1-55963-476-6, pp. xx and 392.
DeGrandi-Hoffman, G. and Martin, J. H. 1993. The size and distribution of the honey bee (*Apis mellifera* L.) cross-pollinating population on male sterile sunflowers (*Helianthus annuus* L.). *J. Apic. Res.* 32, 135–142.
DeGrandi-Hoffman, G. and Watkins, J. C. 2000. The foraging activity of honey bees *Apis mellifera* and non-*Apis* bees on hybrid sunflowers (*Helianthus annuus*) and its influence on cross-pollination and seed set. *J. Apic. Res.* 39, 37–45.
Delaplane, K. S. and Mayer, D. F. 2000. *Crop Pollination by Bees*. CABI Publishing, New York, USA.
Deodikar, G. B. and Suryanarayana, M. C. 1972. Crop yields and bee pollination. *Indian Bee J.* 34, 53–64.
Deshmukh, A. K., Mohana Rao, G., and Karve, A. D. 1985a. Studies on the effect of honeybee pollination on the yield of safflower. *Indian Bee J.* 33(3–4), 35–38.
Deshmukh, A. K., Rao, G. M., and Karve, A. D. 1985b. Studies on the effect of honey bee pollination on the yield of safflower. *Indian Bee J.* 47, 1–2.
Dhileepan, K. 1994. Variation in popuations of the introduced pollinating weevil (*Elaeidobius kamerunicus*) (Coleptera; Curculionidae) and its impact on fruit set of oil palm (*Elaeis guineensis*) in India. *Bull. Entomol. Res.* 84, 477–485.

Dillman, A. C. 1938. Natural crossing in flax. *J. Am. Soc. Agron.* 30, 279–286.

Dillman, A. C. and Goar, L. G. 1937. Flaxseed production in the far western states. U.S. Dept. Agr. Farmers' Bul. 1792, p. 22.

Downey, R. K. and Bolton, J. L. 1961. Production of [Polish and Argentine] rape in Western Canada. Canada Dept. Agr. Res. Br. Pub. 1021, p. 19.

Engel, M. S. and Dingemans-Bakels, F. 1980. Nectar and pollen resources for stingless bees (Meliponinae, Hymenoptera) in Surinam (South America). *Apidologie* 11, 341–350.

Erickson, E. H., Berger, G. A., Shannon, J. G., and Robbins, J. M. 1978. Honey bee pollination increases soybean yields in the Mississippi Delta region of Arkansas and Missouri. *J. Econ. Entomol.* 71, 601–603.

Ernst, W. R., Pearce, P. A., and Pollock, T. L. 1989. Environmental effects of Fenitrothion use in forestry. Conservation and Protection, Environment Canada, Atlantic Region, Canada.

Free, J. B. 1970. *Insect Pollination of Crops.* Academic Press, New York.

Free, J. B. 1993. *Insect Pollination of Crops* (2nd ed.). Academic Press, Harcourt Brace Jovanovich Publication, London, 684 p.

Free, J. B. 1999. Pollination in the tropics. *Beekeeping and Development* 51, 6–7.

Free, J. B. and Ferguson, A. W. 1983. Foraging behaviour of honeybees on oilseed rape. *Bee World* 64, 22–24.

Free, J. B. and Ferguson, A. W. 1986. Foraging of bees on oil-seed rape (*Brassica napus* L.) in relation to the flowering of the crop and pest control. *Journal of Agricultural Science, Cambridge* 94: 151–154.

Free, J. B., Raw, A., and Williams, I. H. 1975. Pollination of coconut *(Cocos nucifera* L.) in Jamaica by honeybees ahd wasps. *Appl. Anim. Ethol.* 1, 213–223.

Free, J. B. and Simpson, J. 1964. The pollination requirements of sunflowers *(Helianthus annuus* L.). *Emp. J. Exp. Agric.* 32, 340–342.

Free, J. B. and Williams, I. H. 1976. Pollination as a factor limiting the yield of field bean *(Vicia faba* L.). *J. Agric. Sci., Camb.* 87, 395–399.

Free, J. B. and Williams, I. H. 1977. *The Pollination of Crops by Bees.* Apimondia and International Bee Research Organisation, Bucharest, Romania.

Furgala, B. 1954a. The effect of the honey bee *Apis mellifera* (L.) on the seed set, yield and hybridisation of the cultivated sunflower, *Helianthus annus* L. Ph.D. thesis. Manitoba University, Canada.

Furgala, B. 1954b. Honey bees increase seed yields of cultivated sunflowers. *Glean. Bee Cult.* 82, 532–534.

Furgala, B., Noetzel, D. M., and Robinson, R. G. 1979. Observations on the pollination of hybrid sunflowers. *Proc. IVth Int. Symp. on Pollination, Md Agric Exp Sta Spec Misc Publ* 1, 45–48.

Gibbons, R. W. and Tattersfield, J. R. 1969. Out-crossing trials with groundnuts (arachis hypogaea L.). *Rhodesia Zambia Malawi Jour. Agr. Res.* 7, 71–75.

Girardeau, J. H. and Leuck, D. B. 1967. Effect of mechanical and bee tripping on yield of the peanut. *J. Econ. Ent.* 60, 1454–1455.

Goyal, N. P. and Atwal, A. S. 1973a. Ecology of sunflower pollination. *Indian Oilseed J.* 3(2), 30–37.

Goyal, N. P. and Atwal, A. S. 1973b. Role of honeybees in sunflower pollination. *Indian Bee J.* 35, 41–47.

Graham, E. H. 1941. *Legumes for Erosion Control and Wildlife.* U.S. Department of Agriculture Misc. Pub. 412, Washington, DC, p. 153.

Greathead, D. J. 1983. The multi-million dollar weevil that pollinates oil palms. *Antenna* 7, 105.

Greenleaf, S. A. and Kremen, C. 2006b. Wild bees enhance honey bees' pollination of hybrid sunflower. *Proc. Nat. Acad. Sci. USA*, 103, 13890–13895. doi: 10.1073/pnas.0600929103

Greenleaf, S. S. and Kremen, C. 2006. Wild bees enhance honeybees' pollination of hybrid sunflower. *Proc. Natl. Acad Sci. USA* 103, 13890–13895.

Griggs, W. H., Hartmann, H. T., Bradley, M. V., Iwakiri, B. T., and Whisler, J. E. 1975. Olive pollination in California. *Bul. Calif. Agr. Exp. Sta.* 869.

Gubin, A. F. 1945. Cross pollination of fibre flax. *Bee Wld.* 26, 30–31.

Guidry, N. P. 1964. Graphic summary of world agriculture. *U. S. Dept. Agr. Misc. Pub.* 705, 64.

Hammer, O. 1963. Summer rape as a competitor affecting the pollination of clovers. *Dansk Froavl* (14), 7.

Hammer, O. 1966. Some problems of competition between summer rape and clover, in relation to pollination. In *2nd Int. Symp. Pollination*, London, 1964. *Bee World* 47(Suppl. 1), 99–106.

Hammons, R. O., Krombein, K. V., and Leuck, D. B. 1963. Some bees (Apoidea) associated with peanut flowering. *J. Econ. Ent.* 56, 905.

Hartmann, H. T. and Opitz, K. W. 1966. Olive production in California. *California Agricultural Experimental Station and Extension Service Circular* 540.

Hassanein, M. H. 1955. The value of pollinating insects to flax seed production in Egypt. *Ann. Agric. Sci. Univ. A. Shams*, 773–784.

Heard, T. A. 1999. The role of stingless bees in crop pollination. *Ann. Rev. Entomol.* 44, 183–206.

Hegde, D. M. 2000. "Oilseeds Situation", 2000, Directorate of Oilseeds Research, ICAR, Hyderabad.

Hegde, D. M. 2004. *Safflower, Packages of Practices for Increasing Production* (ed. D. M. Hegde). DOR, Hyderabad.

Henderson, A. 1986. A review of pollination studies in the palmae. *Botanical Review* 52, 221–259.

Hoffmann, M. 1994. Observações sobre a polinização entomófila de *Helianthus annuus* L. em Viamão, Rio Grande do Sul. *An Soc. Entomol. Bras, Londrina* 23(3), 391–397. [In Portuguese].

Howard, A., Howard, G. L. C., and Abdur Rahman, K. 1916. Studies in Indian oil-seeds. I: Safflower and mustard. *Mem. Dep. Agric. India. Bot. Ser.* 7, 237–272.

Howard, A., Howard, G. L. C., and Rahman, K. A. 1919. Studies on the pollination of Indian crops. *Mem. Dept. Agric. Indian Bot. Ser.* 10, 195–220.

Hoyle, M., Hayter, K., and Cresswell, J. E. 2007. Effect of pollinator abundance on self-fertilization and gene flow: Application to GM Canola. *Ecol. Appl.* 17, 2123–2135.

Huggins, H. D. 1928. Pollination and crop production (concluded). *Agricultural Journal of British Guiana*, 1(90–94), 164–169.

Jay, S. C. 1974. Nectar and pollen collection by honeybees from coconut flowers. *Bee World* 55, 105–111.

Jyoti, J. and Brewer, G. J. 1999. Effect of honeybee (Hymenoptera: Apidae) pollination on sunflower hybrids. Proc. 21st Sunflower Research Workshop. Directorate of Oilseed Research, Nat Sunflower Assoc., Hyderabad, India, January 14–15, pp. 103–107.

Kadam, B. S. and Patankar, V. K. 1942. Natural cross-pollination in safflower. *Indian Jour. Genet and Plant Breed* 2, 69–70.

Kalmer, F. 1996. Sunflower pollinatrion in Czech Republic. Proc.7th International Symposiuym on Pollination, Lethbridge, pp. 407–412.

Kapil, R. P., Grewal, G. S., Kumar, S., and Atwal, A. S. 1971. Insect pollinators of rape-seed mustard. *Indian Journal of Entomology* 33, 61–66.

Kevan, P. G. 1989. Beekeeping for pollination of coconut and honey production in Republic of Maldives. Report on pollination experiments Ariyadhu. Unpublished report (TCP/MDV/8851), A. FAO, Rome.

Kevan, P. G., Hussein, M. Y., Hussey, N., and Wahid, M. B. 1986. Modelling the use of *Elaeidobius kamerunicus* for pollination of oil palm. *Planter* 62, 89–99.

Khan, R. B. and Dethe, M. D. 2004. Median Lethal Time of new pesticides to foragers of honeybee. *Pestology* 28(1), 28–29.

Kisselhagen, S. 1977. Seed yield is increased if there are bees on the rape. *Tidsskrift for Biavl*, 111, 66–67.

Klein, A. M., Steffan-Dewenter, I., and Tscharntke, T. 2003. Fruit set of highland coffee increases with the diversity of pollinating bees. *Proc R Soc B* 270, 955–961.

Klein, A. M., Vaissie' re, B. E., Cane, J. H., Steffan-Dewenter, I., Cunningham, S. A., Kremen, C., and Tscharntke, T. 2007. Importance of pollinators in changing landscapes for world crops. *Proc. R. Soc. B Biol. Sci.* 274, 303–313.

Kluser, S. and Peduzzi, P. 2007. *Global Pollinator Decline: A Literature Review.* UNEP/GRID, Geneva.

Koelling, P. D., Kenworthy, W. J., and Caron, D. M. 1981. Pollination of male-sterile soybeans, *Glycine max*, in caged plots. *Crop. Sci.* 21, 559–561.

Krantz, G. W. and Poinar, G. O. 2004. Mites, nematodes and the multimillion dollar weevil. *J. Nat. Hist.* 38, 135–141.

Kremen, C. et al. 2007. Pollination and other ecosystem services produced by mobile organisms: A conceptual framework for the effects of land-use change. *Ecology Letters* 10, 299–314.

Kremen, C., Williams, N. M., and Thorp, R. W. 2002. Crop pollination from native bees at risk from agricultural intensification. *PNAS* 99, 16812–16816.

Kubisova, S., Nedbalova, V., and Plesnik, R. 1980. Cinnost vcely medonosne *(Apis mellifera* L.) na repce *(Brassica napus* L. var. *napus). Pol'nohospoddrstvo* 26, 744–754.

Kumari, A. and Pandey, A. K. 2005. Pollination ecology of safflower (*Carthamus tinctorius* linn). *J. Ind. Bot. Soc.* 82, 56–61.

Langridge, D. F. and Goodman, R. D. 1974. A study on pollination of sunflowers (*Helianthus annuus*). *Aust. J. Exp. Agric. Anim. Husb.* 14, 201–204.

Langridge, D. F. and Goodman, R. D. 1975. A study on pollination of oilseed rape (*Brassica campestris*). *Austral. J. Exptl. Agric. Anim. Husband.* 15, 285–288.

Latif, A., Qayyum, A., and Abbas, M. 1960. The role of *Apis indica* in the pollination of toria and sarson (*Brassica campestris* var. Toria and dichotoma). *Bee World* 41, 283–286.

Lavee, S. and Datt, Z. 1978. The necessity of cross-pollination after fruit set of Manzanilla olives. *J. Hon. Sci.* 53, 261–266.

León, J. 2000. *Botánica de los Cultivos Tropicales*, 3rd edn. Agroaméricana, Instituto Interamericano de Cooperación para la Agricultura, San José, Costa Rica, pp. 370–378.

Lesik, F. 1953. Microfertilization increases the nectar secretion of honey plants. *Pchelovodstvo, Mosk.* 30, 42–45.

Leuck, D. B. and Hammons, R. O. 1965a. Pollen-collecting activities of bees among peanut flowers. *J. Econ. Ent.* 58, 1028–1030.

Leuck, D. B. and Hammons, R. O. 1965b. Further evaluation of the role of bees in natural cross-pollination of the peanut, *Arachis lypogaea* L. *Agron. J.* 57, 94.

Leuck, D. B. and Hammons, R. O. 1969. Occurrence of atypical flowers of some associated bees *(Apoidea)* in the peanut, *Arachis hypogaea* L. *Agron. J.* 61, 958–960.

Levin, M. D. 1984. Value of honeybee pollination to United States agriculture. *American Bee Journal* 124, 184–186.

Lingappa, S., Viraktamath, S. A., Vastrad, A. S., and Williams, R. 1999. Utilization of honey bee *Apis cerana* Fab. for pollination of watermelon and safflower. Proceedings of Apimondia, Vancouver, Canada, p. 235.

Luttso, V. P. 1957. The pollination of flax by honeybees. *Dokl. TSKhA* 30, 327–331.

Macfarlane, R. P. 1995. Applied pollination on tropical oceanic islands. In: Roubik, D. W. (ed.) *Pollination of Cultivated Plants in the Tropics*. FAO, Rome, pp. 40–49.

Manning, R. and Boland, J. 2000. A preliminary investigation into honey bee (*Apis mellifera*) pollination of canola (*Brassica napus* cv. Karoo) in Western Australia. *Aust. J. Exp. Agr.* 40, 439–442.

Mardan, M. and Osman, M. S. 1982. Beekeeping in coconut small holdings in Pontian, Johor, West Malaysia. Proc. 2nd Int. Conf. Apic. Trop. Climates, New Delhi, pp. 118–126.

Masuo, Y. 1958. On natural crossing in flax. *Crop Sci. Soc. Japan Proc.* 27, 321–323. [In Japanese.] *Abstract in Plant Breed.* 29(4), 817–818. Oct. 1959.

Mayer, D. F. 2003. Bee kills from pesticides and the expert witness. *American Bee Journal* 143(7), 563.

Mayfield, M. M. 2005. The importance of nearby forest to known and potential pollinators of oil palm (*Elaeis guineensis* Jacq.; Arecaceae) in Southern Costa Rica. *Econ. Bot.* 59, 190–196.

McGregor, S. E. 1976. Insect pollination of cultivated crop plants. In: *Agriculture Handbook*. United States Department of Agriculture, Agricultural Research Service, Washington, DC, p. 496.

Meléndez-Ramírez, V., Parra-Tabla, V., Kevan, P. G., Ramírez-Morillo, I., Harries, H., Fernández-Barrera, M., and Zizumbo-Villareal, D. 2004. Mixed mating strategies and pollination by insects and wind in coconut palm (*Cocos nucifera* L. (Arecaceae)): Importance in production and selection. *Agr. Forest Entomol.* 6, 155–163.

Melnichenko, A. N. 1976. Role of insect pollinators in increasing yields of agricultural plants. In: Kozin, R. B. (ed.) *Pollination of Entomophilous Agricultural Crops by Bees*. Amerind Publishing Co. Pvt. Ltd., New Delhi, p. 331.

Mishra, R. C., Kumar, J., and Gupta, J. K. 1988. The effect of mode of pollination on yield and oil potential of *Brassica campestris* L. var. *sarson* with observation on insect pollinators. *J. Apic. Res.* 27, 186–189.

Mitchener, A. V. 1950. Honeybees and sunflowers. *Ann. Con. Man. Agron.*, 27.

Mohammed, A. 1935. Pollination studies in tara *(Brassica napus* L. var. dichotsma Prain) and sarson *(Brassica campestris* L. var. sarson, Prain). *Indian J. Agric. Sci.* 5, 125–154.

Montgomery, B. E. 1958. Preliminary studies of the composition of some Indian nectars. *Proc. Indiana Acad. Sci.* 68, 159–163.

Morandin, L. A. and Winston, M. L. 2005. Wild bee abundance and seed production in conventional organic, and genetically modified canola. *Ecol. Appl.* 15, 871–881.

Moreti, A. C. C. C., Silva, E. C. A., Alves, M. L. T. M. F. et al. 1998. Observações sobre a polinização entomófila da cultura da soja (*Glycine max* Merril). *Boletim da Indústria Animal* 55(1), 91–94.

Moreti, A. C. de C. C., da Silva, R. M. B., da Silva, E. C. A. et al. 1996. Aumento na produção de sementes de girassol (Helianthus annuus) Pela ação de insetos polinizadores. *Sci. Agric.* 53, 280–284.

Morettini, A. 1950. *Oliviocoltura, Ramo Editoriale*. Degli Agricoltori, Roma, Italy.

Mound, L. A. 1962. Extra-floral nectaries of cotton and their secretions. *Emp. Cott. Grow. Rev.* 39, 254–261.

Murrell, D. C. and Nash, W. T. 1981. Nectar secretion by toria *(Brassica campestris L. v. toria)* and foraging behaviour of three *Apis* species on toria in Bangladesh. *J. Apic. Res.* 20, 34–38.

Nderitu, J., Nyamasyo, G., Kasina, M., and Oronje, M. L. 2008. Diversity of sunflower pollinators and their effect on seed yield in Makueni District, Eastern Kenya. *Spanish Journal of Agricultural Research* 6(2), 271–278.

Nogueira-Couto, R. H., Pereira, J. M. S., and De Jong, D. 1998. Pollination of *Glycine wightii*, a perennial soybean, by Africanized honey bees. *Journal of Apicultural Research* 3(4), 289–291.

Ochse, J. J., Soule, M. J., Dijkman, M. J., and Wehlburg, C. 1961. *Tropical and Subtropical Agriculture*, Vol. 2. The Macmillan Co, New York, USA.

Ohler, J. G. (ed.). 1999. *Modern Coconut Management: Palm Cultivation and Products*. FAO, The University of Wisconsin, Madison, UK.

Ohler, J. G. 1999a. Part I. The coconut palm and its environment: The coconut palm. In: Ohler, J. G. (ed.) *Modern Coconut Management: Palm Cultivation and Products*. FAO, The University of Wisconsin, Madison, UK, pp. 11–34.

Osborne, J. L. and Williams, I. H. 1996. Bumble bees as pollinators of crops and wild flowers. In: Matheson, A. (ed.) *Bumble Bees for Pleasure and Profit*. International Bee Research Association, Cardiff, UK, pp. 24–32.

Panchabhavi, K. S, Deviah, M., and Basavanna, G. P. C. 1976. The effect of keeping Apis cerana indica F. colonies on the seed set of sunflower, Helianthus annuus Linn. *Mysore J. Agric. Sci.* 10, 631–636.

Panda, P., Sontakke, B. K., and Sarangi, P. K. 1988. Preliminary studies on the effect of bee *(Apis cerana indica* Fabr) pollination on yield of sesamum and niger. *Indian Bee J.* 50, 63–64.

Pandey, A. K. and Dhakal, M. R. 2004. Some aspects of pollen biology and pollination ecology of Niger (*Guizotia abyssinica* Cass.). In: Srivastava, P. C. (ed.) *Vistas in Palaeobotany and Plant Morphology: Evolutionary and Environmental Perspectives*. Print House, Lucknow, pp. 415–428.

Pardede, D., Randriani, E., and Marfuati, A. 1986. Penyerbukan bunga kelapa di Sub Balai Penelitian Kelapa pakuwon. *Pemberitaan Penelitian. Tanaman Industri* 12, 26–35.

Parkash, R. and Kumaraswami, T. 1984. Toxicity of some insecticide spraying on hymenopertrans pollinators in new Brunswick. *Canadian Entomologist.* 112, 259–269.

Parker, F. D. 1981. How efficient are bees in pollinating sunflowers? *Journal of the Kansas Entomological Society* 54(1), 61–67.

Patil, J. A. and Chavan, V. M. 1948. Selfing methods in Safflower. *Ind. Oilseeds Jour.* 2, 10–12.

Pellett, F. C. 1947. *American Honey Plants*, 4th edn. Orange Judd Publ. Co., Inc., New York.

Phadke, K. G., Pant, N. C., and Sinha, R. B. P. 1967. The foraging behaviour of *Apis indica* F ((Hymenoptera: Apidae) visiting flowers of sesame *(Sesamum orientate* L.). *Indian J. Ent.* 29, 157–163.

PIBA. 2001. *Commodity Report: Positioning Australian Soybeans in a World Market*. Primary Industry Bank of Australia Limited, Canbbera, Australia.

Pimental, D., and 11 others. 1980. Environment and pesticides: A preliminary assessment. *Oikos.* 34(2), 126–140.

Pordel, M. R., Hatami, B., Mobli, M., and Ebadi, R. 2007. Identification of insect pollinators of three different cultivars of winter canola and their effect on seed yield in Isfahan. *J. Sci. Technol. Agric. Natur. Resour.* 10(4B), 413–426.

Porter, P. and Crompton, D. 2008. Canola. Varietal Trials Results. Minnesota Agricultural Experiment Station, University of Minnesota, Saint Paul, Minnesota, USA, pp. 52–53. Available at www.maes.umn.edu/08varietaltrials/canola.pdf

Potts, S. G. et al. 2010. Declines of managed honeybees and beekeepers in Europe? *J. Apic. Res.* 49, 15–22.

Prasad, D., Hameed, S. F., Singh, R., Yazdani, S. S., and Singh, B. 1989. Effect of bee pollination on the quantity and quality of rai crop (Brassica juncea Coss). *Indian Bee Journal* 51(2), 45–47.

Pritsch, G. 1965. Untersuchungen uber die Steigerung der Olfruchtertrage durch Honigbienen. *VM Prace vyzkum. Ustav. vcelaf. CSAZV* 4, 157–163.

Purseglove, J. W. 1968. *Tropical Crops: Dicotyledons*. Longmans, London.

Radaeva, E. N. 1954. Bee pollination increases the yield of sunflower seeds *(Helianthus annuus). Pchelovodstvo, Mosk.* 31, 33–38.

Radchenko, T. H. 1964. The influence of pollination on the crop and the quality of seed of winter rape. *Bdzhil'nitstvo* 1, 68–74.

Radford, B., Nelson, R., and Rhodes, J. 1979. Agents of pollination in sunflower crops in the central Darlingdowns, Queensland. *Aust. J. Exp. Agric. Anim. Husb.* 19, 565–569.

Rahman, K. A. 1940. Insect pollinators of toria (*Brassica napus* L. *var dichotoma) and Sarson* (brassica campestris L. var Sarson) at Lyallpur. *Indian J. Agric Sci.* 10, 422–447.

4 Brassicas

4.1 INTRODUCTION

Brassicas are one of the most popular oilseed crops of the world, which mainly comprises approximately 100 species, including species *Brassica napus* L., spp. oleifera, commonly known as oilseed rape, rapeseed, or canola. Brassica is a big group containing many species, such as rape (*Brassica napus*), sarson (*Brassica campestris* var. sarson), toria (*Brassica campestris* var. toria), Indian mustard or broad-leaved mustard or rai (*Brassica juncea*), white mustard (*Brassica alba*), and black mustard (*Brassica nigra*), which are grown widely as oilseed crops for different purposes and in different geographical locations throughout the world. The Brassicaceae family (formerly Cruciferae) consists of approximately 375 genera and 3,200 species of plants, the majority of which are found in the Northern Hemisphere. It is evident from the published literature of Downey (1983) and Prakash and Hinata (1980) that humans have used vegetable oil pressed from the seeds of the *Brassica* plants since thousands of years. Although it is age-old crops, three decades ago, Lamb (1989) has described that oilseed *Brassica* has become internationally important during the last 30 years. There is no exaggeration in present-day context to say oilseeds are still very important and remain until the civilization exits on planet Earth.

Under the name *rapeseed* and *mustard*, seven important annual oilseeds belonging to the Brassicaceae (Cruciferae) are grown in India. They are Indian mustard (*Brassica juncea* [L.] Czern. and Coss.), commonly called rai (raya or laha); the three ecotypes of Indian rape, *B. campestris* L. ssp. oleifera viz, toria, brown sarson (lotni and tora types), and yellow sarson; Swede rape or gobhi sarson (*B. napus* L.); Ethiopian mustard or karan rai (*B. carinata* Braun.); and taramira or tara (*Eruca sativa* Mill.). On the Indian subcontinent, *B. juncea* is the dominant species grown and, along with *B. rapa* (syn. *B. campestris* L.) and *B. napus* L., is an important source of edible oil in India. These species are regarded as of Asiatic origin. Besides, taramira, believed to be a native of southern Europe and North Africa, is grown in the drier parts of northwest India. Of late, two other exotics, namely, gobhi sarson and karan rai, are becoming popular with the farmers in the areas where winter spell is longer. The rapeseed-mustard groups of crops are largely grown under the hardy conditions of rainfed agriculture with low input management during *rabi* season but have a good inherent potential to convert natural resources into usable biological energy.

4.1.1 Global Scenario

Rapeseed and mustard crops are grown in 53 countries across all six continents worldwide. The global production of rapeseed and its oil ranges from approximately 38 to 42 million tons and 12 to 14 million tons, respectively. China (11–12 million tons), EU (10–13 million tons), Canada (6–7 million tons), and India (4–7 million tons) are the world's major producers; 4.5–6, 1.5, and 2.5 million tons of seed, oil, and meal are traded globally. Canada (2.5–3.5 million tons) and Australia (1.5 million tons) are major exporters, while Japan (2 million tons) and Mexico (1 million tons) are the major importers of seed. Canada is the major exporter of oil, while the United States is the major importer of oil.

Oilseed rape is now one of the major global sources of vegetable oil and the major crop grown in Europe for the production of vegetable oil. Worldwide, rapeseed production is experiencing a rapid growth. According to the FAO, there was a production of 36 million tons in the 2003–4 season, which increased to 46 million tons in 2004–5. In Europe, rapeseed cultivation is mainly focused on animal feed due to its high lipid and medium protein content. This makes it a popular choice for Europeans seeking to avoid the importation of genetically modified organisms (GMOs). Although

DOI: 10.1201/9781032656724-5

the Asian continent accounts for 59.1% of the total cultivated area, its contribution to global production is only 48.6%. In contrast, Europe contributes 29.7% to global production, despite its share being only 16.2% of the total global cultivated area. The yield of rapeseed varies significantly, ranging from 411 kg/ha in the Russian Federation to 3,528 kg/ha in France. Among the seven Asian countries listed in Table 4.1, China and India combined make up 95.4% of the total cultivated area and 96.7% of the rapeseed-mustard production in Asia. Rapeseed-mustard is cultivated on approximately 13% of arable land in India. The global production of rapeseed-mustard has been steadily rising in various countries, primarily driven by the growing demand for edible oils and related products. It is estimated that the demand for oilseeds in India will reach approximately 34 million tons by 2020, with rapeseed-mustard accounting for around 14 million tons (41%) of the total. Hence, it is evident that increasing the productivity potential of rapeseed-mustard genotypes is crucial for meeting these requirements.

4.1.2 Indian Scenario

The oleiferous *Brassicas*, including *B. juncea* (L) Czern (L.) *B. zrapa* (syn. *B. campestris* L.) and *B. napus* L., are the important sources of edible oil in India. India produces 4–7, 1–2, 2–4 million tons of seed, oil, and mustard and does not export or import seed and oil due to heavy domestic consumption. However, meal is exported. Rapeseed and mustard are cultivated in the rabi season (October to November, February to March). Production is highly dependent on the weather. The major growing states are Rajasthan (40%), Uttar Pradesh (16%), and Haryana and Punjab (15%).

4.2 ECONOMIC IMPORTANCE

Rapeseed canola is an oleaginous crop, the oil of which is the most valuable component of its seeds (Amjad 2010). The oil has a reduced amount of erucic acid (Mandal et al. 2002; Aytaç et al. 2006) and is regarded as one of the top-quality oils suitable for human consumption (Porter and Crompton 2008). The oil percentage ranges from 36% to 42%, with an average oil yield of 34% to 35%. The remaining is used as rapeseed meal cake. In the Indo-Gangetic plains of India, sarson and toria (*Brassica campestris* L. sarson; var. toria; Cruciferae) are important oilseed crops, and their leaves and soft shoots are eaten as vegetable. The crops are grown in multiples of half or full acres and quite often as distant rows as intercropping in wheat or chickpea crops.

4.3 PLANT CHARACTERISTICS

Like other Cruciferae, sarson and toria plants have four sepals, four petals, and four long stamens around the ovary and two small ones in an outer whorl. An ovary contains two fused carpels. Sarson and toria crops have distinct growing periods under irrigated conditions in India. Toria is sown in September and produces flowers from early October to the end of November, the peak of blooming being from mid-October to mid-November. Sarson and raya (*Brassica juncea*) are sown, often side by side, in October and November. They produce flowers from December to February. In favorable weather of cold nights and sunny, warm days, flowering continues sequentially and the spikelets gain considerable length, and practically all the pods are full of seed. Out of 21 m ha, under all oilseed crops, 5 million is under Brassica alone.

4.4 FLORAL BIOLOGY OF *BRASSICAS*

Most of rape (*Brassica napus*), sarson (*Brassica campestris* var. sarson), toria (*Brassica campestris* var. toria), Indian mustard or rai (*Brassica juncea*), white mustard (*Brassica alba*), and black mustard (*Brassica nigra*) crops bloom during February to March for over a month. The long terminal racemes produce fragrant flowers that are vibrant yellow in color. Typically, these crops are

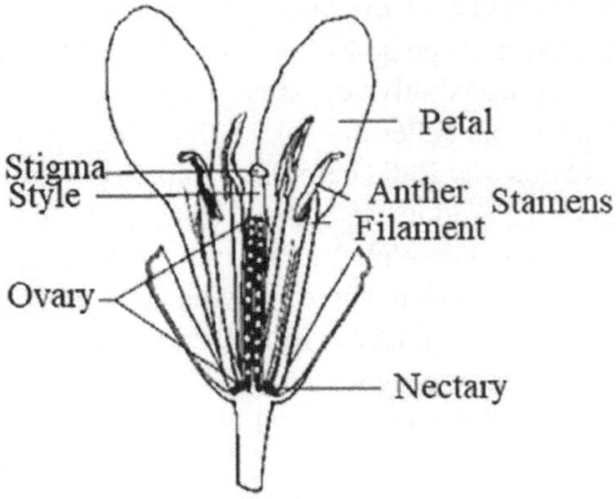

FIGURE 4.1 General design of crucifer flowers, *Brassica oleracea.*

self-pollinated, although some cross-pollination occurs in *Brassica campestris.* The flowers are rich in nectar and pollen, making them highly appealing to honeybees and other natural insect pollinators. Honeybee-mediated cross-pollination enhances seed yield, quality, and oil content. In the afternoon, the average sugar concentration ranges from 40% to 45%, while in the morning, it is only 16% (Sharma 1958; Murrell and Nash 1981). *Brassica rapa* L., a species of mustard from the Cruciferae family, holds significant agricultural importance in Asia (Rubatzkky and Yamaguchi 2000). In *B. rapa*, the flowers emerge as the stem elongates, following the cessation of leaf growth. The stem features multiple branches, small leaves, and vibrant-yellow flowers. Each flower possesses four petals and six stamens, two of which are shorter than the style, while the remaining four are longer (Figure 4.1).

A single stigma is located at the tip of the style. Nectar is secreted at the bases of the shorter stamens and the ovary (Lesik 1953). The flowers remain open for three days. Approximately 95% of Crucifer species require cross-pollination, with some cauliflower varieties being self-fertile. Wind is not an effective pollinator for *Brassica* spp., and bees play a crucial role in pollen transfer. When cross-pollination is absent, self-fertilization generally leads to reduced seed yield, smaller seed size, and diminished yield in subsequent generations (Delaplane and Mayer 2000). Self-incompatible plants rely on the transfer of pollen between individual plants (Wallace et al. 2002). Among Cruciferae, two species, *B. rapa* (syn. campestris) and *B. oleracea*, have been identified as exhibiting sporophytic self-incompatibility (SSI) (Takayama and Isogai 2005).

Self-incompatible plants require pollen transfer from plant to plant (Wallace et al. 2002). Earlier studies on the effects of pollinator insects to seed set of Cruciferae had been reported. In male sterile oilseed rape (*B. napus*), yields of the species were increased by honeybee pollination (Westcott and Nelson 2001; Steffan-Dewenter et al. 2002).

In sarson (*B. campestris*), the presence of insects for pollination enhances seed production, resulting in larger, well-formed grains and a higher number of viable seeds (Khan and Chaudory 1995). Bees, when involved in the pollination process of rapeseed, contribute to increased productivity, improving the overall yield and promoting uniformity and early pod development (Abrol 2007).

According to Sabbahi et al. (2005), this results from the abundant production of flowers during the flowering period, which ensures proper pollination and thus a high and relatively early production of siliques. With respect to the lack of siliqua formation in the exclusion cages, the same authors indicate that although rapeseed is self-fertilizing, it does not produce a large number of mature pods in the absence of insect pollination. Therefore, the absence of cross-pollination, which occurs when free plants in the presence of pollinators interchange pollen, is responsible for the low production of well-developed siliqua.

4.5 POLLINATION REQUIREMENTS FOR BRASSICA

Evidently, pollination is one of the most important natural factors enhancing crop production. Among the various pollinating agents, insect pollinators play a predominant role in increasing the yield of oilseed crop. Almost one-third of total cropped area under oilseeds has been reported to be entomophilous (Mishra et al. 1988). The yield of rapeseed can be doubled through pollination by insects. Pollinators not only enhance the yield of the crop but also contribute uniform and early pod setting. Being a cross-pollinated crop, rapeseed attracts a large number of insect pollinators.

Brassica napus can be self-pollinated or cross-pollinated. In cultivated fields, cross-pollination rates of about 20–30% have been reported (Rakow and Woods 1987). The frequency of cross-pollination is influenced by weather, availability of insect pollinators, and cultivar.

Brassica rapa varieties are generally self-incompatible. Both species are primarily pollinated by honeybees (Williams et al. 1987); however, wind pollination is possible over distances of up to 2.5 km (Timmons et al. 1995). Pollination interventions in the form of planned pollination have been found to greatly benefit quality and quantity of seed production in crucifers. For instance, in male sterile oilseed rape (*B. napus*), yields of the species were increased by honeybee pollination (Westcott and Nelson 2001). Similarly, in sarson (*B. campestris*), pollination by insects increased the seed yield, caused formation of well-shaped, larger grain and more viable seed (Abrol 2009). Impact of insect pollination on production of different oil crop is presented in Table 4.1. *B. rapa* is a diploid with chromosome number of 2n = 20, and *B. napus* is an allotetraploid with a chromosome number of 2n = 38 (Hosaka et al. 1984).

4.6 DIVERSITY OF POLLINATORS

Insects help in pollination of these crops. Crops belonging to family Brassicaceae are predominantly dependent on insect pollination (Entomophilous) (Figure 4.2).

FIGURE 4.2 (a) *Apis dorsata*, (b) *Apis mellifera*, (c) *Apis cerana*, (d) bumblebee foraging on mustard flowers.

FIGURE 4.2 (Continued)

Increase in seed quality and quantity is possible through pollination (Abrol 2007; Shakeel and Mian Inayatullah 2013, 2015). The role of cross-pollination between cruciferous crops is majorly played by honeybees. Absence of cross-pollination generally reduces seeds number, seeds size, and viability of seeds, which can lead to decrease in yield (Delaplane and Mayer 2000). Pollination by insect not only increases crop yield but also improves physiochemical properties of the fruits (Bashir et al. 2018). Crops of family Brassicaceae are very attractive for insect pollinators for a good source of pollens and nectar (Masierowska 2003).

Atmowidi et al. (2007) analyzed the diversity of pollinator insects and its effect to seed set of mustard (*Brassica rapa*) planted in agricultural ecosystem in West Java. At least 19 species of insects pollinated the mustard, and three species, that is, *Apis cerana, Ceratina* sp., and *Apis dorsata*, showed a high abundance. The higher abundance and species richness of pollinators occurred at 0830–1030 h, and the diversity was related to the number of flowering plants. Insect pollinations increased the number of pods, seeds per pod, seed weights per plant, and seed germination.

Brassica rapa L. (Cruciferae) is an important crop in Asia (Rubatzkky and Yamaguchi 2000). Approximately 95% of species of Crucifers require cross-pollination, although some cauliflower varieties are self-fruitful. Wind is not a good pollinator in *Brassica* spp., and bees play an important role in pollen vectors. Selfing in the absence of cross-pollination generally reduces seed yield, seed size, and yield in subsequent generation (Delaplane and Mayer 2000). Self-incompatible plants require pollen transfer from plant to plant (Wallace et al. 2002). In male sterile oilseed rape (*B. napus*), yields of the species were increased by honeybee pollination (Westcott and Nelson 2001; Steffan-Dewenter 2003). Similarly, in sarson (*B. campestris*), insects pollination increase the seed yield, caused formation of well-shaped, larger grain and more viable seed (Khan and Chaudory 1995). The vast natural distribution is covering a wide range of climatic zones and therefore is responsible for the diverse associations between honeybees and flora of wild plants and cultivated crops (Sihag and Mishra 1995).

Bajiya and Abrol (2017) studied flower-visiting insect pollinators of mustard (*Brassica napus*) in Jammu region, India. The blooming crop of the mustard was visited by 15 species of insects belonging to four orders and seven families of class insects. Hymenopterans were the most dominant visitors, constituting 87.48, 88.18% of the insect pollinators, followed by other insect pollinators (12.52, 11.82%) in 2014–15 and 2015–16, respectively. Among the hymenopterans species, *Apis mellifera* L. was highest in number (28.09, 28.31%) of the visitors of mustard flowers, followed by *Apis cerana* F. (25.10, 25.48%), *Apis dorsata* F. (18.00, 18.09%), *Apis florea* F. (8.53, 7.90%), *Xylocopa fenestrata* (5.55, 5.71%), and other insect pollinators (12.52, 11.82%). The foraging activity of honeybees increased with temperature and sunshine and decreased with relative humidity, wind speed, and rainfall. However, the species differences in the population dynamics of bees were evident, as of all the honeybees, *Apis mellifera* was most abundant, followed by *A. cerana*, *A. dorsata*, *A. florae*, and *X. fenestrata*. The foraging population of *Apis mellifera* was highly significant and positively correlated with maximum temperature and sunshine hours and negatively with relative humidity in the evening hours. The same trend was observed in case of *Apis florea*. However, the foraging population of *Apis cerana* was significant and positively correlated with maximum temperature and minimum temperature. In the case of *X. fenestrata*, non-significant observations were recorded with all-weather parameters (Table 4.1).

Insect pollinators observed belonged to four orders, that is, Hymenoptera, Diptera, Lepidoptera, and Coleoptera. A total of 20 major species of insect pollinators was recorded. The highest abundance of pollinator species belonged to Hymenoptera. The most prominent insect pollinator species were *Apis mellifera*, followed by other three honeybee species of *A. cerana*, *A. florea*, and *A. dorsata*, respectively. Some species of solitary bees were also recorded. From Diptera, four species of syrphid fly and one species from Muscidae family were also recorded. Insect pollinators recorded from order Lepidoptera were *Pieris brassicae*, *Vanessa cardui*, and *Papilio demoleus*. Ladybird beetle *Coccinella septempunctata* was recorded from Coleoptera order as occasional visitor. It was noticed that *E. sativa* attracted more insect pollinators than *B. rapa*, which may be attributed to

TABLE 4.1
Insect Visitors and Their Percentage Proportion on Mustard Bloom during 2014–15 and 2015–16 in Jammu Region

Order	Family	Species	Percentage Composition		Total	
			2014–15	2015–16	2014–15	2015–16
Hymenoptera	Apidae	*Apis mellifera*	28.09	28.31	85.27	85.49
		A. cerana	25.10	25.48		
		A. dorsata	18.00	18.09		
		A. florea	8.53	7.90		
		Xylocopa fenestrata	5.55	5.71		
	Andrenidae	*Andrena leaena*	1.66	1.90	2.21	2.69
		A. ilerda	0.55	0.79		
Lepidoptera	Pieridae	*Pieris rapae*	1.36	1.54	1.36	1.54
	Danaidae	*Danaus plexippus*	1.87	1.84	1.87	1.84
Diptera	Muscidae	*Musca* spp.	3.02	1.93	3.02	1.93
	Syrphidae	*Eristalis* spp.	0.99	1.41	3.76	4.08
		Episyrphus balteatus	1.30	1.25		
		Metasyrphus corollae	1.47	1.42		
Coleoptera	Coccinellidae	*Coccinella septempunctata*	0.87	0.88	2.51	2.43
		C. sexmaculata	1.64	1.55		

different amounts and chemical properties of nectar, with number of pollen grains, and flower canopy of both crops. Further studies are needed to confirm the reasons for higher pollinator visitation to *E. sativa* than *B. rapa* through chemical analysis of nectar, amount of pollens, flower physiology, and phenology of both crops.

4.7 YIELD ENHANCEMENT AND OIL POTENTIAL OF RAPESEED, MUSTARD, AND CANOLA (*BRASSICA* SPP.)

Mohammed (1935) in West Punjab at Lyallpur recorded 117 visiting species of insects belonging to seven orders; *Andrena ilerda*, *Apis florea*, and *Halictus* spp. comprised 82% of insects and were the most important in this order. He also observed that stigma of the crops remained receptive for two days after opening, and in five days, their fertility was lost, whereas pollen remained viable for seven days. It was observed in Australia that among *Apis mellifera* foragers, 72% collected nectar only, 25% both nectar and pollen, and 3% collected pollen only (Langridge and Goodman 1975). The pollen was collected between 1000 and 1400 h.

Auto-fertilization of these crops was studied by Mohammed (1935) for six years by enclosing individual plants in muslin bags; in toria, seed set was 12.3%; in brown sarson, 20.3%; and in yellow-seeded sarson, 91.0%. Then he compared self-pollination to cross-pollination in toria. He bagged flowers for cross-pollination and hand-pollinated them; there was 100% pod formation. Sihag (1986) studied seed set in toria plots caged and not caged and left open for natural cross-pollination; there developed 25 and 1,060 pods per plant, 6.3 and 18.2 seeds per pod, and 4 and 385 seeds per plant. Similar observations on sarson gave 59 and 95% pods containing 3.5 and 12.7 seeds per pod (Mishra et al. 1988). Observations made in field plots of both sarson and toria showed that when crop was caged to exclude insects, yield was 68 g; when caged with *Apis cerana*, it gave 219 g, and in open field near honeybee colonies, where other insect pollinators were also present, yield was 244 g (Latif et al. 1960). In general, there is quite a high degree of self-incompatibility in these crops, and good yields are obtained by cross-pollination with the aid of natural fauna in most places where good weather prevails, and probably yield can be further increased with the help of honeybee colonies to the extent of 10–30% (Rahman 1940; Singh 1954), covering for inclement weather. Singh and Singh (1992) observed that bee-pollinated plants in field cages as compared with self-pollinated plants in individual bags produced 3 times heavier pods, 4 times more seed per pod, 50 times more seeds, and 84 times more seeds yield per plant.

Mishra et al. (1988) found that percent pod setting was significantly higher in plants which had access to insect pollination (95.21%) than those which were net-caged (70.27%) or muslin-bagged (58.81%). Number of seeds per pod was also significantly higher in open-pollinated flowers (912.62 seeds/pod) in comparison to nylon-caged (5.55 seeds/pod) and muslin-bagged (3.05 seeds/pod). Pods obtained from open-pollinated flowers had significantly more healthy seeds (84.78%) than those from net-caged (41.45%) or muslin-bagged (48.37%) flowers; the latter two, however, did not differ from each other. The average weight of 100 healthy dry seeds from muslin-bagged flowers was significantly higher (542.20 mg) than the weight of seeds from open-pollinated (478.50 mg) and net-caged flowers (459.40 mg). Percent oil content of healthy seeds from muslin-bagged (41.64%) flowers was significantly higher than that from open-pollinated (39.02%) flowers. However, oil content of seeds from muslin-bagged flowers was equal to that from net-caged flowers. When all the yield parameters were taken into consideration for calculating oil content (mg/pod), it was found that increase in total oil yield from open-pollinated and net-caged plants was 9.76 and 1.55 times more, respectively, than from muslin-bagged plants.

Varieties of *Brasica campestris* L. are generally considered to require insect pollination for seed production (Free 1970; McGregor 1976). Mohammed (1935) and Rahman (1940) concluded that *Andrena ilerda* Cam., *Halictus* spp., and *Apis florea* were the most important pollinators, whereas Latif et al. (1960); Kapil et al. (1971) and Bhalla et al. (1983) reported *Apis cerana indica* to be the principal pollinator in *B. campestris* L. *var. sarson*. In a similar study, Singh et al. (2000) found that there was significant improvement in qualitative and quantitative parameters of rape when pollinated by *Apis cerana Himalaya*.

Pollination by bees in rapeseed produces an increase in productivity, improving the yield and contributing to the uniformity and early establishment of the pod (Abrol 2007). According to Sabbahi et al. (2005), this results from the abundant production of flowers during the flowering period, which ensures proper pollination and thus a high and relatively early production of siliques. With respect to the lack of silique formation in the exclusion cages, the same authors indicate that although rapeseed is self-fertilizing, it does not produce a large number of mature pods in the absence of insect pollination. Therefore, the absence of cross-pollination, which occurs when free plants in the presence of pollinators interchange pollen, is responsible for the low production of well-developed siliques.

Chhuneja et al. (2007) studied the role of *Apis mellifera* in the seed production of *Brassica campestris* var. toria and found that the proportion of pollen-foraging bees was higher up to 50–75 m, from the bee colony and was also higher in the morning and noontime than in the evening. Likewise, the mean number of siliquae per plant, number of seeds per 20 siliquae, thousand seed weight, seed yield per plant, and oil content were significantly higher up to 50 and 75 m distances. One colony of *Apis mellifera* with ten-bee strength was sufficient for effective pollination of 1 ha of crop area. Araneda et al. (2010) evaluated the yield achieved by *B. napus* cv. Artus pollinated by *A. mellifera* in Chile. He showed that the parameter least affected by bee intervention was the grains per silique variable. In contrast, siliques per plant and 1,000 grain weight parameters presented significant differences, contributing to a yield greater than 5 t ha^{-1}, which represented a figure 50.34% higher than in the treatment without bees. He further concluded that the inclusion of bees in crops is fully justified as a production tool.

Deodikar and Suryanarayana (1972) found enhanced yield in oilseed crops when pollinated by bees. Langridge and Goodman (1975) also found significant differences in production of rapeseed in open and caged plots. Plots receiving unrestrained visits had significantly higher plant density, seed yield, seed/plant, weight of 1,000 seeds, weight of seeds per plant, germination, and percentage of oil content as compared to those isolated from insect visits. Abrol (2009) studied that pollination requirement of different oilseed crops and expected increase in yield due to bee pollination were 20 to 222% in different rapeseed-mustard crops and 21–3,400% in sunflower.

The yields of many field, fruit, and seed crops are greatly enhanced by bumblebee visitation (Corbet et al. 1991; Osborne and Williams 1996). For example, field beans are largely dependent on pollination by longer-tongued species, such as *B. pascuorum* and *B. hortorum*, and without them, yields are poor (Free and Williams 1976). Bumblebees are acknowledged to be more reliable pollinators than honeybees, particularly because they will continue foraging even when it is cold and wet. In a poor spring, bumblebee queens (and perhaps also solitary bees, such as *Osmia rufa*) may be the only insects that remain active enough to pollinate early flowering crops, such as hard fruits.

Rapeseed canola is an oleaginous crop and considered to be among the best-quality oils for human consumption (Porter and Crompton 2008; Amjad 2010). Pollination is a basic ecological process, essential for the maintenance of viability and diversity (Potts et al. 2006; Klein et al. 2007), and agriculture today has become even more dependent on pollinators (Aizen et al. 2009), with about 35% of global plant food production, depending on plants which require these agents. One-third of the human diet is constituted by vegetables, legumes, and fruits pollinated by insects, of which more than 80% are honeybees (*Apis mellifera* L.) (Klein et al. 2007; Hu et al. 2008), the most prominent and efficient pollinators (Danforth et al. 2006; Evans and Spivak 2006; Thapa 2006). For rapeseed, although the wind is the principal vector in terms of the distances over which pollen is transported (Hoyle et al. 2007), bees are the principal pollinators, being the most abundant insects in the cultivars and varieties of rapeseed used for seed production (Westcott and Nelson 2001; Pordel et al. 2007). Thus, the introduction of hives of *A. mellifera* in controlled pollination helps increase the production of crops such as rapeseed (*Brassica napus*) (Sabbahi et al. 2006), since this plant is characterized by the production of abundant pollen and good-quality nectar at relatively high concentrations of sugar, in flowers with a color and structure which are attractive to insects, particularly bees (Smith 2002; Sabbahi et al. 2006).

Most of the oilseed crops are cross-pollinated, and adequate pollination is vital for any significant seed production. An increase in seed yield as a result of insect pollination has been reported in mustard (Mohammed 1935; Latif et al. 1960). Latif et al. (1960) found that *Apis cerana* colonies near sarson and toria fields increased seed yield by 60%. Similar increase in field has been reported in oilseed rape (Langridge and Goodman 1975; Kisselhagen 1977) and *Brassica campestris* var Jambuck (Kubisova et al. 1980). Bisht et al. (1980) found that flowers of rapeseed visited by *Apis* species had higher pod set, more number of seeds per pod, and weight of seed was also higher than those deprived of pollinator visit. Mishra et al. (1988) found that in *Brassica campestris*, percent pod setting, number of pods per plant, and proportion of healthy seeds were significantly higher in open-pollinated flowers than in net-caged and muslin-bagged ones. Similarly, average weight of seeds and oil content was higher in open-pollinated flowers. *Apis cerana* was the most common pollinating species. The other pollinators observed included *A. mellifera*, syrphid flies, etc. In a similar study, Prasad et al. (1989) found that pollination of *Brassica juncea* by *A. cerana* resulted in more siliqua setting, increased length of siliqua, seed weight increased yield, and a pronounced effect on oil contents and germination.

Kevan and Eisikowitch (1990) reported that the presence of pollinators, for example, honeybees, on canola (cv. OAC Triton) increases the germinability of resulting seeds from 83 to 96%. Mishra and Kaushik (1992) reported that yield and oil content were higher in honeybee-pollinated crops in comparison to self-pollinated. The results showed that higher percentage of pod setting was obtained in open-pollinated crops of *Brassica nigra*, *B. campestris*, *B. carinata*, *B. napus*, and *Eruca sativa*, having 80, 84.42, 71.90, and 70.30%, respectively. Average weight of 1,000 healthy seeds varied between 3 and 5.1 g under open pollination, whereas it was 1.8 to 3.7 g under self-pollination. Langridge and Goodman (1982), in a trial on oilseed rape (*Brassica napus*) cv. Midas in northern Victoria, found that no increase in yield of seed, oil content, or percentage germination was obtained from plots where bees and larger insects had access compared with plots enclosed in cages of 2.5 mm mesh to exclude these insects. The weight of 1,000 seeds was slightly greater in the enclosed plots. Bees and a Syrphid, *Melangyna viridiceps*, were the predominant insects visiting the flowers. The bees stored some surplus honey and pollen and built up colony population. Self-pollination and possibly wind pollination appear to be the norm for this cultivar.

Evidently, pollination is one of the most important natural factors enhancing crop production. Among the various pollinating agents, insect pollinators play a predominant role in increasing the yield of oilseed crop. Almost one-third of total cropped area under oilseeds has been reported to be entomophilous (Mishra and Sharma 1998; Mishra et al. 1988). The yield of rapeseed can be doubled through pollination by insects. Pollinators not only enhance the yield of the crop but also contributes uniform and early pod setting. Being a cross-pollinated crop, rapeseed attracts a large number of insect pollinators.

From the plant breeder's point of view, the methods employed for the improvement of a crop depend to a large extent on the mode of pollination of the crop concerned (Mand et al. 2010). A study of pollination methods is extremely necessary, as in light of these studies, the right lines of effecting improvement in a crop can be decided upon and adopted. Thus avoiding considerable time and labor. The embryo which gives rise to an adult plant results from the fusion of two male and female gametes or reproductive cells. The constitution of which may be identical or different. In case these two gametes have identical constitution, the plant is self-fertilized. On the other hand, if two gametes fusing together are different in constitution, a large amount of cross-fertilization occurs in the plant, the one generation being different from the other. Cresswell et al. (2004) reported that the flower of *Brassica napus* L. appears to be typically zoophilous (suited to animal pollination) because of its visually attractive petals, robust stigma, and nectaries. Pollination by wind is feasible, however, and its likely effectiveness is not immediately foreseeable because of the complexity of interactions between objects and windborne particles. The flower's petals handicap wind pollination by reducing the target efficiency of the upwind-facing stigma. For downwind-facing flowers, pollen reception was negligible. Several aspects of the plant's architecture (floral structure, pollen cohesiveness, inflorescence structure) are uncompromisingly zoophilous (Cresswell et al. 2002; Cresswell and Osborne 2004). Estimates of the amount of wind pollination suggest that it is unlikely to be important for the long-distance dispersal of *B. napus* genes, such as those from genetically modified varieties. This study illustrates how computational fluid dynamics (CFD) may become a powerful tool in future analyses of wind pollination. However, considering the importance of these crops, insufficient data is known about their pollination requirements. Oilseed rape plants have been reported to benefit greatly from bee pollination. Appreciable differences in the seed yield attributes were noticed when plots with bee pollination were compared with caged plots.

4.7.1 Raya (*Brassica juncea* L., Cruciferae)

It is primarily a self-pollinated crop. Howard et al. (1916) described how a honeybee works in a flower. It lands on a flower and pushes its tongue between the long and short stamens to reach nectar that has sugar content of 22–65% (Sharma 1958). Its head touches long anthers and stigma, and its thorax touches anthers of short stamens. As it moves over to the other side of a flower, its thorax touches the stigma, transferring pollen. The exact role of bees in pollination has not been determined.

It was observed in UK that plots of Indian mustard (*B. juncea*) caged without bees or with bees produced the same yield as obtained in the open (Free and Spencer-Booth 1963). In India, plants inside cages devoid of pollinating insects produced 14.4 seeds per pod, 581 pods per plant, and average yield of 193 g seeds per plant as compared with the figures 17.2, 1,090, and 349 from the exposed plants visited by all insects (Sihag 1986). It was suggested by Free (1993) that in order to eliminate caging effect, foraging bees should be released inside another cage and yield of plants be observed.

Surveys carried out, from 1966–68, in Punjab (including Haryana and Himachal Pradesh) showed that in 59 samples of 30 nets each, the average density of insect visitors (2.3/net) was about the same as recorded on toria and sarson. *Apis florea* was by far the most common bee (75.4% of total), and the other common bees were *Andrena ilerda* (9.3%), *Andrena leaena* (4.1%), *Andrena* spp. (2.1%), *Apis dorsata* (3–8%), *Halictus* sp. (3.7%), and all *Halictines* (4.1%). All the rest of visitors comprised 2.1% of the population, and almost all the groups were represented as recorded on toria and sarson. It was interesting to note that after *A. florea*, *Andrena* was the conspicuous group (15.5%), and bees of family Halictidae (7.8%) were also active on this crop as on toria and sarson (Atwal 1970; Kapil et al. 1969, 1971). Megachile, even though a small group, was observed on toria and sarson, but not on raya.

It was observed that among *Apis dorsata* foragers, 18% collected pollen. The nectar collectors also touched stamens and stigmas. They started foraging when the temperature was at least 15°C (Kapil and Kumar 1975). The reason for Andrena group preferring raya above sarson was not clear.

4.7.2 Taramira (*Eurica sativa*, Cruciferae)

This oilseed crop is grown under unirrigated conditions by itself or mixed with chickpeas. Its flowers open in the morning and remain so for three days. Anthers dehisce as soon as the flower opens, and nectar is available sometime later on; therefore, pollen gatherers visit first and nectar gatherers follow them. It has been reported that taramira is self-sterile and seed is produced with the help of outside agencies (Cobley 1956).

Insect visitors were surveyed (58 samples of 30 sweeps each) in Punjab (including Haryana and Himachal Pradesh) for three seasons and recorded. Among all the visitors, *Apis florea* (50.8%) was the most abundant, followed by *Apis dorsata* (21.2%). *Andrena ilerda* and allied species (18.5%), *Colletes* spp. (2.2%), and all the rest of bees (6.9%) and Halictidae were also recorded, but the presence of *Colletes* on this crop was unique. On the whole, population density was less than half of that recorded on other oilseed Cruciferae. In the drier southern district, *A. florea* held first position, and *Andrena* spp. the second position. However, in central districts, *A. dorsata* was the most abundant, followed by *A. florea* and *Andrena* (Atwal 1970; Kapil et al. 1971). It was reported that *Andrena* in West Punjab had a comparatively higher population, perhaps owing to the drier climate there. The overall density of population on taramira was 0.9 insect per net sweep in East Punjab.

4.7.3 Seed Cauliflower (*Brassica oleracea* L., var. botrytis, Cruciferae)

Other vegetables that belong to this species are cabbage, brussels sprout, broccoli, etc. Cauliflower seed is produced from plants of the previous season, grown in isolated fields at higher altitudes. Flowers are self-pollinated also, but for seed production, cross-pollination is considered desirable (Vesely 1962). Among the insect pollinators, honeybees (*Apis cerana* and *Apis mellifera*) are considered most useful. The wild honeybees, Andrena, Syrphidae, and other Diptera also play an important role (Kakkar 1980, 1981).

The pollination requirements of cauliflower were studied in Himachal Pradesh (Rauala 1972). It was observed that plants in cages with honeybees (*Apis cerana*) had 68% seed set, 9% seeds per pod, and 627 seeds per 100 flowers, whereas in the cages without honeybees, there was 9% seed set, 2.3 seeds per pod, and 28 seeds per 100 flowers. In another experiment it was found that when plants were bagged, there was 19% seed set and 2.3 seeds per pod as against 45% and 4 for plants in cages of wire netting that excluded large insects; the figures for exposed plants in the open were 93% and 6.4 (Verma and Joshi 1983). Sihag (1986) also observed big differences in seed set and yield in plants caged and not caged.

Insect pollinators, including honeybees, are active from morning till evening, but the highest seed set is obtained by their activity between 1100 and 1300 h (Dhaliwal and Bhalla 1981). It was observed that *A. cerana* colonies enhanced pollination: in the open fields up to 50 m, seed set was 94.5%, number of seeds per pod was 21, and there were 84 seeds per plant; at 200 m, where honeybees were not visiting, the figures were 81%, 15, and 41. At 200 m, *Apis dorsata*, *A. florea*, *Andrena*, and other insects were active in the field (Tewari and Singh 1983). It appears foraging range of *A.*

cerana on cauliflower bloom is limited. Within 200 m, 53%, and within 400 m, 86%, of the honeybees foraged, the maximum distance at which they were seen on this crop.

REFERENCES

Abrol, D. P. 2007. Foraging behaviour of *Apis mellifera* and *Apis cerana* as determined by the energetics of nectar production in different cultivars of *Brassica campestris* var. *toria*. *J. Apicultural Sci.* 51, 5–10.

Abrol, D. P. 2009. *Bees and Beekeeping in India*, 2nd edn. Kalyani Publishers, New Delhi, p. 720.

Aizen, M., Garibaldi, L., Cunningham, S., and Klein, A. 2009. How much does agriculture depend on pollinators? Lessons from long-term trends in crop production. *Ann. Bot. (London)* 103, 1579–1588.

Amjad, M. 2010. Canola variety guide in WA 2010. Canola Research Officer, Department of Agriculture and Food, Western Australia. Available at http://agric.wa.gov.au/objtwr/imported_assets/content/fcp/gmcrops/fn406_canolavariety.pdf

Araneda, X. D., Ulloa, R. B., Carrillo, J. A., Contreras, J. L., and Bastidas, M. T. 2010. Evaluation of yield component traits of honeybee pollinated (*Apis mellifera* L.) rapeseed canola (*Brassica napus* L.). *Chilean Journal of Agricultural Research* 70(2), 309–314.

Atmowidi, T., Buchori, D., Manuwoto, S., Suryobroto, B., and Hidayat, P. 2007. Diversity of pollinator insects in relation to seed set of mustard (*Brassica rapa* L.: Cruciferae). *HAYATI Journal of Biosciences* 14(4), 155–161.

Atwal, A. S. 1970. *Insect Pollinators of Crops*. Punjab Agricultural University Press, Ludhiana.

Aytaç, S., Gizlenci, S., Acar, M., and Ustun, A. 2006. Changes in erucic acid concentration of rape seeds in advanced generations. *Journal of Biological Sciences* 6, 60–62.

Bajiya, M. R. and Abrol, D. P. 2017. Flower-visiting insect pollinators of mustard (*Brassica napus*) in Jammu Region. *Journal of Pharmacognosy and Phytochemistry* 6(5), 2380–2386.

Bashir, M. A., Alvi, A. M., Khan, K. A., Rehmani, M. I. A., Ansari, M. J., Atta, S., et al. 2018. Role of pollination in yield and physicochemical properties of tomatoes (Lycopersicon esculentum). *Saudi J. Biol. Sci.* 25, 1291–1297. doi: 10.1016/j.sjbs.2017.10.006

Bhalla, O. P., Verma, A. K., and Dhaliuwal, H. S. 1983. Insect visitors of mustard bloom (*Brassica campestris var* sarson), their number and foraging behaviour under mid-hill conditions. *J. Ento Res.* 7, 15–17.

Bisht, D. D., Naim, M., and Mehrotra, K. N. 1980. Studies on the role of honeybees in rapeseed production. In: *Proc. 2nd Int. Conf. Apiculture in Tropical Climate*. Indian Agricultural Research Institute, New Delhi, pp. 491–496.

Chhuneja, P. K., Singh, J., Gatoria, G. S., and Blossom, S. 2007. Assessing the role of honey bee (*Apis mellifera* L.) in the seed production of *Brassica campestris* var. toria in the Punjab. *Indian Journal of Crop Science* 2(2), 34–40.

Cobley, L. S. 1956. *An Introduction to the Botany of Tropical Crops*. Longmans, London.

Corbet, S. A., Williams, I. H., and Osborne, J. L. 1991. Bees and the pollination of crops and wild flowers in the European Community. *Bee World* 72, 47–59.

Cresswell, J. E., Davies, T. W., Patrick, M. A., Russell, F., Pennel, C., Vicot, M., and Lahoubi, M. 2004. Aerodynamics of wind pollination in a zoophilous flower, *Brassica napus*. *Functional Ecology* 18, 861–866.

Cresswell, J. E. and Osborne, J. L. 2004. The effect of patch size and separation on bumblebee foraging in oilseed rape: Implications for gene flow. *Journal of Applied Ecology* 41, 539–546.

Cresswell, J. E., Osborne, J. L., and Bell, S. A. 2002. A model of pollinator-mediated gene flow between plant populations with numerical solutions for bumblebees pollinating oilseed rape. *Oikos* 98, 375–384.

Danforth, B. J., Sipes, S., Fang, J., and Brady, S. 2006. The history of early bee diversification based on five genes plus morphology. *Proc. Natl. Acad. Sci. USA* 103, 15118–15123.

Delaplane, K. S. and Mayer, D. F. 2000. *Crop Pollination by Bees*. CABI Publishing, New York, USA.

Deodikar, G. B. and Suryanarayana, M. C. 1972. Crop yields and bee pollination. *Indian Bee J.* 34, 53–64.

Dhaliwal, H. S. and Bhalla, O. P. 1981. *Factors Affecting Optimal Pollination of Cauliflower.* Apimondia Publishing House, Acapulco, Mexico, pp. 386–387.

Downey, R. K. 1983. The origin and description of the *Brassica* oilseed crops. In: Kramer, J. K. G., Saner, F. D., and Pigden, W. J. (eds.) *High and Low Erucic Acid Rapeseed Oils: Production, Usage, Chemistry, and Toxicological Evaluation*. Academic, New York, pp. 1–20, 582.

Evans, E. C. and Spivak, M. 2006. Effects of honey bee (Hymenoptera: Apidae) and bumble bee (Hymeoptera: Apida) presence on cranberry (Ericales: Ericacaeae) pollination. *J. Econ. Entomol.* 99, 614–620.

Free, J. B. 1970. *Insect Pollination of Crops*. Academic Press, New York.

Free, J. B. 1993. *Insect Pollination of Crops*, 2nd edn. Academic Press, Harcourt Brace Jovanovich Publ., London, p. 684.

Free, J. B. and Y. Spencer-Booth. 1963. The foraging areas of honey-bee colonies in fruit orchards. *Journal of Horticultural Science* 38, 129–137.

Free, J. B. and Williams, I. H. 1976. Pollination as a factor limiting the yield of field bean *(Vicia faba* L.). *J. Agric. Sci. Camb.* 87, 395–399.

Hosaka, K., Ogihara, Y., Matsubayashi, M., and Tsunewaki, K. 1984. Phylogenetic relationship between the tuberous Solanum species as revealed by restriction endonuclease analysis of chloroplast DNA. *Jpn J Genet.* 59(4), 349–369.

Howard, A., Howard, G. L. C., and Abdur Rahman, K. 1916. Studies in Indian oil-seeds. I. Safflower and Mustard. *Mem. Dep. Agric. India. Bot. Ser.* 7, 237–272.

Hoyle, M., Hayter, K., and Cresswell, J. E. 2007. Effect of pollinator abundance on self-fertilization and gene flow: Application to GM Canola. *Ecol. Appl.* 17, 2123–2135.

Hu, S., Dilcher, D., Jarzen, D., and Winship, D. 2008. Early steps of angiosperm-pollinator coevolution. *Proc. Natl. Acad. Sci. USA* 105, 240–245.

Kakkar, K. L. 1980. Prospects of bees as pollinating agents of cauliflower. In: *Proc. 2nd Int. Conf. Apiculture in Tropical Climate*, New Delhi. pp. 545–556.

Kakkar, K. L. 1981. Foraging behaviour of insect pollinators of cauliflower bloom. *Indian Journal of Ecology* 8(1), 126–130.

Kapil, R. P., Grewal, G. S., Kumar, S., and Atwal, A. S. 1969. Insect pollinators of rape seed and mustard (Abstract). *56th Indian Sci. Cong. Proc.*, pt. 3, p. 509, Bombay, India.

Kapil, R. P., Grewal, G. S., Kumar, S., and Atwal, A. S. 1971. Insect pollinators of rape-seed mustard. *Indian Journal of Entomology* 33, 61–66.

Kapil, R. P. and Kumar, S. 1975. Foraging activity of *Apis dorsata* Fab. on Brassica juncea Hook and Thomas. *J. Bombay Nat. Hist. Soc.* 71(2), 327–331.

Kevan, P. G. and Eisikowitch, D. 1990. The effects of insect pollination on canola *(Brassicanapus* L. cv. O.A.C. Triton) seed germination. *Euphytica* 45, 39–41.

Khan, B. M. and Chaudory, M. I. 1995. Comparative assessment of honey bee and other insects with self-pollination of sarson (Brassica campestris) in Peshawar, Pakistan. In: Kevan, P. G. (ed.) *The Asiatic Hive Bee: Apiculture, Biology and Role in Sustainable Development in Tropical and Subtropical Asia*. Enviroquest Ltd., Ontario, pp. 147–150.

Kisselhagen, S. 1977. Seed yield is increased if there are bees on the rape. *Tidsskrift for Biavl* 111, 66–67.

Klein, A. M., Vaissie' re, B. E., Cane, J. H., Steffan-Dewenter, I., Cunningham, S. A., Kremen, C., and Tscharntke, T. 2007. Importance of pollinators in changing landscapes for world crops. *Proc. R. Soc. B Biol. Sci.* 274, 303–313.

Kubisova, S., Nedbalova, V., and Plesnik, R. 1980. Cinnost vcely medonosne (*Apis mellifera* L.) na repce (*Brassica napus* L. var. *napus*). *Pol'nohospoddrstvo* 26, 744–754.

Lamb, R. J. 1989. Entomology of oilseed Brassica crops. *Annual Reviews of Entomology* 34, 211–229.

Langridge, D. F. and Goodman, R. D. 1975. A study on pollination of oilseed rape (*Brassica campestris*). *Austral. J. Exptl. Agric. Anim. Husband.* 15, 285–288.

Langridge, D. F. and Goodman, R. D. 1982. Honeybee pollination of oilseed rape, cultivar Midas. *Australian Journal of Experimental Agriculture* 22(115), 124–126.

Latif, A., Qayyum, A., and Abbas, M. 1960. The role of *Apis indica* in the pollination of toria and sarson (*Brassica campestris* var. Toria and dichotoma). *Bee World* 41, 283–286.

Lesik, F. 1953. Microfertilization increases the nectar secretion of honey plants. *Pchelovodstvo, Mosk.*, 30, 42–45.

Mänd, M., Williams, I. H., Viik, E., and Karise, R. 2010. Oilseed rape, bees and integrated pest management. In: Williams, I. H. (ed.) *Biocontrol-Based Integrated Management of Oilseed Rape Pests*. Springer Science, Springer, Berlin.

Mandal, S., Yadav, S., Singh, R., Begum, G., Suneja, P., and Singh, M. 2002. Correlation studies on oil content and fatty acid profile of some Cruciferous species. *Genetic Resources and Crop Evolution* 49, 551–556.

Masierowska, M. 2003. Floral nectaries and nectar production in brown mustard (*Brassica juncea*) and white mustard (*Sinapis alba*) (Brassicaceae). *Plant Syst. Evol.* 238, 97–107.

McGregor, S. E. 1976. Insect pollination of cultivated crop plants. In: *Agriculture Handbook*. United States Department of Agriculture, Agricultural Research Service, Washington, DC, p. 496.

Mishra, R. C. and Kaushik, H. D. 1992. Effect of cross-pollination on yield and oil content of *Brassica campestris* L. var. sarson with pollination efficiency of honeybees, *Apis* spp. *Ann. Entomol.* 10, 33–37.
Mishra, R. C., Kumar, J., and Gupta, J. K. 1988. The effect of mode of pollination on yield and oil potential of *Brassica campestris* L. var. *sarson* with observation on insect pollinators. *J. Apic. Res.* 27, 186–189.
Mishra, R. C. and Sharma, S. K. 1998. Technology for management of Apis mellifera in India. In: Mishra, R. C. (ed.) *Perspectives in Indian Apiculture.* Agro Botanica, Bikaner (India), pp. 131–147, 412.
Mohammed, A. 1935. Pollination studies in tara *(Brassica napus* L. var. dichotsma Prain) and sarson *(Brassica campestris* L. var. sarson, Prain). *Indian J. Agric. Sci.* 5, 125–154.
Murrell, D. C. and Nash, W. T. 1981. Nectar secretion by toria *(Brassica campestris L. v. toria)* and foraging behaviour of three *Apis* species on toria in Bangladesh. *J. Apic. Res.* 20, 34–38.
Osborne, J. L. and Williams, I. H. 1996. Bumble bees as pollinators of crops and wild flowers. In: Matheson, A. (ed.) *Bumble Bees for Pleasure and Profit.* International Bee Research Association, Cardiff, UK, pp. 24–32.
Pordel, M. R., Hatami, B., Mobli, M., and Ebadi, R. 2007. Identification of insect pollinators of three different cultivars of winter canola and their effect on seed yield in Isfahan. *J. Sci. Technol. Agric. Natur. Resour.* 10(4B), 413–426.
Porter, P. and Crompton, D. 2008. Canola. In: *Varietal Trials Results.* Minnesota Agricultural Experiment Station, University of Minnesota, Saint Paul, Minnesota, USA, pp. 52–53. Available at www.maes.umn.edu/08varietaltrials/canola.pdf
Potts, S., Petanidou, T., Roberts, S., O'Toole, C., Hulbert, A., and Willmer, P. 2006. Plant-pollinator biodiversity and pollination services in a complex Mediterranean landscape. *Biol. Conserv.* 129, 519–529.
Prakash, S. and Hinata, K. 1980. Taxonomy, cytogenetics and origin of crop Brassicas, a review. *Opera Botanica* 55, 3–57.
Prasad, D., Hameed, S. F., Singh, R., Yazdani, S. S., and Singh, B. 1989. Effect of bee pollination on the quantity and quality of rai crop (Brassica juncea Coss). *Indian Bee Journal* 51(2), 45–47.
Rahman, K. A. 1940. Insect pollinators of toria (*Brassica napus* L. *var dichotoma) and Sarson* (brassica campestris L. var Sarson) at Lyallpur. *Indian J. Agric. Sci.* 10, 422–447.
Rakow, T. and Woods, D. L. 1987. Outcrossing in rape and mustard under saskatchewan prairie conditions. *Can. J. Plant Sci.* 67, 147–151.
Rauala, T. S. 1972. Pollination studies in cauliflower (Brassica oleracea var. botrytis L.). *Journal of Research Punjab Agricultural University* 9, 580–585.
Rubatzkky, V. E. and Yamaguchi, M. 2000. *Sayuran Dunia: Prinsip, Produksi dan Gizi.* Penerbit ITB, Bandung.
Sabbahi, R., De Oliveira, D., Marceau, J. and Houle, E. 2006. Floraison du canola (Crucifera: Brassicaceae) et le rendement des colonies d'abeilles domestiques (Hymenoptera: Apidae). Avalaible at www.agrireseau.qc.ca/apiculture/documents/Rendem_apicole-Rachid_revise.pdf
Sabbahi, R., Oliveira, D. and Marceau, J. 2005. Influence of honey bee (Hymenoptera: Apidae) density on production of Canola (Crucifera: Brassicae). *Journal of EconomicEntomology* 98, 367–372.
Shakeel, M. and Mian Inayatullah, H. A. 2013. Checklist of insect pollinators and their relative abundance on two canola (Brassica napus) cultivars in Peshawar, Pakistan. *Journal of Entomology and Zoology Studies* 326(36), 326–330.
Shakeel, M. and Mian Inayatullah, H. A. 2015. Impact of insect pollinators on the yield of canola (Brassica napus) in Peshawar. *Pakistan J. Agri Urban Entomol* 29, 1–5.
Sharma, P. L. 1958. Sugar concentration of nectar of some Punjab honey plants. *Indian Bee J.* 20(7), 81–91.
Sihag, R. C. 1986. Insect pollination increases seed production in cruciferous and umbelliferous crops. *J. Agric. Res.* 25, 121–126.
Sihag, R. C. and Mishra, R. C. 1995. Crop pollination and *Apis cerana.* In: Kevan, P. G. (ed.) *The Asiatic Hive Bee: Apiculture, Biology, and Role in Sustainable Development in Tropical and Subtropical Asia.* Enviroquest Ltd., Ontario, pp. 129–134.
Singh, M. P., Singh, K. I., and Devi, C. S. 2000. Role of Apis cerena Himalaya pollination on yield and quality of rape seed and sunflower crops. Proceedings of 4th Asian Apicultural Association, International Conference. March 23–28, Kathmandu, Nepal, 1998.
Singh, R. P. and Singh, P. N. 1992. Impact of bee pollination on seed yield, carbohydrate composition and lipid composition of mustard seed. *Journal of Apicultural Research* 31, 128–133.
Singh, S. 1954. Insect pollination and breeding of fruit varieties. *Indian J. Hort. Sci.* 11, 123–125.
Smith, D. R. 2002. Genetic diversity in Turkish honey bees. *Uludag Bee J.*, 2, 10–17.

Steffan-Dewenter, I., Munzenberg, U., Burger, C., Thies, C., and Tscharntke, T. 2002. Scale-dependent effects of landscape context on three pollinator guilds. *Ecology* 83, 1421–1432.

Takayama, S. and Isogai, A. 2005. A self-incompatibility in plants. *Annu. Rev. Plant Biol.* 56, 467–489.

Tewari, G. N. and Singh, K. 1983. Studies on insect pollinators in relation to seed production in cauliflower (*Brassica oleracea* var. *botrytis* L.). *Indian Bee J.* 456, 54–55.

Thapa, R. B. 2006. Honeybees and other insect pollinators of cultivated plants: A review. *J. Inst. Agric. Anim. Sci.* 27, 1–23.

Timmons, A. M., O'Brien, E. T., Charters, Y. M., Dubbels, S. J., and Wilkinson, M. J. 1995. Assessing the risks of wind pollination from fields of Brassica napus ssp. oleifera. *Euphytica* 85, 417–423.

Verma, S. K. and Joshi, N. K. 1983. Studies on the role of honeybees in the pollination of cauliflower (*Brassica oleracea* var. *botrytis*). *Indian Bee Journal* 45(2/3), 52–53.

Vesely, V. 1962. The economic effectiveness of bee pollination on winter rape (*Brassica napus* l., var. Oleifera metz.). Min. Zemedel. Lesn. A vodniho hospodar. *Ust. Vedtech. Inform. Zemedel. Ekon.* 8, 659–673.

Wallace, H. M., Maynard, G. V., and Trueman, S. J. 2002. Insect flower visitors, foraging behaviour and their effectiveness as pollinators of *Persoonia virgata* R. Br. (Proteaceae). *Austral J. Entomol.* 41, 55–59.

Westcott, L. and Nelson, D. 2001. Canola pollination: An update. *Bee World* 82, 115–129.

Williams, I. H., Martin, A. P., and White, R. P. 1987. The effect of insect pollination on plant development and seed production in winter oil-seed rape (*Brassica napus* L.). *J. Agric. Sci., Camb.* 109, 135–139.

5 Sunflower

5.1 INTRODUCTION

Sunflower (*Helianthus annuus* L., Compositae) is an important oilseed crop prized for its oil of low-cholesterol quality and is grown throughout the world (Free 1993; Abrol and Shankar 2012). Sunflower, native to North America, is the state flower of Kansas (US), wherein oilseed sunflower has been an economically important crop in the United States since 1966. The cultivated sunflower is one of the 67 species in the genus *Helianthus.* All are native to the Americas, belong to the Compositae family, and have a typical composite flower (Berglund 2007). Sunflower has a long history as an economic plant, but the exact time and place of its first cultivation are uncertain. Early English and French explorers have spread along the trade routes to Italy, Egypt, Afghanistan, India, China, and Russia. Sunflower developed as a premier oilseed crop in Russia and has found wide acceptance throughout Europe. Sunflower is grown over an area of more than 25 million hectares in about 70 countries, with a world average yield of 1,637 kg/ha. Russia, Ukraine, Romania, China, Kazakhstan, Spain, Bulgaria, France, India, Turkey, South Africa, USA, and Myanmar are the major sunflower-growing countries, whereas Russia and Ukraine have the largest share of about 50% in total sunflower production in the world. China, Turkey, Bulgaria, and France are the highest-yielding countries, with an average yield of more than 2 tons/ha as against less than 1 ton/ha by India.

Sunflower is a major source of vegetable oil in the world. It is a photo-insensitive and short-duration crop and can be grown over a wide range of climatic conditions in many parts of India. In comparison to world scenario, it is a new introduction to India and fits well in many crop rotations practically throughout the country. It can cover the unsown fields after the *rabi* and *kharif* sowings. It particularly suits the fields after potato harvest in December or early spring, wherein the crop favorably utilizes the leftover nutrients in the soil. In recent years, it has become very popular in Central and Northern India, and its cultivation is being encouraged by government price support policy.

5.2 FLORAL BIOLOGY OF SUNFLOWER

Similar to the other Compositae flowers, the sunflower head is surrounded by green bract and, when it opens, has a single row of conspicuous ray golden-yellow florets. From the outer periphery inward, the rings of concentric disc florets open at a rate of two to four per day and continue for six to ten days, and the crop flowers for three to five weeks (Figure 5.1 and 5.2). Normally, a flower head ranges from 1,000 to 2,000 in the number of florets. Each disc floret has a single ovary containing one ovule which, on fertilization, produces one sunflower seed. A floret on opening develops in two stages, first the male, and then the female. Even though there is mechanical possibility for the achievement of self-pollination, intra-floral pollination does not occur in nature. However, some of the varieties may be partially self-fertile.

The role of honeybees and other insects is essential for inter-floral cross-pollination. It was demonstrated that bumblebees achieved 35% seed set, whereas there was only 1% seed set in flower heads covered with cloth bags (Figure 5.3). Cross-pollination takes place by the transfer of pollen (mediated by insects) from another floret in the same flower head or from floret on other flower heads (Free and Simpson 1964). There are important advantages of cross-pollination, high percentage of seed set and crop yield; as a result of hybrid vigor, seeds are stouter and raise healthy plants, and greater percentage of oil is obtained from hybrid seed (Furgala 1954a, 1954b)

DOI: 10.1201/9781032656724-6

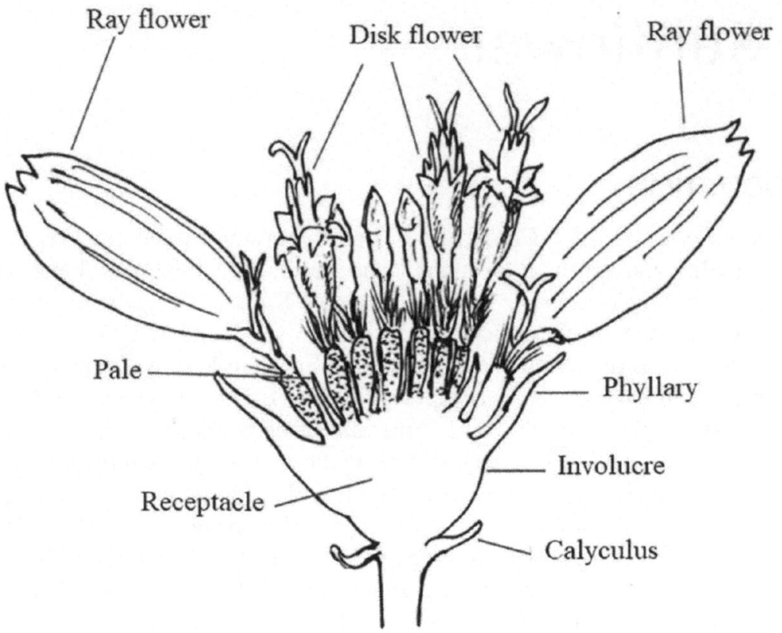

FIGURE 5.1 Floral biology of sunflower.

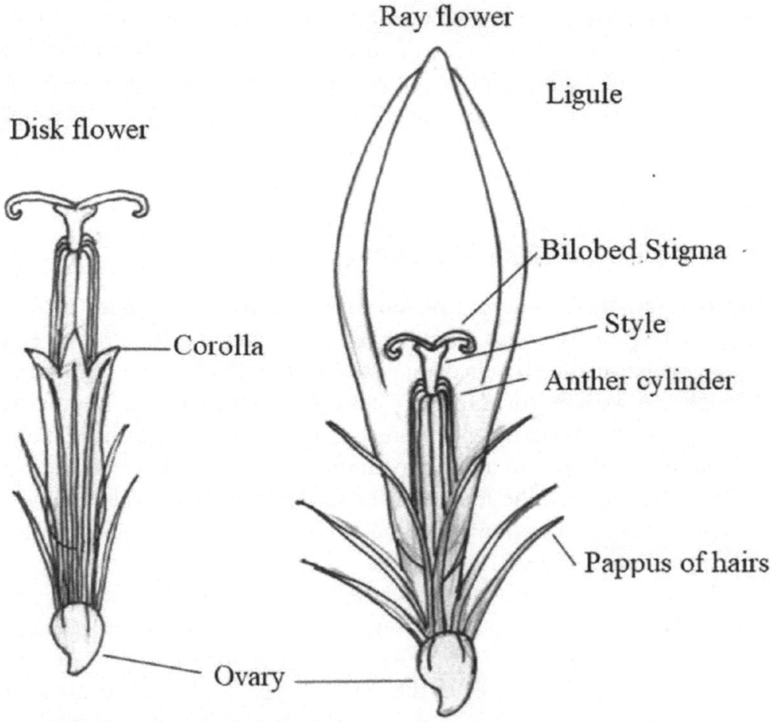

FIGURE 5.2 Different parts of disc and ray flowers.

5.3 DIVERSITY OF POLLINATORS

Honeybees play a crucial role in cross-pollinating the disc florets, leading to final seed set and seed yield (Free 1970; Rao et al. 1995). Bee pollination not only enhances the quantity but also improves the quality of sunflower seeds (Wakhle et al. 1978). Ramya et al. 2014 observed both *Apis* and non-*Apis* bees (97.3%) foraging actively on sunflower heads. Among all major *Apis* bee pollinators observed were rock bee, *Apis dorsata* F. (46.8%), and Indian bee, *A. cerana indica* F. (10.7%). The non-*Apis* pollinators mainly lured to sunflower were stingless bees, *Trigona* (21.4%), and small carpenter bee, *Ceratina* (18.4%).

Hevia et al. (2016) portrayed that a functioning livestock drove road serves as an efficient reservoir of wild bee pollinators in a highly transformed agricultural matrix. They suggested that the drove road enhances wild bee visitation to sunflower heads, with a resulting increase in seed set in areas closer to the drove road, even against a background of high honeybee abundance. They used pan traps to assess bee abundance and richness at 13 sites along the drove road and in adjacent sunflower fields at 10, 75, and 150 m from the drove road. Wild bee visitation to sunflower heads was higher at 10 m compared to 75 and 150 m from the DR, but differences in honeybee visitation were non-significant.

The sunflower seed set was significantly higher at a distance of 10 m compared to 75 and 150 m. This increase in seed set was closely linked to the abundance of wild bees, but not honeybees. Most species of wild sunflowers, such as Helianthus, naturally cannot self-pollinate and depend on cross-pollination (Heiser et al. 1969; Hurd et al. 1980). Therefore, the transfer of pollen between sunflower plants is crucial for successful fruit development. Furthermore, even when sunflower florets are perfect (containing both male and female reproductive organs), they have mechanisms that discourage self-pollination, such as protandry and secondary pollen presentation (Leins and Erbar 1989; Seiler 1997). Consequently, insect visits are necessary to transfer pollen from the male florets to the female florets within a single sunflower head (capitulum). In the past, sunflower and early sunflower varieties required insect pollination for viable seed production and high yields.

Sufficient number of pollinators is required to ensure adequate seed set. Although self-compatible sunflower hybrids usually out-produce self-incompatible cultivars, modern hybrids benefit from insect pollination (Brewer et al 2023). Parker (1981a, 1981b, 1981c) recorded 22 species and/or groups of insects (mostly bees) visiting commercial sunflowers.

Various bee species exhibited consistent abundance throughout each season (Figure 5.3–5.10). The study focused on multiple aspects, including the daily visitation pattern, quantity of pollen

FIGURE 5.3 *Apis dorsata* foraging on sunflower.

FIGURE 5.4 *Apis mellifera* visiting disc flowers of sunflower.

FIGURE 5.5 *Apis cerana indica* visiting disc flowers of sunflower.

FIGURE 5.6 Ceratina bee foraging on sunflower.

FIGURE 5.7 *Xylocopa* bee collecting pollen and nectar on sunflower.

FIGURE 5.8 Large carpenter bee, *Xylocopa* bee, foraging on disc flower.

FIGURE 5.9 *Megachile bicolor* foraging on sunflower.

FIGURE 5.10 Different *Apis* bees foraging on same capitulum.

FIGURE 5.11 *Pithitis smaragdula* foraging on sunflower.

carried on bees' body hairs, behavior of bees during visitation, pollen movement within the plots, nectar volume, sugar concentration, and the impact of bees on seed yields. Furthermore, the author extensively examined and presented strategies for managing native bees as prospective pollinators in commercial sunflower plantations, specifically emphasizing their relevance in the United States.

Breeders have put their sincere efforts to develop self-fertile sunflowers and obtained hybrids which showed high levels of auto-compatibility (80–90%) (Fick and Zimmer 1976; Vranceanu et al. 1988) and reported free from pollinator dependency.

However, empirical research has demonstrated that cross-pollination facilitated by insects present in sunflower fields (Parker 1981a, 1981b, 1981c; Griffiths and Erickson 1983; Jones 1988; Virupakshappa et al. 1992; Aslan and Yavuksuz 2010) benefits sunflower hybrids. The presence of pollinators not only enhances grain yield (Fick 1979; Robinson 1980; Krause and Wilson 1981; Langridge and Goodman 1981; Freund and Furgala 1982; Low and Pistillo 1986; DeGrandi-Hoffman and Watkins 2000; Singh et al. 2000; DeGrandi-Hoffman and Chambers 2006; Chambó et al. 2011) but also increases the oil concentration in the grains (Furgala et al. 1979; Mahmood and Furgala 1983; Kleinschmidt 1986; Calmasur and Ozbek 1999; Kasina et al. 2007). Similar findings have been reported by Ryan and Brugnoni (1986), Astiz et al. (2011), and Chamer (2012).

5.4 POLLINATION REQUIREMENTS AND YIELD ENHANCEMENT

Sunflower crops are visited by a number of insects, including Hymenoptera, Lepidoptera, Coleoptera, and Diptera. Honeybees, wild bees, bumblebees, and solitary bees are the most important groups foraging on sunflower. According to an estimate by Goyal and Atwal (1973), *Apis mellifera* visited a flower 261 times, *A. dorsata* visited 11 times, and a moth, a butterfly, a housefly, and a beetle each visited once only in a day. Honeybees, among all the insects, acquire a pre-eminent position because the colonies can be moved to the desired fields for pollination, and at the same time, surplus honey can be gathered.

Benedek et al. (1972) revealed that fertilization in sunflower fields seems to be dependent on the density of honeybees visiting the crop. However, density of honeybees is influenced by the effect of competitor flora and also by other factors to a much greater degree than by the number of colonies moved to the field. Further, they advised increasing the number of colonies of honeybees on sunflower fields for sufficient pollination on the crop in bloom. Although native wild bees are often better pollinators of sunflower than honeybees, the honeybee is the only managed pollinator of sunflower available. However, if pollen sources other than sunflower are nearby, the honeybee will forage sunflower primarily for nectar and will not transfer sunflower pollen efficiently. Honeybee colonies are placed in seed production fields at a rate of one hive per 1–2 ac. A bee density of more than 20 bees per 100 heads in bloom is needed to transfer sufficient pollen from the male line to the female sterile line. Placement of honeybee colonies will depend upon proximity and acreage of competing nectar and pollen sources. With no competition, all honeybee colonies are placed at one end of the target field. With competing nectar and pollen sources, placement of honeybee colonies at 800 ft intervals may be necessary (Brewer 2007).

Poor soil fertility, water stress, damage due to insect pests, and diseases are some of the grave concerns for the low productivity and successful cultivation of sunflower. Simultaneously, it is evident that insufficient pollination can significantly minimize the yields (Free 1999). Sunflower benefits from insects that visit its flowers for pollen or nectar as reward. Honeybees are known as the main pollinators of sunflower in most parts of the world. Honeybee pollination, especially by *Apis mellifera*, increases the seed yield by 30% and oil content by more than 6% in hybrid varieties (Furgala et al. 1979; Jyoti and Brewer 1999). Hoffmann (1994) also recorded that 96% of the sunflower insect visitors were *A. mellifera* in Brazil. Other bees, especially the non-*Apis* bees, are also known to be frequent visitors of sunflower but are largely perceived as unreliable and ineffective pollinators mainly due to their low activity (Radford et al. 1979). However, DeGrandi-Hoffman and Watkins (2000) noted a possible indirect role of non-*Apis* bees, shown by their ability to enhance pollination by *Apis mellifera*.

Evidently, Greenleaf and Kremen (2006) showed that non-*Apis* bees are beneficial in sunflower pollination in their behavioral interactions with honeybees. Honeybees are forced to move to many flowers when several non-*Apis* bees forage at the same time and in the same floral heads. This improves efficiency of honeybees in pollinating sunflower. Other than bees, different insect species visit flowers of sunflower. For example, among the 20 insect species observed by Arya et al. (1994) visiting sunflower in India, only 12 were bee species, implying that sunflower supports high diversity of flower visitors. Foraging time per bout differs with the flower visitor. Up to 2,089 sec per head have been recorded for honeybees (Fell 1986). The time taken by the bee on a flower is crucial, as it indicates its pollinating potential.

However, there are additional factors that affect this process, including the availability of an adequate reward, such as pollen or nectar, and the presence of competing individuals. The role of wind in the cross-pollination of sunflowers is typically insignificant due to the sticky and heavy nature of the pollen, which has evolved for animal pollination.

Pollinators play a crucial role in providing essential ecosystem services to both natural and agricultural ecosystems. Previous studies conducted at the landscape level have demonstrated positive correlations between the richness and abundance of floral resources and the diversity and activity of pollinators (Steffan-Dewenter et al. 2002; Klein et al. 2003).

On sunflowers, wild bees may contribute more pollination services through enhancement than that of direct pollination, because they concentrate on male flowers rather than female ones. Bees visiting only male flowers make no direct contribution to its pollination but do contribute through enhanced pollination (Greenleaf and Kremen 2006). Intensive studies on sunflower pollination requirements and increase of yield have been reviewed by Rudnev (1941a, 1941b), Goyal and Atwal (1973), and Parker (1981a, 1981b, 1981c).

In an attempt to determine the effect of honeybee visits on sunflower seed production in Manitoba, Furgala (1954b) placed 2.5 colonies per hectare along the east border of sunflower fields. The seed production per hectare in these two crops was as much as recorded in four fields located 4.8 km away from the honey colonies. Furthermore, he noted that seed yield in the farmers' fields decreased progressively as the distance increased from the east to the west. Some observations were made by Rudnev (1941a, 1941b) in Russia in plots 100 m^2 at distances of 500–1,250 m from the honeybee colonies. The number of bees per day was counted throughout the cropping season, and later on, the seed yield was recorded. Within that distance, the number of bees decreased from 100 to 61, and the yield decreased from 6,000 g to 3,700 g. In general, the greater the number of bee visits per floret, the higher the seed yield. Rundev (1941a, 1941b) further observed that when the bee visits per floret were 1, 1.4, 3.4, and 6.1, the average seed harvest per million florets was 53, 76, 133, and 210 kg, respectively. He further calculated that for efficiently pollinating all florets in one hectare of sunflowers, 2,000 bees per day are needed for three weeks of flowering period. A strong colony of *Apis mellifera* can meet the need. Nectar secretion of 15 sunflower cultivars was determined at an average of 18.1 mg/50 flowers. The research indicated that a higher production of sugar resulted in an increased number of bee visits. As a result, there was an average seed yield increase of 26.0%, with the highest increase reaching up to 101% (Kalmer 1996).

According to an estimate by Montgomery (1958), depending upon weather conditions, 21–75 kg of honey can be extracted from 1 ha. In Manitoba, an actual gain of 47 kg was recorded from colonies by placing them in sunflower fields for 15 days (Mitchener 1950). Thus, honeybees can play a vital role in sunflower seed production and also in collecting surplus honey. There was variation in sugar concentration in different cultivars and places; that range was 49–51% (Lesik 1953) and 13–26% (Parker 1981a, 1981b, 1981c). Total sugar produced per flower ranged from 0.4 to 0.6 mg. In USA honeybee colonies taken to sunflower field produced on an average 36 kg of honey, and in USSR from 14–47 kg/colony. It is generally believed that honeybees visit this crop for nectar only, but there are records indicating that 60–85% of the pollen collected was from sunflower. Availability of alternative sources of pollen is the deciding factor.

Panchabhavi et al. (1976) in Karnataka kept 67 *Apis cerana* colonies in a sunflower field of 20 ha and observed that there was 72% seed set, whereas in two other fields open to natural pollinators but without colonies, there were 55 and 57% seed set. Honeybees are vital in the process of hybrid seed production utilizing cytoplasmic male sterile lines, as they are responsible for transferring pollen from the male lines to the female (male sterile) lines. The significance of honeybees in this context has been studied in Pakistan, specifically regarding insect pollination of sunflower (Manzoor-ul-Haq and Fiaz 1980). Nderitu et al. (2008) studied the diversity of sunflower (*Helianthus annuus* L.) pollinators and their influence on seed yield in Eastern Kenya and found that individuals belonging to 14 insect species visited sunflower floral heads, which included six Lepidopteran, five Hymenopteran, two Dipteran, and one Coleopteran species. *Apis mellifera* L. was the most frequent visitor and had the highest pollination efficiency. Plots where insect visitors had access produced on average 53% more seed yield compared with plots where insect visitors were excluded.

Sihag and Kaur (1994) reviewed the floral biology and pollination ecology of sunflower (*Hilianthus annuus*). Majority of research on sunflower pollination indicate that honeybees are the primary pollinating agents and that colonies should be provided to the field for higher seed yield (Barbier and Abid 1966; Sihag and Kaur 1994).

The pollinators of sunflower were documented by Panchabhavi and Devaiah (1977). According to Khalifman (1959), the presence of honeybees in sunflower fields not only enhanced seed production but also reduced damage caused by the sunflower moth (*Homoeosoma nebulella*), a phenomenon known as hysteresis. Sunflower attracted honeybees and other wild bees as its insect visitors (Vaish et al. 1978; Swaminathan and Bhardwaj 1982). *A. cerana* indica and other insect pollinators were found to be most active in the morning (Deodikar et al. 1976; Panchabhavi and Devaiah 1977; Vaish et al. 1978). Their visits were positively influenced by relative humidity (RH) and temperature (Deshmukh 1977; Vaish et al. 1978; Swaminathan and Bhardwaj 1982). Shrivastava and Shrivastava (1986) reported that *A. cerana* indica and *Xylocopa* spp. were the predominant pollinators of sunflower.

Crops like mung bean, sesame, and niger, when grown in the vicinity, reduced the visits of pollinators to sunflower to negligible. Several studies made in India clearly reveal the role of honeybees in the seed production of sunflower, and in their absence, seed yield was low (Sihag and Kaur 1994). Dhoble and Shinde (1982) reported that seed set in sunflower covered with nylon net was markedly lower than that of open-pollinated crops.

Deodikar et al. (1976) and Thakar (1974) found that plots with *A. cerana* colonies produced significantly higher seed yield and weights compared to plots without bees. Panchabhavi et al. (1976) and Channabasavanna (1979) observed a 27% increase in seed set when bee colonies were introduced in the field. Shrivastava and Shrivastava (1986) reported that pollinated plants had 658 grains that were fully developed. Bhattacharya et al. (1982) demonstrated that honeybees were the primary pollinators of sunflower, and when floral heads were bagged to exclude honeybees, there was a reduction in seed yield, seed weight, seed viability, and oil contents. Wakhle et al. (1978) also reported an increase in oil contents due to bee pollination in sunflower. Abrol (1996) identified *Bombus haemorrhoidalis*, *Apis cerana*, *Xylocopa aestuans*, *Halictus* sp., *Syrphus* sp., *Formica* sp., butterflies, and moths as important pollinators of sunflower.

Parker and Frohlich (1983) evaluated the efficacy of a small population of the sunflower leafcutting bee, *Eumegachile* (Sayapis) *pugnata* (Say), as a domesticated pollinator. For these purposes, artificial wooden nesting sites were made available for the population, obtained from adjacent forest and range lands. After release, both females and males visited and effectively pollinated most hybrid cultivars.

Ramya et al. (2014) highlighted the need of intervention of pollinators to ensure productivity as self-fertility varies widely among sunflower varieties. For this purpose, they have selected a less self-fertile variety, *viz*, Morden, and another highly self-fertile variety, *viz*, CO5, to ascertain the role of pollinators in enhancing the productivity of sunflower.

A paired-plot experiment was performed to investigate the impact of pollinators on two different varieties: open pollination and pollinator exclusion. The results indicated that open-pollinated plots had a higher seed yield per head compared to plots where pollinators were excluded. In the case of the less self-fertile Morden variety, the seed set was doubled in open pollination. Additionally, the seed oil yield was 13% higher in open pollination compared to pollinator exclusion. Similarly, for the highly self-fertile CO5 variety, the seed set and seed oil content were 11.4% and 5.8% higher, respectively, in open pollination compared to the pollinator exclusion treatments.

Yield difference between the treatment effects was significantly higher in less self-fertile Morden than in highly self-fertile CO5. Seed quality (100 seed weight, oil content, germination percentage, and vigor index) was significantly more in the presence of pollinators in both the varieties.

Seed set per head nearly doubled in open pollination than in pollinator exclusion treatments in less self-fertile Morden, and the difference in seed set percent between treatments was found to be 36.5, whereas in highly self-fertile CO5, it was found only 11.4% (Ramya et al. 2014). Since it is very hard to get percent self-fertile varieties, even highly self-fertile varieties also require some amount of alien pollen grains for pollination. The percentage weight of set seeds in Morden variety was significantly more in open pollination (91.8%) than in pollinator-excluded plots (62.9%). But in CO5, the difference was only minimal. Seeds produced in open pollination were heavier and plumper, with fewer number of chaffy seeds. Hence, the filled seeds weighed more and recorded more weight/volume percentage in the Morden variety, while in the case of Morden, the yield gap due to pollinator absence was more when compared with CO5 because of its less self-fertile nature. There was a marked increase in oil content to an extent of 13.7% in open pollination in the Morden variety, whereas it was only 6.8% in CO5 (Ramya et al. 2014).

Chamer et al. (2015) assessed the importance of entomophilous pollination and their beneficial effect on sunflower pollination. The results demonstrated that pollen amount and pollen quality that arrive to the stigma have a joint effect on sunflower grain set, some yield components, and positive effect of the pollen redistribution across the head.

5.5 NIGER (*GUIZOTIA ABYSSINICA* CASS)

It is another important oilseed crop belonging to family Compositae. Niger is a comparatively new crop in South India which provides forage to honeybees during the dearth period. It covers an area of 4.8 lakh hectares, and seed production is 1.2 lakh hectare tons. It is also grown in Madhya Pradesh, Maharashtra, Bihar, Orissa, and Tamil Nadu.

5.5.1 Floral Biology

The composite head has 40–60 tubular hermaphrodite flowers and a row of marginal florets. Spread over seven to eight days, the flowers in the outer rows open and dehisce in the morning, and the style lobes of stigma grow during the day and curl backward but do not touch their own pollen; therefore, cross-pollination is common through external agencies like insects.

5.5.2 Pollination and Seed Set

In an experiment in Maharashtra, 40 flower heads in two plots were caged, one with *Apis cerana* and the other without bees; in the former, on an average, 40 seeds were produced, but in the latter, there were only 15 (Bhambure 1958). In another experiment, yield was compared under different situations; in measured plots without *A. cerana* bees, it was 0.9 kg; those caged with bees, 1.2 kg; and in the open with natural pollinators, was 1.1 kg (Panda et al. 1988). In the open, bees were observed at 1.8–8.1/m^2 per minute. It indicates that there is some self-fertilization, but the crop benefits from pollinators that bring about cross-pollination.

5.6 RECOMMENDATIONS FOR SUNFLOWER

Bees play a major role in the dispersal of pollen grains, thereby augmenting sunflower yield both qualitatively and quantitatively. Inadequate pollination is one of the major causes resulting into chaffy seeds in sunflower heads. Therefore, it is imperative to put bee colonies nearby sunflower fields to ensures the quantitative and qualitative characters of sunflowers (Ramya et al. 2014). Hevia et al. (2016) emphasized that for effective crop pollination, the conservation of natural or semi-natural habitats (such as drove roads) within intensive agricultural landscapes should be prioritized in future agricultural policies. Similarly, Perrot et al. (2019) also emphasized to maintain the suitable conditions for pollinators in sunflower production at field scale. Development of early maturing sunflower hybrids, low-input-requiring genotypes, resistance for biotic stresses such as *Alternaria*, and head borer in sunflower is of dire need to ameliorate the productivity level of national scenario in India. Apart from these, it may be introduced in new or non-traditional areas, like the states of Odisha, Bihar, Bengal, and Chhattisgarh, to increase the cropped area in our country. Further, the scenario will be improved with the aid of technological backup or interventions to the farming community.

REFERENCES

Abrol, D. P. 1996. Sunflower pollination: Abundance and diversity of pollinating insects and their effect on seed yield. *Indian Bee J.* 58(2), 60–63.

Abrol, D. P. and Shankar, U. 2012. Pollination in oil crops-recent advances and future strategies. In: Gupta, S. K. (ed.) *Technological Innovations in Major World Oil Crops, Volume 2. 2011.* Springer Verlag, pp. 221–267. ISBN 978-1-4614-0826-0.

Arya, D. R., Sihag, R. C., and Yadav, P. R. 1994. Diversity, abundance and foraging activity of insect pollinators of sunflower (*Helianthus annuus* L.) at Hisar (India). *Indian Bee J.* 56, 172–178.

Aslan, M. M. and Yavuksuz, C. 2010. Effect of honey bee (*Apis mellifera* L.) and bumblebee (*Bombus terrestris* L.) pollinators on yield and yield factors in Sunflower (*Helianthus annuus* L.) production areas. *J. Anim. Vet. Adv.* 9, 332–335.

Astiz, V., Iriarte, L., Flemmer, A., and Hernández, L. 2011. Self-compatibility in modern hybrids of sunflower (*Helianthus annuus* L.) fruit set in open and self-pollinated (bag isolated) plants grown in two different locations. *Helia* 34, 129–138.

Barbier, E. C. and Abid, M. 1966. Pollination and seed formation in sunflower. *Al Awamia* 21, 63–83.

Benedek, P., Manningear, S., and Nagy, B. 1972. The number of colonies and the density of honeybees in sunflower fields in relation to the pollination of the crop. *Z. Ang. Ent.* 71, 385–389.

Berglund, D. R. (ed.). 2007. *Sunflower Production: Extension Publication A-1331 (EB-25 revised) September 2007.* Agricultural Experiment Station, North Dakota, p. 124.

Bhambure, C. S. 1958. Effect of honey bee activity on niger seed production. *Indian Bee J.* 20, 189–191.

Bhattacharya, P., Samuli, R. C., Ghos, M. R., Dasgupta, S. K. and Roy, A. 1982. Sunflower seed yield as influenced by pollination and insect pests. Proc. 10th Int. Sunflower Conference, Sulfers Paradise, pp. 132–234.

Brewer, G. J., Miwa, K., and Hanford, K. 2023. Measuring bee effects on seed traits of hybrid sunflower. *Plants* 12, 2662.

Calmasur, O. and Ozbek, H. 1999. Pollinator bees (Hymenoptera, Apoidea) on sunflower (*Helianthus annuus* L.) and their effect on seed setting in Erzurum region. *Turk. J. Biol.* 23, 73–79.

Chambó, E. D., Garcia, R. C., Oliveira, N. T. E., and Duarte-Junior, J. B. 2011. Honey bee visitation to sunflower: Effects on pollination and plant genotype. *Sci. Agric.* 68, 647–651.

Chamer, A. M. 2012. Influencia del servicio de polinización sobre el rendimiento delgirasol (*Helianthus annuus* L.) en la Argentina. Facultad de Ciencias Exactas yNaturales. Universidad de Buenos Aires, Buenos Aires, p. 110.

Chamer, A. M., Medan, D., Mantese, A. I., and Bartoloni, N. J. 2015. Impact of pollination on sunflower yield: Is pollen amount or pollenquality what matters? *Field Crops Research* 176, 61–70.

Channabasavanna, G. P. 1979. Role of insect pollinators in sunflower production. *Curr. Res.* 8(1), 1–3.

DeGrandi-Hoffman, G. and Chambers, M. 2006. Effects of honey bee (Hymenoptera: Apidae) foraging on seed set in self-fertile sunflowers (*Helianthus annuus* L.). *Environ. Entomol.* 35, 1103–1108.

DeGrandi-Hoffman, G. and Watkins, J. C. 2000. The foraging activity of honey bees *Apis mellifera* and non-*Apis* bees on hybrid sunflowers (*Helianthus annuus*) and its influence on cross-pollination and seed set. *J. Apicult. Res.* 39, 37–45.

Deodikar, G. B., Seethalakshmi, V. S., and Suryanarayana, M. C. 1976. Floral biology of sunflower with special reference of honeybees. *Journal of Palynology* 12(11), 115–125.

Deshmukh, S. D. 1977. Effects of bee visits and climate on sunflower pollination. *Sc. & Cult.* 43, 543–545.

Dhoble, Y. S. and Shinde, Y. M. 1982. Role of insect pollinators for effective seed setting in sunflower. *J Maharashtra Agric. Univ.* 7(1), 103.

Fell, R. D. 1986. Foraging behaviors of *Apis mellifera* L. and *Bombus* spp. on oilseed sunflower (*Helianthus annuus* L.). *J. Kans. Entomol. Soc.* 59, 72–81.

Fick, G. N. 1979. Some factors to consider in the selection of a hybrid. *Sunflower* 5, 26.

Fick, G. N. and Zimmer, D. E. 1976. Yield stability of sunflower hybrids and open polli-nated varieties. Proc. 7th Int. Sunflower Conf, 27 July, Krasnodar, Russia, pp. 253–258.

Free, J. B. 1970. *Insect Pollination of Crops.* Acad. Press, London, p. 544.

Free, J. B. 1993. *Insect Pollination of Crops* (2nd ed.). Academic Press, Harcourt Brace Jovanovich Publication, London, 684 p.

Free, J. B. 1999. Pollination in the tropics. *Beekeeping and Development* 51, 6–7.

Free, J. B. and Simpson, J. 1964. The pollination requirements of sunflower *(Helianthus annuus). Emp. J. exp. Agric.* 32, 340–342.

Freund, D. E. and Furgala, B. 1982. Effect of pollination by insects on seed set and yield. *Am. Bee J.* 122, 648–652.

Furgala, B. 1954a. Honey bees increase seed yields of cultivated sunflowers. *Glean. Bee Cult.* 82, 532–534.

Furgala, B. 1954b. The effect of the honey bee *Apis mellifera* (L.) on the seed set, yield and hybridisation of the cultivated sunflower, *Helianthus annus* L. Ph.D. Thesis, Manitoba University, Canada.

Furgala, B., Noetzel, D. M., and Robinson, R. G. 1979. Observations on the pollination ofhybrid sunflowers. Proc. 4th Int. Symposium on Pollination. Maryland, USA, pp. 45–48.

Goyal, N. P. and Atwal, A. S. 1973. Role of honey bees in sunflower pollination. *Indian Bee Journal* 35, 40–47.

Greenleaf, S. S. and Kremen, C. 2006. Wildbees enhance honeybees' pollination of hybrid sunflower. *Proceedings of the National Academy of Sciences of the United States of America* 103(37), 13890–13895. http://dx.doi.org/10.1073/pnas.0600929103

Griffiths, W. A. and Erickson, Jr., E. H. 1983. Hybrid sunflowers. In: Jones, C. E. and Little, R. J. (eds.) *Handbook of Experimental Pollination Biology.* Van Nostrand Reinhold, New York, NY, pp. 522–535.

Heiser, Jr., C. B., Smith, D. M., Clevenger, S. B., and Martin, Jr., W. C. 1969. The North American sunflowers (*Helianthus*). *Mem. Torrey Bot. Club* 22, 1–218.

Hevia, V., Bosch, J., Azcárate, F. M., Fernández, E., Rodrigo, A., Barril-Graells, H., and González, J. A. 2016. Bee diversity and abundance in a livestock drove road and its impact on pollination and seed set in adjacent sunflower fields. *Agriculture, Ecosystems and Environment* 232, 336–344. http://dx.doi.org/10.1016/j.agee.2016.08.021

Hoffmann, M. 1994. Observações sobre a polinização entomófila de *Helianthus annuus* L. em Viamão, Rio Grande do Sul. *An. Soc. Entomol. Bras* Londrina 23(3), 391–397. [In Portuguese].

Hurd, Jr., P. D., La-Berge, W. E., and Linsley, E. G. 1980. Principal sunflower bees of North America with emphasis on the southeastern United States (Hymenoptera: Apoidea). *Smithson. Contrib. Zool.* 310, 1–157.

Jones, W. A. 1988. A pollination study into relative self-compatibility levels of 25 sunflower oilseed cultivars. Proc. 2nd Australia and International Beekeeping Congress, Gold Coast, Australia.

Jyoti, J. and Brewer, G. J. 1999. Effect of honeybee (Hymenoptera: Apidae) pollination on sunflower hybrids: Proc. 21st sunflower research workshop. *Nat Sunflower Assoc.* January 14–15, 103–107.

Kalmer, F. 1996. Sunflower pollinatrion in Czech Republic. Proc. 7th International Symposium on Pollination, Lethbridge, pp. 407–412.

Kasina, J. M., Nderitu, J., Nyamasyo, G., and Oronje, M. L. 2007. Sunflower pollinatorsin Kenya: Does diversity influence seed yield? *Afr. Crop Sci. Conf. Proc.* 8, 1149–1153.

Khalifman, I. A. 1959. Heterosis in plants as the after effects of pollination by bees (Hysteresis). *Bee World* 40, 303–313.

Klein, A. M., Steffan-Dewenter, I., and Tscharntke, T. 2003. Fruit set of highland coffee increases with the diversity of pollinating bees. *Proc. R. Soc. B* 270, 955–996.

Kleinschmidt, G. J. 1986. Pollination requirement of a range of sunflower cultivars. *Aust. Beekeep.* 6, 253–254.

Krause, G. L. and Wilson, W. T. 1981. Honey bee pollination and visitation patterns onhybrid oilseed sunflower in central Wyoming (Hymenoptera: Apidae). *J. Kansas Entomol. Soc.* 54, 75–82.

Langridge, D. F. and Goodman, R. D. 1981. Honeybee pollination of sunflower cultivars Hysun 30 and Sunfola. *Aust. J. Exp. Agric. Anim. Husbandry* 21, 435–438.

Leins, P. and Erbar, C. 1989. On the mechanisms of secondary pollen presentation in the campanulales-asterales-complex. *Bot. Acta* 103, 87–92.

Lesik, F. 1953. Microfertilization increases the nectar secretion of honey plants. *Pchelovodstvo, Mosk.* 30, 42–45.

Low, A. and Pistillo, G. 1986. The self fertility status of some sunflower cultivars in Australia. *Field Crops Res.* 14, 233–245.

Mahmood, A. N. and Furgala, B. 1983. Effect of pollination by insects on seed oil percentage of oilseed sunflower. *Am. Bee J.* 123, 663–667.

Manzoor-ul-Haq and Fiaz, M. 1980. Insect pollination of sunflower *(Helianthus annuus)* in Pakistan. *J. Apic. Res.* 19, 83–88.

Mitchener, A. V. 1950. Honeybees and sunflowers. *Ann. Con. Man. Agron.* 27.

Montgomery, B. E. 1958. Preliminary studies of the composition of some Indian nectars. *Proc. Indiana Acad. Sci.* 68, 159–163.

Nderitu, J., Nyamasyo, G., Kasina, M., and Oronje, M. L. 2008. Diversity of sunflower pollinators and their effect on seed yield in Makueni District, Eastern Kenya. *Spanish Journal of Agricultural Research* 6(2), 271–278.

Panchabhavi, K. S., and Devaiah, M. A. 1977. A note on the activities of insect pollinators on sunflower during winter and summer seasons at Bangalore. *Current Science* 6, 88–89.

Panchabhavi, K. S., Devaiah, M., and Basavanna, G. P. C. 1976. The effect of keeping Apis cerana indica F. colonies on the seed set of sunflower, Helianthus annuus Linn. *Mysore Journal of Agricultural Sciences* 10, 631–636.

Panda, P., Sontakke, B. K., and Sarangi, P. K. 1988. Preliminary studies on the effect of bee (*Apis cerana indica*) pollination on yield of sesamum and niger. *India Bee J.* 50, 63–64.

Parker, F. D. 1981a. A candidate red clover pollinator *Osmia coerulescens* (L.). *J. Apic. Res.* 20, 62–65.

Parker, F. D. 1981b. How efficient are bees in pollinating sunflowers? *J. Kans. Entomol. Soc.* 54, 61–67.

Parker, F. D. 1981c. Sunflower pollination: Abundance, diversity and seasonality of bees and their effect on seed yields. *J. Apic. Res.* 20(1), 49–61. doi: 10.1080/00218839.1981.11100473

Parker, F. D. and Frohlich, D. R. 1983. Hybrid sunflower pollination by a manageable composite specialist: The sunflower leafcutter bee (Hymenoptera: Megachilidae). *Environ. Entomol.* 12, 576–581.

Perrot, T., Gaba, S., Roncoroni, M., et al. 2019. Experimental quantification of insect pollination on sunflower yield reconciling plants and field scale estimates. *Basic Appl. Ecol.* 34, 75–84.

Radford, B., Nelson, R., and Rhodes, J. 1979. Agents of pollination in sunflower crops in the central Darlingdowns, Queensland. *Aust. J. Exp. Agric. Anim. Husb.* 19, 565–569.

Ramya, M., Muthuraman, M., and Saravanan, P. A. 2014. Yield enhancement in self-fertile sunflower (*Helianthus annuus* L.) by honeybees. *Madras Agric. J.* 101(7–9), 269–272.

Rao, M. G., Nadre, K. R., and Suryanarayana, M. C. 1995. Studies on pollination of male sterile line for foundation seed production in hybrid sunflower. *Ind. Bee J.* 57(4), 170–173.

Robinson, R. G. 1980. Artifact autogamy in sunflower. *Crop Sci.* 20, 814–815.

Rudnev, V. Z. 1941a. The effect of pollination by bees on seed yield of sunfl ower seeds. *Soc. Zern. Hox.* 2, 134–140.

Rudnev, V. Z. 1941b. The technique of utilization of bees for the pollination of sunflower. *Soc. Zern. Hox.* 3, 77–84.

Ryan, G. S. and Brugnoni, L. F. 1986. El cultivo de girasol. Serie Agricultura. In: *Oleaginosas No. 4802.* INTA, Buenos Aires, p. 40.

Seiler, G. J. 1997. Anatomy and morphology of sunflower. In: Schneiter, A. A. (ed.) *Sunflower Technology and Production.* Agronomy Monograph No. 35. ASA, CSSA, SSSA, Madison, WI, pp. 67–111.

Shrivastava, O. S. and Shrivastava, G. P. 1986. Nocturnal pollination in sunflower *Helianthus annaus* L. In: Kapil, R. P. (ed.) *Pollination Biology: An Analysis.* Inter-India Publication, New Delhi, pp. 59–65.

Sihag, R. C. and Kaur, G. 1994. Colony foraging: An overview. *Indian Bee J.* 56.

Singh, M. P., Singh, K. I., and Devi, C. S. 2000. Role of *Apis cerana* pollination on yield andquality of rapeseed and sunflower crops. In: Matsuka, M., Verma, L. R., Wongsiri, S., Shrestha, K. K., and Partap, U. (eds.) *Asian Bees and Beekeeping in Asia: Progress of Research and Development, Proc. 4th AAA Int. Conf.* Oxford and IBH Publishing Co. Pvt. Ltd, Kathmandu, New Delhi, pp. 186–189, March 23–28, 1998.

Steffan-Dewenter, I., et al. 2002. Scale-dependent effects of landscape context on three pollinator guilds. *Ecology* 83, 1421–1432.

Swaminathan, R. and Bhardwaj, S. C. 1982. Bee pollinators of sunflower and their foraging behaviour. *Indian Bee Journal* 44(2), 32–34.

Thakar, C. V. 1974. Experiments on pollination of sunflower. *Indian Bee J.* 36(1–4), 36–38.

Vaish, O., Agrawal, S., and Joshi, M. 1978. Frequency of insect visitors for pollen foraging on sunflower in relation to daily temperature and humidity. Proc. 8th International Sunflower Conference, United States, pp. 148–157.

Virupakshappa, K., Gowda, J., and Ravikumar, R. L. 1992. Autogamy and self-incompatibility as influenced by genotypes and planting date in sunflower. Proc. 13th Int. Sunflower Conf. September 7–11, Pisa, Italy, pp. 1281–1290.

Vranceanu, A. V., Stoenescu, F. M., and Pirvu, N. 1988. Genetics progress in sunflowerbreeding in Romania. 24 Proc. 12th Int. Sunflower Conf. July 25–29, Novi Sad, Yugoslavia, pp. 404–410.

Wakhle, D. M., Nair, S. K., and Phadke, R. P. 1978. Effect of bee-pollination on the oil and protein content in the seeds of sunflower *(Helianthus annuus* L.). *Indian Bee J.* 40, 1–2.

6 Sesamum

6.1 INTRODUCTION

Sesame (*Sesamum indicum* L., Pedaliaceae), which originated in Africa, is probably the oldest oilseed crop cultivated in many parts of the world. Sesame seed found in an excavation at Harappa civilization is dated to circa 2000 BC. It is one of the most important oilseed crops, having been grown in the Near East and Africa for more than 5,000 years for cooking and medicinal needs (Joshi 1961; Weiss 1971; Mabberley 1997; Purseglove 1968). In the context of world scenario, China, India, and Myanmar are the world's largest producers of sesame, followed by Sudan, Nigeria, Pakistan, Bangladesh, Ethiopia, Thailand, Turkey, and Mexico (Desai 2004).

A fair record has been mentioned through the Indian subcontinent and throughout Mesopotamia and Eastern Anatolia since time immemorial. Sesame has been spread to China by the second century BCE. Today, China is the world's largest consumer and one of the primary producers of sesame. According to an estimate, about 7.8 million hectares of the total world cropped area are under sesame cultivation, and it ranks sixth in the world among vegetable oils. Asia and Africa collectively contributed 3.67 million tons to the global production of sesame, out of the total 3.83 million tons. Specifically, Asia accounted for 2.29 million tons, while Africa produced 1.38 million tons. Myanmar stands first as producer of sesame in the world, followed by India, China, Ethiopia, Sudan, Uganda, Nigeria, Bangladesh, Egypt, and Pakistan (FAOSTAT 2012).

Sesame is a herbaceous annual plant cultivated for its nutritious and flavored edible seeds. It is commonly known as til in Hindi vernacular and has been grown under diverse kinds of soils and fits well into different crop rotations, with low management input by small holders and marginal farmers (Bisht et al. 1980). Sesame is thinly spread throughout India as a kharif crop in the North and a rabi crop in the South, covering an area of 2.3 ha.

6.2 ECONOMIC IMPORTANCE

Sesame has a long history of traditional use as food, medicine, and lamp oil for illumination. It is also known as gingelly, til, and benne seed and popularly hailed as the "queen of oilseeds." Yet it is neglected crop and primarily cultivated by resource-poor and marginal farmers as a major commodity in most of the regions where it is grown. Despite its medicinal and nutritional advantages, it is less cultivated, as it requires intensive and costly labor inputs during harvesting. This contrast presents a paradox, requiring further insights to work toward improvements in agricultural productivity to ensure food and nutritional security and poverty alleviation for vulnerable small and marginal farmers.

6.2.1 As Food and Nutritional Benefits

Sesame seeds possess an attractive, nutty flavor that is highly appealing. When the seeds are crushed, they yield an aromatic edible oil with a distinct taste. When crushed, seeds can be transformed into a creamy paste, adding another dimension to their culinary versatility. The seeds of sesame contain a fair amount of oil (ca. 50%); unsaturated fatty acids, primarily oleic and linoleic; protein (ca. 25%), including the significant limiting amino acid methionine; and dietary fiber and nutrients, including minerals, lignans, tocopherol, and phytosterol. Usually, white-seeded varieties are edible and used for culinary purposes, whereas the oil obtained from black-seeded varieties of sesame is used for the manufacture of soaps and perfumes. Sesame seeds are incredibly versatile, serving as food, flavor enhancers, and a valuable oil source. These seeds provide a rich protein

DOI: 10.1201/9781032656724-7

source of exceptional quality, along with their oil content (Bedigian 2000; Bedigian and Harlan 1983; Dirar 1993; Mkamilo and Bedigian 2007).

Rich in polyunsaturated fats, its oil is remarkably stable because of its natural antioxidants sesamin and sesamolin, which are unique phenylpropanoid lignans that occur in few other plant groups (Bedigian 1984, 1988, 2003a; Bedigian et al. 1985). The oil possesses a low cholesterol content and a significant proportion of polyunsaturated fat. It finds application as a salad dressing, cooking oil, shortening, and margarine. Sesame seeds, paste, and oil have great value in a diverse range of edible products, whether consumed in their raw or roasted form. Whole sesame seeds are incorporated into breads, pastries, crackers, and confectionary items (Bedigian 2000, 2004a) and are sprinkled on breads, rolls, and cookies before baking. Salad oil is derived from the crude sesame oil pressed from the seeds, while the "refined" oil extracted from toasted seeds is used as a flavoring agent in cooked dishes.

6.2.2 As Medicinal Value and Industrial Use

Sesame seeds are utilized to treat conditions such as piles, dysentery, and urinary issues. Additionally, they can be applied as a poultice on ulcers. The decoction of pedalium fruit is effective in addressing urinary complaints, spermatorrhea, and impotency. Furthermore, the infusion of leaves and stems is beneficial in managing venereal diseases like gonorrhea and dysuria.

The roots have antibilious properties. Sesame oil has pharmaceutical uses and offers medicinal and curative properties (Duhoon and Tripathi 2005). It has also been used in making liniments and ointments. Massaging sesame oil into the scalp once a week is an excellent way to nourish the scalp and restore the natural balance and luster of hair.

In the regions of eastern and southern Africa, leaves are utilized for treating snakebites and malaria, while in India and China, they are employed in the treatment of cancers. Sesame oil is also employed as a remedy for cough and earache, as well as an emmenagogue and abortifacient. Additionally, sesame seeds are highly regarded for their beneficial laxative properties. Throughout history, sesame oil has been extensively utilized in the production of soaps, lubricants, and cosmetics, as well as serving as a source of lamp oil.

Sesame oil is a pharmaceutical solvent, and as an insecticide, it is exceptionally synergistic with pyrethrin, which is derived from *Pyrethrum cinerarifolium* Trev. (Bedigian et al. 1985; Haller et al. 1942). It is a carrier for medicines and perfumes. Sesame oil is highly versatile and can be used as a substitute for olive oil. It is also utilized in the production of penicillin injections containing oil and wax according to the USP. Additionally, it is an ingredient in ammonia liniment NF. Apart from that, sesame oil contains various other unsaponifiable substances, such as sterols, triterpenes, pigments, and tocopherols. In India, sesame oil is an essential component of vegetable ghee and is commonly used for anointing hair and skin. Lower-quality oils are employed in the manufacturing of soaps, paints, and lubricants, while in the past, sesame oil was used as a lamp oil. Furthermore, the ash obtained from burned sesame stems possesses medicinal properties and is used as a salt (Bedigian and Harlan 1983).

6.2.3 As Ornamentals

Ceratotheca triloba is domestically grown as an ornamental plant to a limited extent.

6.2.4 Sesame Seed Constituents

Per 100 g edible portion, dry decorticated sesame seed contains water (3.8 g), energy (2,640 kJ; 631 kcal), protein (20.5 g), fat (60.2 g), carbohydrate (11.7 g), dietary fiber (11.6 g), calcium (60 mg), magnesium (345 mg), phosphorous (667 mg), iron (6.4 mg), zinc (6.7 mg), vitamin A (9 IU), thiamin (0.70 mg), riboflavin (0.09 mg), niacin (5.80 mg), and folate (115 µg) (Hu et al. 2007).

6.2.5 Sesame and Cultural Survival

For centuries, the Chinese have burned sesame oil to make soot for their fine ink blocks and used as India ink. Sesame has served as a source of inspiration in various domains of world culture, such as literature, art, music, and intellectual pursuits (Bedigian 1984, 2000, 2004a; Batra et al. 2001; Bedigian and Harlan 1986). Alongside other seeds, leaves, and flowers, sesame holds a significant importance in religious ceremonies (Bedigian 2004; Bedigian and Harlan 1986; Mehra 1967, 2000). For thousands of years, subsistence farmers in Asia and Africa have cultivated sesame, often alongside other vital cereal and legume species, as a staple crop.

Smallholders have safeguarded and created new biological diversity, availing themselves of environmental influences for selection, managing soil and water needs, and maintaining their livelihoods with the help of sesame, even under harsh conditions. Sesame is a resilient crop that thrives in environments where few others can, requiring minimal fertilizer, pesticide, fossil fuel usage, or intensive labor. Climate change concerns have finally brought attention to the world's harshest dry, marginal lands. Drought and high temperatures as well as increasing population pressures combine to impact indigenous farmers' livelihood, as Yardley (2009) shows about farmers in Andhra Pradesh.

6.3 FLORAL BIOLOGY

Sesame is an erect tropical herbaceous annual herb belonging to the Pedaliaceae family. Depending on cultivar and growth conditions, it grows to a height of 0.5 to 2 m. Sesame genotypes reveal a lot of morphological and genetic diversity (Brar and Ahuja 1979) (Figure 6.1). It comprises about 20 species, most of which are indigenous to tropical Africa, with two disjunct sections found in India (Ihlenfeldt et al. 1979).

Some botanists have opined that the Pedaliaceae family comprises 13 genera (Ihlenfeldt 2001) and an estimated 75 species of annual and perennial herbs. Flowers are hermaphrodite and are borne singly or in group. They have four stamens and an ovary with many ovules. When a flower opens at 0400 h, the stigma is receptive and is covered over with pollen. The flowers are shed

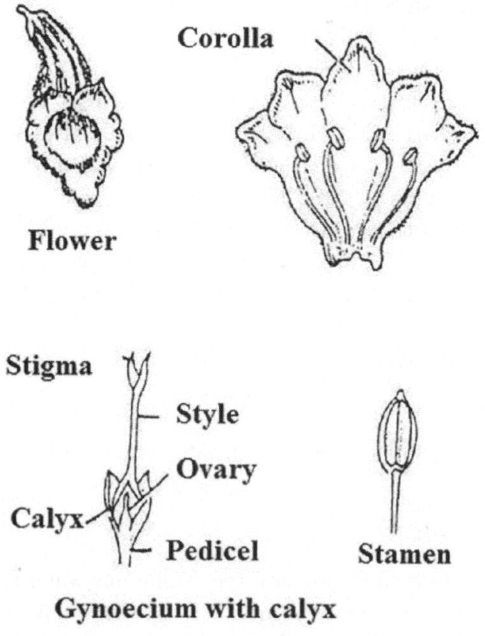

FIGURE 6.1 Floral biology of sesame.

12 h later. In many flowers, anthers do not shed pollen, and insects attracted to extra floral nectaries bring about cross-pollination (Howard et al. 1919). Plants have floral and extrafloral nectaries at the base. Its nectar has 38% sugar concentration, and honeybees exploit it during the dearth period. Despite being commonly perceived as a self-pollinating crop, the sesame blossom's structure promotes cross-pollination.

6.3.1 Ecology of Sesame and Various Wild Relatives

Usually, the sesame crop is more suitable in the tropics and subtropics, but the development of newer cultivars and summer plantings has extended its range into temperate regions too. It occurs mainly between 25° S and 25° N but extends and grows up to 40° N in China, Russia, and the United States, 30° S in Australia, and 35° S in South America. Of course, it is sensitive to low temperatures, and for this reason, it generally grows better from sea level up to 1,500 m. Certain cultivars of sesame, despite being classified as short-day plants, have developed the ability to adjust to different photoperiods. In conditions where the day length is 10 hr, flowering typically commences around 42–45 days after the seeds are sown. However, specific Turkish cultivars exhibit even faster flowering, with some varieties blooming as early as 37 days after sowing (Bedigian 1984). Temperature and moisture have major regulating and modifying effects on the number of days to flowering. Generally, high temperatures are more conducive to optimal growth and production.

Mode of Pollination: entomophilous or mainly mediated by insect pollinators.

6.3.2 Mutual Relationship of Flowering and Pollinators

Sesame flowers attract a wide variety of insects (Figure 6.2–6.10). Flowering of sesame plants and bees have a mutual relationship, in which both benefit from each other. Of total pollination activities,

FIGURE 6.2 Sesame crop with flowers and capsules.

FIGURE 6.3 *Apis dorsata* foraging on sesame flower.

FIGURE 6.4 *Apis dorsata* inside the sesame flower for nectar.

FIGURE 6.5 Bumblebee, *Bombus haemarrgdalis*, visiting sesame flower.

FIGURE 6.6 *Megachile carbonaria* bee foraging on sesame flower.

over 80% are performed by insects, wherein bees contribute with nearly 80% of total insect pollination (Robinson and Morse 1989). Sesame flowers, typically considered self-pollinating, actually have a structure that facilitates cross-pollination. The extent of cross-pollination can range from 0.5 to 65%, depending on factors such as insect activity, flora diversity, and environmental conditions (Kumar and Lenin 2000a, b). According to Ashri (2007), cross-pollination rates in Nigeria ranged from 2.7 to 51.7%. Insect pollinators are most abundant on sesame flowers during the time periods of

FIGURE 6.7 *Megachile anthracina* foraging on sesame flower.

FIGURE 6.8 *Megachile disjuncta* foraging on flower.

FIGURE 6.9 Nomia bee visiting flowers of sesame.

FIGURE 6.10 Clearwing hawk moth foraging on flower.

1100–1300 h and 1300–1500 h. This is because nectar flow is copious in the sesame crop, especially in the middle period of the day; thereafter, the nectar concentration gradually diminishes (Kamel et al. 2013). The floral structure of this crop helps bees by providing them the required landing surface and ample reward in the form of nectar (Ashri 1990).

Panda et al. (1988) have recorded that bee pollination resulted to a significant increment of the seed yield by 22 to 33% rather than pollination without insects. The plants exposed to insect pollinators gave 25% higher yield as compared with those covered with cages to exclude insects. Cages with *A. cerana* inside gave 2.6 kg pods per plot as compared with 1.6 kg in cages with bees (Phadke et al. 1967; Panda et al. 1988). Out in the open, *A. cerana*, *A. dorsata*, and *A. florea* were the most common insects on the crop, and they began foraging at 0600, 0700, 0800 h and visited 6.3, 5, and 1.9 flowers per minute, and flowers pollinated by them gave 51, 53, and 4 seeds per pod, respectively. Seed weight was also heavier as compared with the controls (Rao et al. 1981). While foraging, 72% of *A. cerana* collected nectar, 23% pollen only, and 4% both nectar and pollen. At some point, they engaged in anther contact to facilitate cross-pollination, resulting in enhanced crop yield (Phadke et al. 1967). *Apis mellifera* is recognized for its diligent interaction with flowers. In Punjab, comprehensive research was conducted on the genetics and pollination needs of this particular plant (Sikka and Gupta 1949).

6.4 POLLINATION REQUIREMENTS

Co-evolution of flowering plants and their pollinators started about 225 million years ago (Price 1975; Maiti and Maiti 2011). Of the total pollination activities, over 80% is performed by mediation of insects, especially with bees; that is why they are considered as the best pollinators (Robinson and Morse 1989). Insufficient number of suitable pollinators leads to decline in fruit and seed production (Partap 2001).

Even though sesame is predominantly a self-pollinated crop, honeybees increase cross-pollination (Uzo and Ojiake 1981). Given that emasculation is easy, and pollen abundant, with hand-pollination it is possible to obtain a large number of seeds per capsule.

Flowering initiation is influenced by the duration of daylight (Kotecha et al. 1975) and shows variation among different cultivated varieties. The bell-shaped flowers start forming three to eight weeks after planting and continue to develop for several weeks, sometimes even until the arrival of frost. The flowers open gradually, starting from the bottom of the stem and progressing toward the top. The corolla, which is fused and divided into lobes, displays colors such as white, rose-pink, lavender, or mauve. Many cultivars feature dark-purple markings on the inside, resembling the appearance of foxglove. These flowers are arranged in racemes found in the leaf axils. The lowest lobe is often as much as 1 cm longer than the others and often shaded purple. A yellow streak in the interior contains flavonoids (Bedigian 1984). Sesame possesses a cleistogamous nature, meaning, it self-pollinates prior to the flowers opening, although occasional cross-pollination by insects is possible. The growth pattern of sesame landraces is indeterminate, allowing the plant to continually produce leaves, flowers, and capsules for as long as weather conditions permit. In this regard, sesame contradicts a conventional belief in domestication, which involves a transition from an indeterminate to a determinate growth pattern (Harlan 1992). Flowers arise in leaf axils on the upper stem and branches, and most flowers open at dawn, wilt after midday, and drop at dusk. Pollen shed (anthesis) occurs just before the flowers open (Khidir and El Awad 1972); hence, most fertilization is cleistogamous. The stigma is receptive one day before flower opening and remains receptive for another day. Under natural conditions, pollen remains viable for 24 hr. Flowers are mostly self-pollinated, but cross-pollination is possible and, under exceptional conditions, may reach 68% (Yermanos 1980). Sesame's ability of self-pollination makes sense for situations where there is a low supply of resources available, such as limited soil moisture and nutrients. If fewer pollinators are available, self-compatibility is an evolutionary advantage. Reduced flower size brings anther and stigma in close proximity and is an advantage in dry habitats for such dryland species.

6.5 DIVERSITY OF POLLINATORS AND INCREASE IN YIELD

Sesame is self-pollinating, although differing rates of cross-pollination have been reported by Yermanos (1980), Ashri (2007), and Sarker (2004). The pollination process occurs at the time the flowers open (Kafiriti and Deckers 2001; Langham 2007). Yermanos (1980) found less than 1% when the sesame was fruit set/seed set.

Both open pollination and bee pollination treatments were effective to increase the seed yield of sesame up to 22 to 33% more than that in "pollination without insects" (Panda et al. 1988). In addition to increasing the yield, cross-pollination also helps raise quality through a more unified ripening period and an earlier harvesting time. According to Kumar and Lenin (2000a, b), Apoidea were the predominant flower visitors (96%) of sesame. Among these, *A. mellifera* was the most abundant (44.9%), followed by *A. dorsata* (31.4%) and *A. florea* (19.7%). *A. mellifera* comprised 30 and 32% of the foraging population on sesame crops in Egypt, where species of *Megachile*, *Polistes*, and *Eristalis* were also important (Rashad et al. 1979a).

Number of bees visitation was drastically reduced or almost ceased after 1500–1700 h. These findings can help save the pollinators by applying insecticides late in the afternoon. By protecting the pollinators, high yields can be ensured. Bee pollination not only ensures increase in yields of sesame but also improve its quality. It ensures uniform maturity and early harvest of crop. Provision of bee colonies during the flowering period of crop is a simple but essential input (Rashad et al. 1979b).

Mahfouz et al. (2012) observed a significant positive correlation (R2 = 0.5911 and 5817; P = 0.05) between the mean temperature, relative humidity, and the presence of pollinators visiting sesame flowers. These results suggest that temperature and relative humidity play a significant role in influencing insect activity on sesame flowers. Kamel et al. (2013) surveyed the insect pollinators associated with sesame, *Sesamun indicum* L. (*Pedaliaceae*), at the Agriculture Research Farm, Faculty of Agriculture, University of Suez Canal, during the growing seasons of 2011 and 2012. Sampling was conducted on a weekly basis, with three sampling sessions per day.

Three methods were used to collect and identify insects from the sesame plants (a sweep net, pitfall traps, digital camera, and eye observation). A total of 29 insect species were collected and properly identified during the survey. The insect pollinators observed on the plants were classified into four groups: 18 belonged to Hymenoptera, 7 to Diptera, 3 to Lepidoptera, and 1 to Coleoptera. The findings revealed that the honeybee, *Apis mellifera*, was the most dominant species in the 2011 season and the second most dominant species in the 2012 season. Conversely, the small carpenter bees, *Ceratina tarsata*, held the title of the most dominant species in the 2012 season and the second most dominant species in the 2011 season. The percentage of Hymenoptera was higher in both seasons, accounting for 90.94% and 89.59% of the total, followed by Diptera with 3.93% and 5.38%, Lepidoptera with 3.58% and 3.62%, and finally, Coleoptera with 1.53% and 1.39%, respectively. Pathirana (1994) estimated the degree of natural cross-pollination (NCP) in three sesame genotypes with contrasting characters grown in alternate rows at two locations. The lower portion (consisting of the first-formed five capsules), upper portion (comprising the last five fully developed capsules), and middle portion of capsules were individually collected. The presence of dominant markers in the offspring was used as the measure for non-competitive pollination (NCP). The extent of NCP between different genotypes and locations was comparable, ranging from 4.02% to 5.10% when considering the total number of hybrid seeds per plant. The lower portion of the early (1.03–1.31% NCP) and medium (1.51–2.08%) flowering genotypes and the upper portion of the late-flowering genotype (2.27–2.49% NCP) recorded less NCP than the total plant. The presence of natural cross-pollination (NCP) between the late-flowering genotype and the lower portion of the early-flowering genotype was found to be insignificant, ranging from 0% to 7%. These results emphasize the importance of implementing controlled self-pollination to ensure purity in sesame cultivation. However, in situations where controlling pollination is not feasible, the contamination caused by NCP can be mitigated by selectively collecting seeds. Specifically, seeds should be sampled from the lower capsules of early- and medium-flowering genotypes, as well as from the upper capsules of late-maturing genotypes. This strategy aims to minimize the impact of cross-pollination and maintain seed purity in the absence of controlled pollination practices.

Mahfouz et al. (2012) determined the pollinators visiting sesame (*Sesamum indicum* L.) seed crop with reference to foraging activity of some bee species like *Apis mellifera*, *Anthidium* sp., and *Xylocopa* sp. The findings indicated that the proportion of insects belonging to the Hymenoptera order was prominent, with Lepidoptera, Diptera, and Coleoptera following suit. The peak activity of Hymenopterous fauna was observed during the fourth week of the flowering period, gradually declining in the subsequent weeks. The highest count of pollinators occurred between 9000 and 1100 h, followed by 1100 and 1300 h, 1300 and 1500 h, and 1500 and 1700 h. Among the bee species, *Apis mellifera* exhibited the highest population, trailed by *Xylocopa* sp., and finally *Anthidium* sp., across all time intervals. Additionally, it was evident that temperature, wind, and relative humidity influenced the proportion of insects visiting sesame flowers. An experiment was conducted on studies on bumblebee diversity for crop pollination in the farmers' field across various landscapes, viz, plain areas, rainfed steep landscapes, and at elevated rainfed areas of Jammu region, Jammu, and Kashmir during the months of July to mid-September 2012 to 2014. A total of 17 species of native pollinators belonging to three major families, namely, Apidae, Megachilidae, and Halictidae, of order Hymenoptera, was recorded to visit flowers of *Sesamum indicum* L. in the study areas (Table 6.1). Versatility in foraging insect guild and efficient supply utilization of available resources to establish a strong plant–pollinator interaction in *Sesamum indicum* L. were depicted through the generated data in the present study. In all the areas under study, the presence of three species of bumblebees (family Apidae), namely, *Bombus haemorrhoidalis*, *B. trifasciatus*, and *Bombus* sp. as pollinator was prominent, along with six species of Megachilid bees, namely, *Megachile bicolor*, *M. lanata*, *M. hera*, *M. disjuncta*, *M. anthracina*, and *M. carbonaria*; three species of honeybees, viz, *Apis dorsata*, *Apis mellifera*, and *Apis indica*; four species of other *Apidae*

TABLE 6.1
List of Insect Visitors Recorded from Sesame, *Sesamum indicum*, Flowers from Various Landscapes in Jammu Region

			Landscape Type		
Sl. No.	Insect Morphospecies	Family	High-Altitude, Sub-Temperate Zone	Elevated Land	Plain
1	*Amegilla zonata* L.	Apidae	–	+	+
2	*Amegilla confuse* Smith	-do-	–	+	+
3	*Xylocopa latipes* Drury	-do-	–	+	+
4	*Xylocopa pubescens* Spinola	-do-	–	+	+
5	*Bombus haemorrhoidalis* Smith	-do-	+	+	–
6	*Bombus trifasciatus* Smith	-do-	+	+	–
7	*Bombus* sp.	-do-	+	+	–
8	*Apis dorsata* Fab	-do-	–	+	+
9	*Apis mellifera*	-do-	–	+	+
10	*Apis indica*	-do-	–	+	+
11	*Nomia iridiscens* Smith	Halictidae	+	+	+
12	*Megachile bicolour* Fab.	Megachilidae	+	+	+
13	*Megachile hera* Bing.	-do-	+	+	+
14	*Megachile lanata* Fab.	-do-	+	+	+
15	*Megachile disjuncta* Fab.	-do-	+	+	+
16	*Megachile carbonaria* Smith	-do-	+	+	+
17	*Megachile anthracina* Smith	-do-	+	+	+

bees, viz, *Xylocopa latipes*, *X. pubescence*, *Amegilla zonata*, and *A. confusa*; and one species of Halictidae, namely, *Nomia irridipens*. Diurnal visitation pattern of insect foragers shows peak activity time during forenoon, between 1000 and 1200 h.

Spatial distribution among various solitary bee foragers of *S. indicum* revealed bravura of fact on the availability of pollinating bees in the Jammu region for this important crop. Megachilidae and Halictidae bees were found to visit the flower of *S. indicum* in high-altitude, sub-temperate areas of Sunderbani, Sureinsar, and in the elevated landscape as well as plains of Jammu district. *Amegilla* spp. amongst members of Apidae were found to visit these flowers in both sub-temperate and plains, whereas *Xylocopa* spp. were mostly found to forage on the flowers in plain areas of Jammu. Bumblebees were found confined in the sub-temperate areas as a major flower visitor of sesame.

Striking plant–pollinator interaction was recorded between the study plant and the flower visitor pool. Whitish-pink-tinged color of petals of the flowers with upside-down bell-shaped structure in morphology could easily help in visual navigation cue for entomoforagers. Ample content of nectar, profuse pollen, collectively performed the duty to offer best available rewards to the foragers. Groupwise, foraging behavior, when observed keenly, revealed many remarkable features which helped in separating the proficient pollinators from large flower visitor pool recorded from these plants. Field observation on major insect visitors generated a handful of information regarding exclusive plant–pollinator interaction of major social and solitary bees foraging on sesamum. Large-bodied Megachilid bees, viz, *Megachile bicolor*, *M. lanata*, *M. disjuncta*, *M. anthracina*, and *M. carbonaria*, were found to enter the flower and use their mandible to hold the petals after they landed on the flower to collect rewards (Chaudhary and Jain 1978). Large- to medium-bodied *Apis*, viz, *Xylocopa* spp., *Amegilla* spp., etc., was found to spend less time while probing resources of these flowers. Members of Halyctid bees, viz, *Nomia* sp., were also found to approach flowers, intruding its head part inside corolla tube.

The average length of foraging bout, when calculated, was found higher in Megachilidae and bumblebees, whereas in other cases, moderate foraging length was recorded. Of all these insects, bumblebees and Megachile bees were the most efficient pollinators as they inserted tongues deep in the flower to collect nectar, as compared to honeybees, which mostly collected nectar from the side of flowers, thereby acting as nectar thieves. Sesame flowers were of whitish-pink tinge with upside-down bell-shaped structure in morphology. Bumblebees and Megachile bees were found to enter the flower to collect nectar and pollen rewards, whereas the honeybees acted as nectar thieves, collecting nectar from the sides of the flower. Bumblebees and Megachile bees initiated activities early in the morning, at 0730 h, and maximum abundance was observed between 1100 and 1300 h; thereafter, the population declined and continued up to 1700 h in the evening. Bumblebees, on average, spent 7.12 + 0.28 sec/flower ($n = 20$). In almost all cases of such major insect visitors, while foraging, their body was found to rub the pollen-loaded anthers intensely, resulting in a heap of pollen adhered in the ventral parts, which were brushed against the rough stigma head of the next flower while foraging in the next, ensuring successful and efficient sternotribic (carrying pollen chiefly on the sterna region of body) pollination.

6.6 IMPACT OF PESTICIDES ON THE POLLINATORS

Pollinators could be protected by rational pest management tactics—that is, pesticide application, if needed, should be done in the late afternoon to protect the pollinators for high seed yield or spray at a time of day when crop flowers are closed. In order to obtain optimum yields by controlling insect pests on one hand and providing safety to pollinators on the other, the role of pest management in pollinator conservation is essential, as it discourages the use of pesticides, which poses negative effects on non-target organisms, such as pollinators. There are four different schools of thought in integrated pest management, such as integrated insecticide management, ecologically based pest management, organic pest management, and exploitation of transgenics. All have the clear mandate to reduce pesticide pressure of crops as well as on ecosystem. These pesticide reduction strategies of integrated pest management may have multiple benefits, not just for crop pest parasites and predators, but also for pollinators. A number of factors may be detrimental and limit the efficiency of bees as pollinators, such as the effects of pesticides on pollinators (Desneux et al. 2007; Chauzat et al. 2009; Krupke et al. 2012).

REFERENCES

Ashri, A. 1990. Sesame genetic resources: Collection, evaluation and conservation. In: *Oil Crops: Sesame and Sunflower Subnetworks. Proc, Joint 2nd Workshop, Cairo, Egypt (1989).* International Development Research Centre, Ottawa, pp. 2–9.

Ashri, A. 2007. Sesame (*Sesamun indicum* L.). In: Singh, R. J. (ed.) *Genetic Resources, Chromosome Engineering and Crop Improvement. Vol 4: Oilseed Crops.* CRC Press, Boca Raton, FL. USA, pp. 231–289.

Batra, A., Ali, D. J., Goyal, S., and Ajita. 2001. Queen of oilseeds (*Sesamum indicum* L.): A lesser exploited medicinal herb with immense potentiality. In: Khan, I. A. and Khanum, A. (eds.) *Role of Biotechnology in Medicinal and Aromatic Plants*, Vol. 4. Chapter 14 Ukaaz, Hyderabad.

Bedigian, D. 1984. *Sesamum indicum* L. Crop origin, diversity, chemistry and ethnobotany. PhD dissertation. University of Illinois, Urbana-Champaign. University Microfilms DA8502071, Dissertation Abstracts International 45: 1985: 3410-B.

Bedigian, D. 1988. Sesamum indicum L. (Pedaliaceae): Ethnobotany in Sudan, crop diversity, lignans, origin, and related taxa. In: Goldblatt, P. and Lowry, P. P. (eds.) *Modern Systematic Studies in African Botany (MSB 25)*, Missouri Botanical Garden Press, St. Louis, pp. 315–321.

Bedigian, D. 2000. Sesame. In: Kiple, K. F. and Ornelas-Kiple, C. K. (eds.) *Cambridge World History of Food*, Vol. 1. Cambridge University Press, New York, pp. 411–421.

Bedigian, D. 2003a. Evolution of sesame revisited: Domestication, diversity and prospects. *Genetic Resources and Crop Evolution* 50, 779–787.

Bedigian, D. 2004. History and lore of sesame in Southwest Asia. *Economic Botany* 58(3), 329–353.

Bedigian, D. and Harlan, J. R. 1983. Nuba agriculture and ethnobotany with particular reference to sesame and sorghum. *Economic Botany* 37, 384–395.

Bedigian, D. and Harlan, J. R. 1986. Evidence for cultivation of sesame in the ancient world. *Economic Botany* 40, 137–154.

Bedigian, D., Siegler, D. S., and Harlan J. R. 1985. Sesamin, sesamolin and origin of sesame. *Biochemical Systematics and Ecology* 13(2), 133–139.

Bisht, D. S., Naim, M., and Mehrotra, K. N. 1980. Studies on the role of honeybees in rapeseed production. Proc. 2nd Inl. Conf. Apic Trop Climates, New Delhi, pp. 491–496.

Brar, G. S. and Ahuja, K. C. 1979. Sesame: Its culture, genetics breeding and biochemistry. In: Malik, C. P. (ed.) *Annual Rev. Plant Sciences.* Kalyani Publ., New Delhi, pp. 247–460.

Chaudhary, J. P. and Jain, J. L. 1978. Nesting and foraging behaviour of a mason bee Megachile lanata Lepel (Megachilidae: Hymenoptera). *Indian Journal of Entomology* 40(4), 405–411.

Chauzat, M. P., Carpentierm, P., Martel, A. C., Bougeard, S., Cougoule, N., Porta, P., Lachaize, J., Madec, F., Aubert, M., and Faucon, J. P. 2009. Influence of pesticide residues on honey bee (Hymenoptera: Apidae) colony health in France. *Environmental Entomology* 38, 514–523. doi: 10.1603/022.038.0302

Desai, B. B. 2004. Seeds handbook: Biology, production, processing, and storage. *Science*, 787.

Desneux, N., Decourtye, A., and Delpuech, J. M. 2007. The sublethal effects of pesticides on beneficial arthropods. *Annual Review of Entomology* 52, 81–106. doi: 10.1146/annurev.ento.52.110405.091440

Dirar, H. A. 1993. *The Indigenous Fermented Foods of the Sudan and Nutrition. A Study in African Food and Nutrition.* 1st edn. Cambridge University Press, Cambridge, UK, pp. 169–176.

Duhoon, S. S. and Tripathi, A. K. 2005. Medicinal an curative properties of sesame (*Sesamum indicum* L.). Paper presented in National Seminar on "Strategies for enhancing production and export of sesame and niger", ARS, RAU, Mandor, Jodhpur, p. 56.

Faostat. 2012. Food and agriculture organization of the United Nations. Roma, Italy. Available at http://faostat.fao.org/site/567/DesktopDefault.aspx?PageID=567It

Haller, H. L., LaForge, F. B., and Sullivan, W. N. 1942. Some compounds related to sesamin: their structures and their synergistic effect with pyrethrum insecticides. *J. Org. Chem.* 7(2), 185–188.

Harlan, J. R. 1992. *Crops and Man.* 2nd Edition, American Society of Agronomy and Crop Science Society of America, Madison, USA.

Hu Y. -M., Ye W. -C., Yin Z. -Q. and Zhao S. -X. 2007. Chemical constituents from flos *Sesamum indicum* L. *Yao Xue Xue Bao.* 42, 286–291.

Ihlenfeldt, H. -D. 2001. Fruits in *Pterodiscus* Hook. and a key to the species. *Mitt. Inst. Allg. Bot. Hamburg* 28/29, 5–21.

Ihlenfeldt, H. D. and Grabow-Seidensticker, U. 1979. The genus *Sesamum* L. and the origin of the cultivated sesame. In: Kunkel, G. (ed.) *Taxonomic Aspects of African Economic Botany.* Proceedings of the IX Plenary Meeting of AETFAT. Excmo, Las Palmas de Gran Canaria, pp. 53–60.

Joshi, A. B. 1961. *Sesamum.* Indian Central Oilseed Committee, Hyderabad (now Ind. Council of Agric. Res., New Delhi), p. 109.

Kafiriti, E. M. and Deckers, J. 2001. Sésame: *Sesamum indicum* L. In: *Raemaekers, Agriculture en Afrique tropicale.* Drection Générale de la Coopération Internationale (DGCI), Bruxelles, Belgique, p. 1634.

Kamel, S. M., Blal, A. E. H., Mahfouz, H. M., and Said, M. 2013. Pollinator fauna of sesame crop (*Sesamum indicum* L.) in Ismailia governorate, Egypt. *Cercetări Agronomice în Moldova* 46(2) (154), 53–64.

Khidir, M. Q., and EL Awad S. H. 1972. Studies on floral biology in sesame. *Sudan Agric. Journal* 7, 17–26.

Kotecha A. K., Yermanos D. M. and Shropshire, F. M. 1975. Flowering in cultivars of sesame (Sesamum indicum) differing in photoperiodic sensitivity. *Econ. Bot* 29, 185–191.

Krupke, C. H., Hunt, G. J., Eitzer, B. D., Andino, G., and Given, K. 2012. Multiple routes of pesticide exposure for honey bees living near agricultural fields. *PLoS ONE* 7(1), e29268. doi: 10.1371/journal.pone.0029268

Kumar, R. and Lenin, J. K. 2000a. Insect pollinators and effects of cross pollination on yield attributes of sesame *(Sesamum indicum* L.). *Indian Bee J.* 62(1–2), 75–80.

Kumar, R. and Lenin, J. K. 2000b. Insect pollinators and effects of cross pollination on yield attributes of sesame (*Sesamum indicum* L.). *Indian Bee Journal* 62, 80–88.

Langham, D. R. 2007. Phenology of sesame. In: Janick, J. and Whipkey, A. (eds.) *Reprinted from: Issues in New Crops and New Uses.* ASHS Press, Alexandria, VA, pp. 144–182.

Mabberley, D. J. 1997. *The Plant-Book.* 2nd edn. Cambridge University Press, Cambridge, p. 680.

Mahfouz, H. M., Kamel, S. M., Belal, A. H., and Said, M. 2012. Pollinators visiting sesame (*sesamum indicum* l.) seed crop with reference to foraging activity of some bee species. *Cercetări Agronomice în Moldova* 45(2)150, 49–55.

Maiti, P. K. and Maiti, P. 2011. *Biodiversity: Perception, Peril and Preservation.* Asoke K. Ghosh, New Delhi, India.

Mehra, K. L. 1967. History of sesame in India and its cultural significance. *Vishveshanand Indological Journal* 6, 93–107.

Mehra K. L. 2000. History of sesame in India and its significance. *Asian Agri History* 4, 5–19.

Mkamilo, G. S. and Bedigian, D. 2007. *Sesamum indicum* L. Record from PROTA4U. In: van der Vossen, H. A. M. and Mkamilo, G. S. (eds.) *PROTA (Plant Resources of Tropical Africa/Ressources végétales de l'Afrique tropicale)*, Wageningen, Netherlands.

Panda, P., Sontakke, B. K., and Sarangi, P. K. 1988. Preliminary studies on the effect of bee *(Apis cerana indica Fab.)* pollination on yield of sesamum and niger. *Indian Bee J.* 50, 63–64.

Partap, T. 2001. Mountain agriculture, marginal land and sustainable livelihoods: Challenges and opportunities. International Symposium on Mountain Agriculture in HKH Region (21–24 May 2001). ICIMOD, Kathmandu, Nepal.

Pathirana, R. 1994. Natural cross-pollination in Sesame *(Sesamum indicum* L.). *Plant Breeding* 112, 167–170.

Phadke, K. G., Pant, N. C., and Sinha, R. B. P. 1967. The foraging behaviour of *Apis indica* F ((Hymenoptera: Apidae) visiting flowers of sesame *(Sesamum orientate* L.). *Indian J. Ent.* 29, 157–163.

Price, P. 1975. *Insect Ecology.* John Wiley and Sons, New York, USA.

Purseglove, J. W. 1968. *Tropical Crops. Dicotyledons.* Longmans, London.

Rao, G. M., Lazar, G. M., and Suryanarayana, M. C. 1981. Foraging behavior of honeybees in sesame (Sesamum indicum L.). *Indian Bee J.* 43, 97–100.

Rashad, S. E., Ewies, M. A., and El Rabie, H. G. 1979a. Insect pollinators of sesame (*Sesame indicum* L) with special reference to the role of honeybees. Proceedings IVth International Symposium on pollination, Maryland Agricultural Experiment Station.

Rashad, S. M., Mohamed, M. I., and El-Hefny, A. M. 1979b. Some notes on the biology and behavior of *Nomia unidentata* Oliver. Proceedings of IVth International Symposium on Pollination. Maryland Agricultural Experiment Station Miscellaneous Publication, 1, 313–319.

Robinson, W. E. and Morse, R. A. 1989. The value of honeybees as pollinators of US crops. *American Bee Journal* 129(1), 477–487.

Sarker, A. M. 2004. Effect honeybee pollinisation on the yield of rapeseed, mustard and sesame. *Geobros* (Jodhpur) 31, 49–51.

Sikka, S. M. and Gupta, N. P. 1949. Pollination studies in *Sesamum orientale* L. *Indian J. Genetics* 9, 33–191.

Uzo, J. O. and G. U. Ojiake. 1981. Breeding and selection method for sesamum on the basis of assessment of major Nigerian strains, F1 hybrids and segregating generation. *FAO Plant Prod. Prot. Bull.* 29, 90–96.

Weiss, E. A. 1971. *Castor, Sesame & Safflower.* Leonard Hill Books, London, p. 901.

Yardley, J. 2009. As india grows, rural areas left in the dust. *The International Herald Tribune*. September 5, p. 4.

Yermanos, D. M. 1980. Sesame. In: Fehr, H. and Hadleys, H. (eds.) *Hybridization of Crop Plants*. Agronomy-crop Science Society of America, Madison, WI, pp. 549–563.

7 Safflower

7.1 INTRODUCTION

Safflower (*Carthamus tinctorius* L.) is an oilseed crop that belongs to the family Compositae or Asteraceae (Weiss 2000). The term *Carthamus* is derived from the Arabic word *quartum*, or *gurtum*, which refers to the color of dye obtained from safflower flowers. Cultivated safflower originated from two related species: wild safflower (*C. oxyacanthus*) from Afghanistan and neighboring countries, and saffron thistle (*C. lanatus*) from Ethiopia (Chavan 1961). Safflower has been cultivated in India since ancient times and is referred to as kusumba in ancient scriptures.

Safflower is an underutilized and often neglected minor oilseed crop. In India, it is commonly known as kusum in Hindi and is grown as a winter crop in dry regions, usually under unirrigated conditions. Safflower is a versatile crop cultivated for centuries in India for its orange-red dye (carthamin) extracted from its vibrant flowers and its high-quality oil rich in polyunsaturated fatty acids, particularly linoleic acid (78%). Its oil, containing unsaturated fatty acids, is considered beneficial for dietary purposes. Safflower flowers have medicinal properties and are widely used in Chinese herbal preparations (Dajue and Mündel 1996). The tender leaves, shoots, and thinning of safflower plants are consumed as pot herbs and salads. They are rich in vitamin A, iron, phosphorus, and calcium. Young safflower plants are sold as a green vegetable in Indian markets and neighboring countries (Nimbkar 2002). Safflower can be grazed upon or stored as hay or silage. The forage is palatable and offers feed value and yields similar to or better than oats or alfalfa. Thus, each part of the safflower plant holds value and serves different purposes. Safflower exhibits high adaptability to low-moisture conditions, which is why its production is primarily limited to regions with scarce rainfall worldwide.

The Carthamus genus comprises 25 species, with only *C. tinctorius* being cultivated. Despite its immense potential to be grown under diverse conditions and utilized for various purposes, safflower remains a minor crop globally in terms of cultivation area and production compared to other oilseed crops. However, there has been a renewed interest in this crop in recent years.

1. Huge shortfall in oilseed production due to scanty rainfall.
2. Preference of consumers for healthy oil with less amounts of saturated fats.
3. The medicinal uses of flowers and edible dyes.

Safflower is the most suitable crop to meet the preceding shortcomings in the world scenario. Safflower is a promising oilseed crop for the production of specialty oils in the Mediterranean area. Oil quality traits such as high oleic acid content, high linoleic acid content, high saturated fatty acid content, or high gamma-tocopherol content have been developed in this crop. Velasco et al. (2012) found that cross-fertilization rates varied across three environments, with average rates of 5.7%, 12.1%, and 13.2%. However, they observed higher frequencies of up to 35.9% at the single-plant level and up to 58.3% at the single-head level. India ranks second position in harvested area of safflower seed among Asian countries after Kazakhstan with 150,000 ha. Production of safflower (rabi) in India during 2014–15 was 0.90 lakh tons (anonymous, 2015–2016). In India, productivity levels of Maharashtra and Karnataka states (2014–15) account for 55 and 37% of total safflower area and production, respectively. The other safflower-producing states are Andhra Pradesh, Orissa, Madhya Pradesh, Chhattisgarh, and Bihar. Safflower production in India was mostly confined to rainfed conditions during winter.

DOI: 10.1201/9781032656724-8

7.2 WORLD DISTRIBUTION AND PRODUCTION

Safflower, a member of the Asteraceae family, is an annual plant with a thistle-like appearance. Its florets are clustered together in compact inflorescences known as capitula. In Europe and the Americas, its seeds are harvested for oil and birdseed (Dajue and Mündel 1996). Recently, safflower has become a vehicle for plant molecular farming (PMF), which introduces GM traits for producing useful biomolecules (Horn et al. 2004). Adventitious presence of GM material with medical or industrial applications is a cause of public concern (Gerhard et al. 2005).

For centuries, safflower cultivation has been a tradition spanning from China to the Mediterranean region, extending along the Nile valley all the way to Ethiopia (Weiss (1971). Presently, it is grown commercially in India, the United States, Mexico, Ethiopia, Kazakhstan, Australia, Argentina, Uzbekistan, China, and the Russian Federation. Pakistan, Spain, Turkey, Canada, Iran, and Israel also grow safflower to a limited extent. Safflower acreage and production around the world have witnessed wide fluctuations in the past. Safflower seed production in the world rose from 487,000 MT in the year 1965 to 1,007,000 MT in 1975, and subsequently, it decreased to 921,000 MT in 1985 (anonymous, 2002). Until 1980, Mexico was the largest safflower producer globally. During that time, safflower cultivation in Mexico covered an expansive area of 528,000 ha, and its annual production exceeded 600,000 metric tons in the years 1979–1980.

However, the area and production of safflower in Mexico decreased significantly in later years, becoming only 10% of the area and production recorded for the years 1979–1980 (Cervantes-Martinez et al. 2001). Commercial production of safflower in the United States was started in the 1950s, and the area rapidly increased to 175,000 ha mainly in the states of California, Nebraska, Arizona, and Montana. Presently, it is grown over an area of 100,000 ha (Esendal 2001). Currently, safflower cultivation in China spans an area ranging from 35,000 to 55,000 ha, yielding an annual seed production of 50 to 80 mt. Among the Chinese provinces, Xinjiang holds the title of the largest safflower producer, contributing 80% to the country's total safflower output. Additional safflower-producing provinces in China include Yunnan, Sichuan, Henan, Hebei, Shandong, Jiangsu, and Zhejiang (Zhaomu and Lijie 2001). At present, India holds the position of the world's leading safflower producer, followed by the United States, Mexico, and China. In the years 2004–2005, India had an estimated safflower cultivation area of 387,000 ha, resulting in a seed production of 154,000 mt (anonymous, 2004–2005). Maharashtra and Karnataka states contribute 72% and 24%, respectively, to the safflower cultivation and production in India, with other safflower-producing states including Andhra Pradesh, Orissa, Madhya Pradesh, Chhattisgarh, and Bihar. Safflower production in India is primarily limited to rainfed conditions during winter

7.3 ECONOMIC IMPORTANCE

For ages, safflower cultivation in India has served a dual purpose. Apart from obtaining a vibrant orange-red dye from its beautiful florets, it has been extensively grown to produce top-notch edible oil with abundant polyunsaturated fatty acids. This oil plays a crucial role in lowering cholesterol levels in the bloodstream. Nutritionally speaking, safflower oil closely resembles olive oil, as it boasts high concentrations of linoleic and oleic acid. The monounsaturated fatty acid like oleic acid is also known to reduce low-density lipoprotein (LDL; bad cholesterol) without affecting high-density lipoprotein (HDL; good cholesterol) in blood (Smith 1996). According to Weiss (1971), safflower oil possesses excellent stability and maintains its consistency even in low temperatures, making it a viable option for use in frozen or chilled food products. Kleingarten (1993) suggests that safflower oil is more suitable for hydrogenation in the production of margarine compared to soy or canola oils. Additionally, Smith (1996) notes that safflower oil is nonallergenic, making it a suitable choice for injectable medications. Safflower is considered to be ideal for cosmetics and is used in "Macassar" hair oil and Bombay "sweet oil" (Weiss 1971). Safflower oil is preferred for the paint and varnish industry, owing to its specific properties of absence of linolenic acid, presence of high linoleic acid, low color values, low free fatty acids, low unsaponifiables, and no wax, which make the quality in paints, alkyd resins, and coatings beyond comparison (Smith 1996). However, with the development of cheaper petroleum products and a shift to water-based paints, the use of safflower oil in the paint and varnish industry has been reduced drastically in recent times. In India, safflower oil is also used for lighting and the manufacture of soap and to waterproof leather buckets (Weiss 1971). Furthermore, safflower oil finds application in the preparation of "roghan," a substance employed for leather preservation and as a glass adhesive. In addition, charred safflower oil is utilized for treating sores and aiding in rheumatism, as mentioned by Weiss (1971). Moreover, safflower oil is used in the production of infant foods and liquid nutritional formulations.

7.4 FLORAL BIOLOGY

The safflower plant can be described as a bushy, herbaceous annual possessing several branches, which are categorized as primary, secondary, and tertiary, with each terminating into a globular structure called capitulum. Safflower blooms from March to April and produces 15–150 orange-yellow flowering heads terminating the main axis and branches. The flowering head that terminates the main axis blooms first, then flowering proceeds downward, with those flower heads on the lowest branches opening last. A flower head consists of from 20 to 100 yellow and orange florets surrounded by bracts. Each floret consists of five petals united to form a long corolla tube. The stamens consist of five filaments and five anthers (Figure 7.1). Anthers are united around the style. The pistil consists of a single one-chambered ovary having one ovule. In many self-fertile varieties, anthers release pollen early in the day, and soon afterward, the style elongates and the stigma appears above the top of the anther tube, covered with pollen grains. Thus, self-pollination occurs in self-sterile varieties; the style elongates and passes through the anther tube before anthers release pollen. In such varieties, self-fertilization does not occur, and instead, insects perform cross-pollination. The florets produce abundant nectar and pollen, making the crop a significant honey source in regions

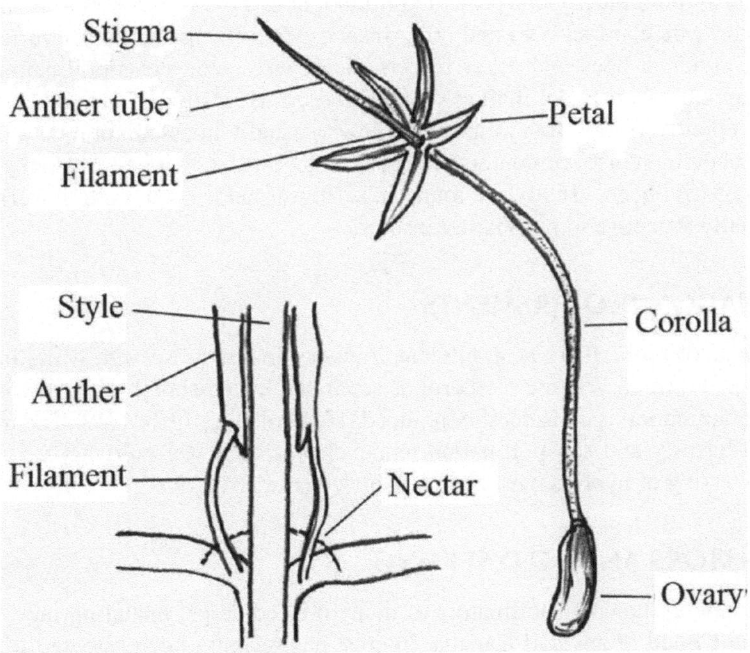

FIGURE 7.1 Floral morphology of safflower.

where it is extensively cultivated. Honeybees play a vital role as pollinators, not only benefiting the seed production in self-sterile varieties but also improving the yield and quality of self-fertile varieties. To ensure effective pollination, it is recommended to place five colonies of *Apis mellifera* and four to six colonies of *Apis cerana* evenly across each hectare when the flowers are at 10–15% blooming stage. The yellow and orange florets are surrounded by bracts that are sharp and pointed. Anthers dehisce in the morning, and soon after the stigma appears above them, catching pollen grains; thus, self-fertilization is possible.

Safflower is a predominantly self-pollinating crop. However, depending on the genetics and environmental conditions, outcrossing may be considerable (Li and Mundel 1996). A diversity of bees, as well as other insects, are attracted to safflower pollen and nectar and thus contribute to outcrossing. Thus, to ensure planned crossing, flowers must be emasculated prior to pollen shedding. Safflower is categorized as a partially allogamous crop, typically exhibiting an outcrossing rate of less than 10% (Knowles 1989). This outcrossing is primarily facilitated by insects, with honeybees being the primary pollinators (Pandey and Kumari 2008; Singh and Nimbkar 2007). However, there have been reports of higher outcrossing rates. Claassen (1950) observed outcrossing rates ranging from 5% to 40%, while rates as high as 59% have been reported in India (Singh and Nimbkar 2007). It is worth noting that some estimations of high outcrossing rates are based on plants growing in close proximity, where contact between neighboring inflorescences may lead to overestimated outcrossing (Christianson et al. 2008). In a study conducted in Germany, Rudolphi et al. (2008) found an outcrossing rate ranging from 29.9% to 63.1% within plots with a plant density of 40 plants per square meter. Additionally, they reported an outcrossing rate between plots ranging from 6.5% to 18.1%.

McPherson et al. (2009) studied pollen flow at different distances at locations in Canada and Chile. They found outcrossing frequencies between 0.5 and 1.7% at plant distances between 0.3 and 3 m and decreasing frequencies at longer distances, resulting in values lower than 0.05% at distances between 50 and 100 m. In addition, their findings revealed that the frequency of outcrossing varied across different spatial locations, suggesting an uneven distribution of pollen and asymmetrical pollination patterns. Safflower is an often cross-pollinated crop and pollinated by honeybees, bumblebees, and syrphid flies Boch (1961); Deshmukh et al. (1985). Flowers have an abundance of pollen and nectar, which attract bees and other insects. Commercial safflower varieties are mostly self-pollinated, although bees and other insects may cause some cross-pollination and increase seed set. The amount of cross-pollination will depend on the safflower variety. Large populations of several insect predators are often found in safflower, usually in association with the green peach aphid. The most common insect predators are lady beetles and lacewing flies. Honeybees, as well as several species of wild bees, usually are found in safflower fields during the flowering period. The flower is especially attractive to pollinating insects.

7.5 POLLINATION REQUIREMENTS

In specific regions of India, there is significant cross-pollination observed, while in other regions, the occurrence is minimal. When a seed crop is separated by a distance of 30 m with another crop in between, contamination is considerably reduced (Howard et al. 1916, 1919; Kadam and Patankar 1942). The self-fertility and self-pollination tendencies of safflower cultivars vary. Experimental production and testing of hybrids have been conducted (Rubis 1970).

7.6 POLLINATORS AND SEED SETTING

Honeybees are among the chief pollinators of many oilseed crops, including rape, sunflower, and safflower. In the United States and Canada, interest has recently been directed to the cultivation of safflower (*Carthamus tintorius*). Here again, honeybees are the predominant insect visitor, and plants from which insects were excluded yielded only half as much seed as those open to insects.

Langridge and Goodman (1980), in a trial conducted on safflower (*Carthamus tinctorius*) cv. Gila in northern Victoria without irrigation, showed no significant increase in seed yield per plant, oil content, percentage germination, or weight of 1,000 seeds when plots were accessible to bees and larger insects compared to plots enclosed in cages to exclude these insects. The predominant insects observed working on the flowers for nectar and pollen were honeybees and native bees. This suggests that cv. Gila is primarily self-pollinated or pollinated by airborne pollen.

Kumari's research focused on the pollen biology of safflower and indicated that safflower is primarily self-pollinated, although bees or other insects are generally necessary for optimal fertilization and maximum yield. The highest germination percentage (73.6 ± 28.2%) was recorded 48 hr after anthesis. Viability is reduced to about 73.76% in TTC after 24 hr of anthesis. If anther dehiscence occurs before style elongation, the stigma becomes coated with pollen and gets self-fertilized. However, if anther dehiscence occurs after style elongation, cross-pollination is required. The absence of pollinators leads to self-pollination. The anthesis of disc florets occurs between 0530 and 0800 hours. The number of pollen grains produced per anther/flower is 236 ± 74/5906 ± 372, respectively. The stigma remains receptive for 32–56 hr after anthesis.

There is about 85.99% seed setting in the open-pollinated heads, followed by 38.15% in muslin cloth, and 35.54% in butter paper-bagged flowers. Khalil et al. (1986) observed 19 insect species of five orders (Hemiptera, Lepidoptera, Coleoptera, Diptera, and Hymenoptera) as visitors to safflower in bloom. The major pollinators belonged to superfamily Apoidea, especially honeybees, which represented 58.15–63.36% of total pollinators. The native bees *Ceratina tarsata* mor., *Megachile* sp., *Nomia* sp. and *Prosopis klugi* Fr. proved to be of great importance for safflower pollination (Boch 1961).

Honeybees (*Apis meltifera* L.) visited safflower (*Carthatmts tinctoius*) for nectar and pollen. About 90% of the honeybees observed visited the crop in the morning. The nectar contained about 25% sugars. Honeybees were by far the predominant visitors, although other bees also visited the flowers. Safflower plots to which insects had full access produced twice as much seed as plots from which insects were excluded during the blooming period. Large numbers of honeybees are attracted by safflower during the flowering period. Bees are especially attracted to safflower because of its abundant pollen and abundant nectar, which they gather readily. The benefits of honeybees or other insect pollinators to commercial varieties depend upon the amount of self-fertility in the variety. Since Frio is lower in self-fertility than Gila, yield increase from cross-pollination is expected to be greater. Experiments with the variety Gila have indicated that honeybees may increase seed set by 5% or more. However, under different environmental conditions, the benefits from pollinators may vary.

Nectar sugar concentration ranges from 13 to 29% at different times of the day, and honeybees, *Apis mellifera*, mostly visit for nectar, and very few for pollen. It was observed near Bombay that *A. cerana*, *A. dorsata*, and *A. florea* began foraging at 1000 h, and pollen collectors worked up to 1400 h, and nectar collector up to 1500 h; the three species were at these jobs for 24 and 21 seconds, 17 and 13 sec, 32 and 33 sec, indicating that *A. florea* worked more thoroughly (Deshmukh et al. 1985). In Karnataka, the presence of honeybees (*A. cerana*) resulted in a 54.8% increase in safflower yield compared to naturally pollinated crops. When two honeybee colonies were present, the yield reached 14.4 q/ha, whereas with only one colony, it was 12.2 q/ha (Lingappa et al. 1999). The success of pollination directly affects seed production in safflower, as the plants tend to self-pollinate in the absence of pollinators (Knowles 1969). Claassen (1950) reported a range of 0 to 100% cross-pollination, with most plants experiencing cross-pollination rates between 5% and 40%. Various studies have demonstrated the contribution of pollinators to different extents in safflower pollination (Kadam and Patankar 1942; Levin and Butler 1966; Butler et al. 1966; Levin et al. 1967). Although safflower is typically considered a self-pollinated crop, insects, particularly bees, play a significant role in its pollination (Boch 1961; Eckert 1962; Rubis et al. 1966). Factors such as temperature and humidity can influence the seed setting of bagged flowers (Patil and Chavan 1958). Kumari and Pandey (2005) found that open-pollinated heads had a seed setting rate of 85.99%, whereas bagged conditions using muslin cloth and butter paper resulted in 38.15% and 35.54% seed setting, respectively.

Safflower is primarily a self-pollinated crop, but bees and other insects are generally necessary for optimal fertilization and maximum yield (Figure 7.2). Without pollinators, self-pollination occurs. Safflower exhibits a unique behavior compared to other members of the Asteraceae family, as most of them are mostly self-incompatible (Kumari and Pandey 2005).

Among the pollinating insects, Hymenoptera (bees), Lepidoptera (butterflies), and Diptera (flies) play a significant role in pollinating safflower crops. The foraging behavior of bees in safflower is similar to that observed in sunflower (Lindsley 1978; Parrish and Bazzaz 1978; Prasad and Rao 1984), dahlia (Heslop-Harrison and Shivanna 1977), niger (Panda et al. 1995; Pyke 1984; Reddy

FIGURE 7.2 Bee pollinators on safflowers: (a) *Apis dorsata*, (b) *A. cerana*, (c) *Bombus haemorrhoidalis*, (d) *Megachile haryanensis*, (e) *Megachile cephalotes*.

FIGURE 7.2 (Continued)

1976; Dhakal and Pandey 2003; Pandey and Dhakal 2004; Neff and Simpson 1990), and safflower itself (Kumari and Pandey 2005). The safflower plant has moderately branched stems, with each branch terminating in a head, forming complex synflorescences. The basic unit of these synflorescences is the dichasm, and the cymes are arranged in a nearly corymbose fashion.

7.7 IMPACT OF PESTICIDES ON POLLINATORS

Several neonicotinoids, that is, imidacloprid, acetamiprid, thiamethoxam, etc., however, showed very strong toxicity to pollinating insects, particularly to the honeybee (*Apis mellifera* L.). They not only triggered chronic sub-lethal effect on honeybee colonies supplemental pollen diet containing multiple brood cycle (Galen and Michael 2015) but also caused many other effects, like behavioral disturbances, orientation difficulties, and impairment of social activities (Decourtye et al. 2004a, 2004b).

Telangre et al. (2018) recorded that in comparison to imidacloprid, acetamiprid was safest to honeybees. The initial impact of spraying acetamiprid 20% SP insecticide at the full dose on honeybee visits was relatively low on the first day after spraying, ranging from 2.66 to 3.66 bees per square meter. However, on the fifth day after spraying, the bee visits increased to a range of 3.00 to 4.66 bees per square meter. The visit of honeybees was initially high in the 14th day after spraying and found at par (in the range of 6–6.66 bees/m^2), with honeybees observed before spraying of acetamiprid 20% SP, followed by 10th day after spraying (in the range of 5–6.77 bees/m^2) honeybees population. They further revealed the comparison of full dose and half dose of acetamiprid, among which at half dose acetamiprid has found more visits of honeybees than the full dose of acetamiprid spraying on safflower. Again, Telangre et al. (2018) compared the effect of neonicotinoids on foraging behavior of honeybees on safflower (*Carthamus tinctorius* L.) at Parbhani during *rabi* 2016–17.

REFERENCES

Boch, R. 1961. Honeybee activity of safflower (*Carthamus tinctorius* L.). *Canad. J. Plant Sci.* 41, 559–562.

Butler, G. D., Werner, E. G., and Leven, M. D. 1966. Native bees associated with safflower in South central Arizona. *Kans Ent. Soc. Jour.* 39(3), 434–436.

Cervantes-Martinez, J. E., Rey-Ponce, M., and Velazquez-Cagal, M. 2001. Evaluation of accessions from world collection of safflower for Alternaria incidence and seed oil content. In: Proceedings of the 5th International Safflower Conference, Williston, ND, and Sidney, MT, July 23–27, 2001.

Chavan, V. M. 1961. *Niger and Safflower.* Indian Central Oilseeds Committee Publication, Hyderabad, pp. 57–150.

Christianson, J., Mcpherson, M., Topinka, D., Hall, L., and Good, A. G. 2008. Detecting and quantifying the adventitious presence of transgenic seeds in safflower, Carthamus tinctorius L. *J. Agric. Food Chem.* 56, 5506–5513.

Dajue, L. and Mündel, H. H. 1996. *Safflower. Carthamus tinctorius L. Promoting the Conservation and Use of Underutilized and Neglected Crops.* 7th edn. Institute of Plant Genetics and Crop Plant Research, Gatersleben/International Plant Genetic, Rome.

Decourtye, A., Armengaud, C., Renou, M., Devillers, J., Cluzeau, S., Gauthier, M. et al. 2004a. Imidacloprid impairs memory and brain metabolism in the honeybee (*Apis mellifera* L.). *Pesticide Biochemistry and Physiology* 78, 83–92.

Decourtye, A., Devillers, J., Cluzeau, S., Charreton, M., and Pham Delegue, M. H. 2004b. Effects of imidacloprid and deltamethrin on associative learning in honeybees under semi-field and laboratory conditions. *Ecotoxicology and Environmental Safety* 57, 410–419.

Deshmukh, A. K., Mohana Rao, G., and Karve, A. D. 1985. Studies on the effect of honeybee pollination on the yield of safflower. *Indian Bee J.* 33(3–4), 35–38.

Dhakal, M. R. and Pandey, A. K. 2003. Changes in pollinator populations during the flowering span of niger (*Guizotia abyssinaca* Cass.). *J. Ind. Bot. Soc.* 82, 74–77.

Eckert, J. E. 1962. The relation of honey bees to safflower. *Am. Bee. J.* 102(9), 349–350.

Esendal, E. 2001. Global adaptability and future potential of safflower. In: Proceedings of the 5th International Safflower Conference, Williston, ND, Sidney, MT, July 23–27, 2001.

Galen, P. D. and Michael, S. E. 2015. Assessment of chronic sub lethal effects of Imidacloprid on honey bee colony health. *Environ. Toxicol. Chem.* 10(3), 748.

Gerhard, F., Andrew, C., and Karen, A. 2005. Animal nutrition with feeds from genetically modified plants. *Arch Anim Nutr.*59, 1–40.

Heslop-Harrison, J. and Shivanna, K. R. 1977. The receptive surface of the angiosperm stigma. *Ann. Bot.* 41, 1233–1258.

Horn, M. E., Woodard, S. L., and Howard, J. A. 2004. Plant molecular farming: systems and products. *Plant Cell Rep* 22, 711–720.

Howard, A., Howard, G. L. C., and Rahman, K. A. 1916. Studies in Indian oilseeds I: Safflower and mustard. *Mem. Dept. Agric. Indian Bot. Ser.* 7, 214–272.

Howard, A., Howard, G. L. C., and Rahman, K. A. 1919. Studies in the pollination of Indian crops. *India Dept. Agr. Mem. Bot. Ser.* 10, 195–220.

Kadam, B. S. and Patankar, V. K. 1942. Natural cross-pollination in safflower. *Indian Jour. Genet. Plant Breed.* 2, 69–70.

Khalil, S. I. Y., El-Shakaa, S. M. A., and Hassan, M. R. 1986. Insect pollinators of safflower (*Carthamus tinctorius* L.) and their effect on its yield. *Zagazig J. Agricult. Res.* 13, 575e591.

Kleingarten, L. 1993. In Notes Safflower Conference, Billings, MT, February 18, 1993. Mundel, H. H. and Braun, J. (eds.). Lethbridge, AB, Canada, p. 5.

Knowles, P. F. 1969. Centers of plant diversity and conservation of crop germplasm, Safflower. *Econ. Bot.* 23, 324–329.

Knowles, P. F. 1989. Safflower. In: Downey, R. K., Robellen, G., and Ashri, A. (eds.) *Oil Crops of the World.* McGraw-Hill, New York, pp. 363–374.

Kumari, A. and Pandey, A. K. 2005. Pollination mechanism and behaviour of pollinators in safflower (*Carthamus tinctorius* L.). *J. Ind. Bot. Soc.* 82, 56–61.

Langridge, D. F. and Goodman, R. D. 1980. A study of pollination of safflower (*Carthamus tinctorius*) cv. Gila. *Australian Journal of Experimental Agriculture and Animal Husbandry* 20(102), 105–107.

Levin, M. and Butler, G. D. 1966. Bees associated with Safflower in South Central Arizona. *J. Eco. Ent.* 59, 654–657.

Levin, M. D., Butler, G. D., and Rubis, D. D. 1967. Pollination of safflower by insects other than honeybees. *Jour. Econ. Ent.* 60, 1481–1482.

Li, D. and H. H. Mundel. 1996. Safflower. Carthamus tinctorius L., Vol. 7, Institute of Plant Genetics and Crop Plant Research, Gatersleben/International Plant Genetic Resources Institute, Rome.

Lindsley, E. G. 1978. Temporal patterns of flower visitation by solitary bees with particular reference to southwestern United States. *J. Kansas. Entomol. Soc.* 51, 531–546.

Lingappa, S., Viraktamath, S. A., Vastrad, A. S., and Williams, R. 1999. Utilization of honey bee *Apis cerana* Fab. for pollination of watermelon and safflower. Proceedings of Apimondia, p. 235.

McPherson, M. A., Good, A. G., Topinka, A. K. C., Yang, R. C., Mckenzie, R. H., Cathcart, R. J., Christianson, J. A., Strobeck, C., and Hall, L. M. 2009. Pollen-mediated gene flow from transgenic safflower (*Carthamus tinctorius* L.) intended for plant molecular farming to conventional safflower. *Environ. Biosafety Res.* 8, 19–32.

Neff, J. L. and Simpson, B. B. 1990. The roles of phenology and reward structure in the pollination biology of wild sunflower (*Helianthus annuus* L.), Asteraceae. *Israel. J. Bot.* 39, 197–216.

Nimbkar, N. 2002. Safflower rediscovered. *Times Agricultural Journal* 2, 32–36.

Panda, P., Rath, L. K., Padhi, J., and Panigrahi, D. 1995. Relative abundance and foraging behaviour of common bee species on niger in Phulbari District, Orissa. *Indian Bee J.* 57(1), 10–14.

Pandey, A. K. and Dhakal, M. R. 2004. Some aspects of pollen biology and pollination ecology of Niger (*Guizotia abyssinica* Cass.), Vistas. In: Srivastava, P. C. (ed.) *Palaeobotany and Plant Morphology: Evolutionary and Environmental Perspectives.* Print House, Lucknow, pp. 415–428.

Pandey, A. K. and Kumari, A. 2008. Pollination ecology of safflower (*Carthamus tinctoris* linn). Paper presented at: Safflower: Unexploited potential and world adaptability, 7th International Safflower Conference, Wagga Wagga, NSW, Australia.

Parrish, J. A. D. and Bazzaz, F. A. 1978. Pollination niche separation in a winter annual community. *Oecologia* 35, 133–140.

Patil, J. A. and Chavan, V. M. 1958. Selfing methods in safflower. *Indian Oilseeds J.* 2, 10–12.

Prasad, R. B. and Subba Rao, M. S. 1984. Safflower in the plateau region of Bihar: Constraints and opportunities. *Oil Seed Production* 159–164.

Pyke, G. H. 1984. Optimum foraging theory: A critical review. *Ann. Rev. Ecol. Syst.* 15, 523–574.

Reddy, C. S. 1976. Floral mechanism, pollen productivity and pollen incidence in *Madhuca indica* with remark on the mode of pollination. *New. Botanist.* 3, 11–16.

Rubis, D. D. 1970. Bee pollination in the production of hybrid safflower. *Rep. Pollination Conf. Univ. Ark.*, 9th MP 127, 43–49.

Rubis, D. D., Levin, M. D., and Mc Gregor, S. E. 1966. Effects of honey bee activity and cages on attributes of thin hull and normal safflower lines. *Crop Sci.* 6, 11–14.

Rudolphi, S., Becker, H. C., and Von Witzke-Ehbrecht, S. 2008. Outcrossing rate of safflower (*Carthamus tinctorius* L.) under the agro climatic conditions of Northern Germany. In: Proc. 7th Int. Safflower Conf., Wagga Wagga, Australia, November 3–6. Available at http://safflower.wsu.edu/Conf2008/Home%20 information/About%20the%20conference.pdf

Singh, V. and Nimbkar, N. 2007. Safflower (Carthamus tinctorius L). In: Singh, R. J. (Ed.), *Genetic Resources, Chromosome Engineering, and Crop Improvement*, Vol. 4. Oilseed Crops. CRC Press, Taylor & Francis Group, Boca Raton, pp. 167–194.

Smith, J. R. 1996. *Safflower.* AOCS Press, Champaign, IL, p. 624.

Telangre, A. H., Matre, Y. B., Latpate, C. B., and Zanwar, P. R. 2018. Effect of neonicotinoids *i.e* acetamiprid 20% SP on foraging behaviour of honey bee on safflower (*Carthamus tinctorius* L.). *International Journal of Chemical Studies* 6(5), 185–188.

Velasco, L., Fischer, M., and Fernandez-Martinez, J. M. 2012. Short communication. Estimation of cross-fertilization rate in safflower (Carthamus tinctorius L.). *Spanish Journal of Agricultural Research* 10(1), 155–159.

Weiss, E. A. 1971. *Castor, Sesame and Safflower.* Barnes and Noble, New York, pp. 529–744.
Weiss, E. A. 2000. *Oilseed Crops.* 2nd edn. Blackwell Science, Malden, pp. 259–273.
Zhaomu, W. and Lijie, D. 2001. Current situation and prospects of safflower products development in China. In: Proc. 5th International Safflower Conference, Williston, ND, and Sidney, MT, July 23–27, 2001, pp. 315–319.

8 Groundnut

8.1 INTRODUCTION

The groundnut (*Arachis hypogaea L.*) belongs to the family *Leguminosae* and subfamily *Papilionoideae* (Isleib et al. 1994). Peanut (groundnut) and the genus *Arachis* originated in central Brazil in South America. Early archaeological evidence suggests that peanut was domesticated in northern Argentina and eastern Bolivia and subsequently grown in Mexico, the Caribbean Basin and throughout Brazil, and the coastal regions of Peru. It is believed that the crop did not reach other parts of the world until after Columbus arrived in America, after which time it was taken from Brazil to Africa and the Far East by the Portuguese. The Spaniards are believed to have taken the crop to the western Pacific, Indonesia, and China early in the sixteenth century.

The groundnut crop is cultivated in 108 countries on about 22.2 million hectares, of which 13.69 million hectares are in Asia (India, 8 million hectares; China, 3.84 million hectares), 7.39 million hectares in sub-Saharan Africa, and 0.7 million hectares in Central and South America. Average pod yields on a global scale increased slightly from 1.08 Mt ha^{-1} in the 1980s to 1.15 Mt ha^{-1} in the 1990s (Carley and Fletcher 1995), and the global production is 29 million tons of pods. India, China, and the United States are the leading producers and grow about 70% of the world's groundnuts (CTA 2001).

Groundnut (*Arachis hypogaea* L.), also known as peanut, is a high-value legume product and important source of farmers' family income and employment generation in developing and less-developed countries. Peanut is used widely throughout the world, being one of the principal oilseeds as well as a high-quality food for human consumption (Figure 8.1). Groundnut oil is a major vegetable oil in many parts of the world, and its sale generates billions of dollars annually to regional and national economies (Florkowski 1994). Besides creating employment and income to local farmers and people engaged in processing and mills crushing, the by-products of groundnut, such as groundnut oil, cake, and meal, play a role in improving the nutritional status of human beings as well as for livestock.

Groundnut is the sixth most important oilseed crop in the world. It contains 48–50% oil and 26–28% protein and is a rich source of dietary fiber, minerals, and vitamins. Groundnut is grown on 26.4 million hectares worldwide, with a total production of 37.1 million metric ton and an average productivity of 1.4 metric ton ha^{-1} (FAO 2003). Over 100 countries worldwide grow groundnut.

Developing nations account for 97% of the global land area and 94% of the global production of this particular crop. The cultivation of groundnut is primarily concentrated in Asia and Africa, representing 56% and 40% of the global land area and 68% and 25% of the global production, respectively. The cultivated groundnut, scientifically known as *Arachis hypogaea* L., belongs to the Arachis genus in the Stylosanthinae subtribe of the Aeschynomeneae tribe, which falls under the Leguminosae family (refer to Figure 8.1). Cultivated groundnut consists of two subspecies, namely, hypogaea and fastigiata, each further divided into two botanical varieties: var. hypogaea and var. hirsuta for the former, and var. fastigiata, var. vulgaris, var. peruviana, and var. aequatoriana for the latter. Each of these botanical varieties possesses distinct plant, pod, and seed characteristics (Krapovickas and Gregory 1994).

However, most of the commercially cultivated varieties belong to var. *hypogaea* (common name/market type: Virginia or runner), var. *fastigiata* (Valencia), and var. *vulgaris* (Spanish) botanical variety groups.

Groundnut ranks 6th among the oilseed crops and 13th among the food crops of the world. Several countries in Africa, Asia, and the Americas participate in the exportation of groundnuts. However, during the 1980s, the primary exporting nations were the USA, Argentina, and China. China and Argentina experienced a boost in production and exports due to an increase in crop yield and cultivated land. While numerous countries with rising yields focus on meeting domestic

DOI: 10.1201/9781032656724-9

FIGURE 8.1 Field view of groundnut crop.

demand, they are inclined to export during periods of global scarcity when prices are high, seizing the opportunity.

Groundnut is the most important oilseed crop in India, covering nearly half of the area under oilseeds. The vegetable oil is rich in omega-3 fatty acid and extensively used for cooking purpose. In addition to providing high-quality edible oil (48–50%), easily digestible protein (26–28%), and nearly half of the 13 essential vitamins and 7 of the 20 essential minerals necessary for normal human growth and maintenance, it produces high-quality fodder for livestock.

The major growing states are Gujarat, Andhra Pradesh, Tamil Nadu, Karnataka, Rajasthan, and Maharashtra. These constituting and contributing around 80% of groundnut area and production. The total area under groundnut is about 8 million hectares and annual production is over 9 million tons of pods. Groundnut plays a significant role in the livelihoods of smallholder farmers of rainfed area.

Groundnut (*Arachis hypogaea* L.) is the most important oilseed crop of semi-arid tropics in general and the Saurashtra region of Gujarat in particular. Gujarat is one of the largest groundnut-producing states of India, with an area of 20 lakh hectares which is mostly grown under kharif.

Groundnut cultivation takes place in India throughout the kharif, rabi, and summer seasons, employing different cropping systems. The average production of rabi groundnut is greater (1,600 kg/ha) compared to kharif (1,000 kg/ha). The primary factor affecting lower productivity is rainfed groundnut cultivation, which is susceptible to various insect pests and diseases. The interplay between weather conditions and pests during a specific season plays a crucial role in determining crop productivity, particularly considering that groundnut and its pests are highly sensitive to extreme weather events.

8.1.1 Plant

The peanut plant, which can grow between 10 and 20 in tall, is an annual plant that can either stand upright or spread out with branches. It is commonly cultivated in rows spaced approximately 1 1/2 to 3 ft apart. The seeds are typically planted during the spring season, once the risk of frost has subsided, and the harvest usually takes place before the arrival of early fall frost. The plant features a main taproot along with less-sturdy lateral roots, allowing for easy extraction of the entire plant from the soil. Its leaves consist of four leaflets measuring around 1 1/2 to 2 1/2 in in length. The seeds develop just below the surface of the soil, but they are attached to the branches near the base of the plant.

8.2 FLORAL BIOLOGY

Groundnut is a self-pollinated tropical annual legume crop. Being a self-pollinated crop, groundnut seeds can be maintained for several generations without losing its genetic purity. In practice, however, progressive deterioration of original stock occurs through admixture with other varieties in farmers' field as well as in the threshing and processing yards. Groundnut or peanut plants have automatic self-pollination. The flowers open at sunrise, and anthers dehisce at about the same time, pollinating the ovary. At locations where bee activity is high, some cross-pollination can occur (Nigam et al. 1983).

8.2.1 Groundnut Flower

Flowers (Figure 8.2) are aerial, but pod development is subterranean due to geotropic movement of the gynophores (pegs). Flowers are borne on inflorescences located in the axils of leaves. Flowering

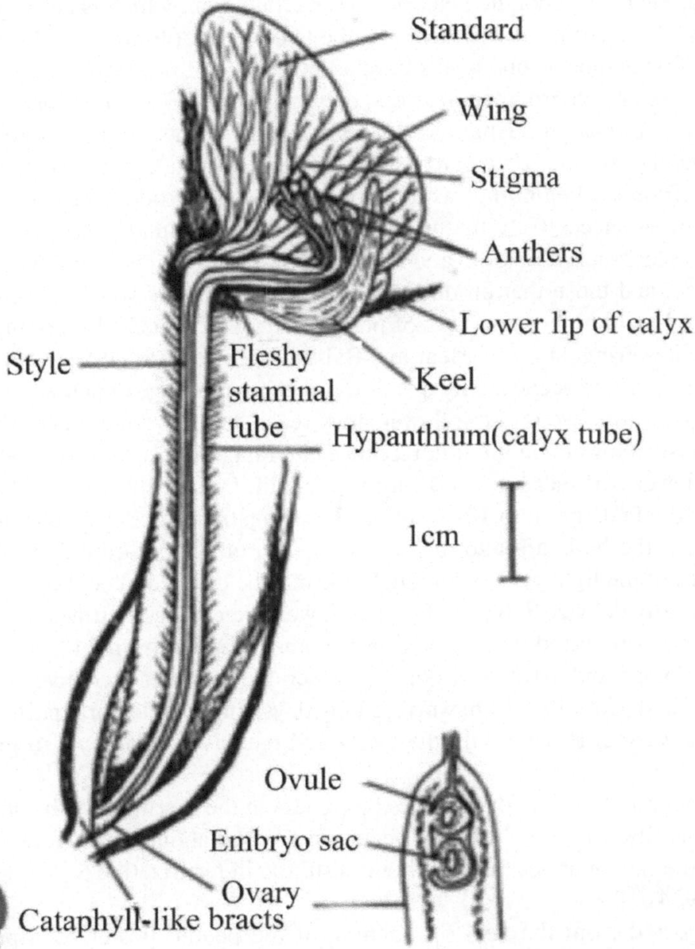

FIGURE 8.2 Longitudinal section of groundnut flower.

is sensitive to light, temperature, and relative humidity. High temperatures and low levels of relative humidity reduce flowering, while temperatures between 22°C and 33°C and soil moisture of 40% are ideal for flowering. A light intensity greater than 45% of full sunlight is necessary for optimum development—flowering is stimulated by light received three days prior to opening of flowers. Generally, only one flower opens at each node at any given time at sunrise.

8.2.2 Flowering

Groundnut exhibits a varying level of growth indeterminacy. The onset of flowering typically takes place approximately 25 days after the emergence, but this timing can be influenced by factors such as genotype, environmental conditions, and temperature. For instance, in sp. *Hypogaea*, a decrease in the mean daily temperature from 30°C to 20°C extends the duration required for the first flowering from 25 to 38 days. Similarly, in sp. *Fastigiata*, this period is reduced from 35 to 24 days. The peak flowering period, characterized by abundant flower production, usually falls within 5 to 11 weeks after planting, contingent upon the cultivar's duration and the prevailing season. Notably, a significant proportion of the initially formed flowers typically develops into mature fruits. Overall, the flowering process can be observed to occur in distinct stages, usually encompassing four to five phases.

Stage 1 is characterized by a low production of flowers, followed by a rapid increase in flowering during stage 2. Stage 3 marks the peak of flower production, which is subsequently followed by a decline in stage 4. In certain cases, there may be two peaks of flowering. However, the flowers that open after the first and/or second peaks have a low chance of producing mature fruits, unless early flowers are affected by environmental stress, preventing them from functioning normally (Bear and Bailey 1973; Mohammad and Khanna 1932). Groundnut plants produce a surplus of flowers compared to the number of pods they can sustain for development. Approximately 40% of the flowers fail to develop from the beginning, while another 40% only produce pegs (Smith 1954). Less than 20% of the flowers successfully mature into fruits under optimal conditions, and sometimes this number drops to less than 15% (Lim and Hamdan 1984). Genotypes that flower early exhibit greater synchronization, and those that produce the majority of flowers within the first two weeks of the flowering period yield higher numbers of pods (Sastry et al. 1985). Removing some flowers daily can extend the flowering duration (Bear and Bailey 1973; Seshadri 1962).

Typically, in an inflorescence, only one flower reaches the stage of anthesis on a given day, except for var. vulgaris, where two or more flowers may reach anthesis simultaneously. The time intervals between successive flowers on the inflorescence can range from one to several days. A day prior to anthesis, the flower bud measures 6–10 mm in length. During the day, the hypanthium elongates gradually, and the buds grow to 10–20 mm. The elongation process speeds up at night, and at the time of anthesis, the buds are approximately 50–70 mm long (Smith 1950). Generally, the buds open at the start of the light period, although this may be delayed in cold or wet weather conditions. The anthers usually dehisce 7–8 hr before the flower opens, or sometimes even later (Bolhuis et al. 1965). While it was reported that the stigma remains receptive from 24 hr before to 12 hr after the flower opens (Hassan and Srivastava 1966), in general, it becomes receptive only a few hours before anthesis (Sastri and Moss 1982). On warm, sunny days, most of the floral parts tend to wither within 5–6 hr after flowering, leaving only the ovary and the base of the style turgid beyond the day of anthesis.

Flowering generally occurs during the early hours in the morning, with some delay on cold/wet/dull days. Generally, anther dehiscence occurs in the bud about 1 hr before flowering. The pollen tube grows at the rate of about 1 cm per hour, resulting in fertilization 5–6 hr after pollination (Lim and Gumpil 1984).

Pollen is ejected from the keel tip opening of the peanut flower by a pistonlike mechanism (Kerner 1897) that involves the wing petals, keel, and ascending stylar hairs. The sensitive system is activated when wing petals are depressed, which in turn causes the conical keel to lower and the enclosed anther group in the constricted keel to become compacted. The stylar hairs below the stigma are then pressed among the dehisced anthers. Further depression of wing petals causes the pollen to become even more compacted among the hairs and the pollen from the anthers, borne by the stylar hairs, to sweep upward. As the keel continues to lower, the stigma first, and then the stylar hairs bearing the pollen, is thrust out of the opening at the keel apex. Pollen is thereby conveniently available to the head of a bee visitor as it positions itself astride the wing petals.

A flower (Figure 8.2) consists of five petals, ten stamens, and a pistil. Five petals are: one yellow to orange standard, two yellow to orange wings, and two petals fused to form a paler yellow keel. Of the ten monadelphous stamens, two are usually not fully developed. Stamens are fused together from one half to two thirds of their length. Eight fertile normally developed anthers consist of four globose, dorsifixed, uniloculate anthers alternating with four adnate, introse, oblong anthers. Pistil consists of an ovary, style, and stigma. Ovary is superior, small, and conical, with a beak-shaped point at the tip, and contains a single sessile carpel with one to six ovules. Style is glabrous throughout its length and covered with bristles near the club-shaped stigma and is enclosed in a filiform hypanthium. Stigma becomes receptive to pollen 24 hr before flower opening and remains so after up to 12 hr, while pollen is shed 1 to 8 hr before the flower opening. Anthesis and pollination occur at sunrise, with self-pollination taking place within the closed keel of the flower. About 40% of the flowers fail to begin pod development, and another 40% abort before pod development.

8.2.2.1 Pollination Requirements

Peanuts are largely self-fertilized; however, there is an increase in the set of seed caused by cross-pollination. Some selections have a structure that doesn't allow self-pollination, necessitating cross-pollination by bees (Reed 1924). Srinivasalu and Chandrasekaran (1958) observed that variations exist among different varieties regarding the protrusion of the stigma from the keel in relation to cross-pollination. Leuck and Hammons (1969) found that certain cultivars possess a floral structure that hinders self-pollination but facilitates cross-pollination through bees. According to Leuck and Hammons (1965a, 1965b), bees were responsible for tripping pollen in approximately 80% of open peanut flowers. Girardeau and Leuck (1967) demonstrated that flowers confined in cages, without any manipulation, yielded 4 to 11% fewer fruits compared to flowers manually flexed or manipulated with water drip. Moreover, their uncovered plots resulted in a significant increase of 6 to 11% in comparison to plots enclosed to exclude bees. Shibuya et al. (1955) linked the presence of only one seed to inadequate pollen on the stigma.

8.2.2.2 Pollinators

While peanuts do not require pollinators, studies in peanut reveal that bees can "trip" ejection of pollen and cause cross-pollination, resulting in increased yield in peanut (Hammons and Leuck 1966; Leuck and Hammons 1967). Peanut flowers bloom from May to harvest in late August. The species visiting peanut flowers comprised of *Melissodes communis* Cresson and *Melissodes bimaculatus* Lepeletier (Apidae)

Leuck and Hammons (1969) reported that the presence of atypical flowers in the peanut, *Arachis hypogaea* L., could facilitate insect cross-pollination by a large bee (Apoidea) species. Six wild bee species were observed visiting peanut flowers. They included *Augochloropsis metallica fulgida* (Smith) (Halictidae), *Chalicodoma* (*Chelostomoides*) *georgica* (Cresson), (Megachilidae) *Megachile* (*Delomegachile*) *addenda* Cresson (Megachilidae), *Melissodes* (*Melissodes*) *comptoides* Robertson (Anthophoridae), *Ceratina* (*Ceratinula*) *cockerelli* Smith (Anthophoridae), and *Ceratina* (*Zadontoemrus*) *dupla floridana* Mitchell (Anthophoridae). In 1967, near Tifton, the bee species that displayed the highest abundance on peanut flowers were *Pyrobombus impatiens* Cresson (Apidae), *Megachile* (Litomegachile) *mendica* Cresson (Megachilidae), *Apis mellifera* L. (Apidae), and *Lassioglossum (Dialictus) versatum* (Robertson) (Halictidae).

Peanut flowers are cross-pollinated, and pollination by bees is most common. Hammons and Leuck (1966) reported that thrips are also vectors of peanut pollen transfer, but halictid bees were the most important visitors to peanut flowers at Tifton, Georgia. Hammons (1963) observed *Lasioglossum*, *Megachile*, *Bombus*, and *Apis* species visiting peanut flowers. *Anthidium* and *Melissodes* species were also observed visiting peanut flowers (Hammons et al. 1963; Leuck and Hammons 1969), but halictids and megachilids were the main pollinators affecting cross-pollination. Diwan and Salvi (1965) reported that *Apis cerana* did not prefer peanut flowers. However, in earlier studies, Heide (1923) had reported *Apis cerana* actively visited peanut flowers between 1900 and 2100 h. *Apis mellifera adansonii*, *Nomia* spp., and *Megachile* spp. were reported as important flower visitors in the Malawi area of Africa (Gibbons and Tattersfield 1969). *Mellisodes spp.* also efficiently tripped and retripped the peanut flower. Sometimes these bees retripped a flower during a single visitation by pumping the wing petals and gathering pollen with each ejection. The honeybee, *Apis mellifera* L., can efficiently trip peanut flowers. The abundance of honeybees in peanut fields was observed in different visitation attitudes. Some approached the flower from the side, between the standard and wing petals; others approached it from the top, alighting head down on the plane of the standard petal and probing at its base. These may be positional attitudes that the honeybee assumes in searching for nectar or pollen in this or other flower types (Leuck and Hammons (1965a, 1965b).

Leuck and Hammons (1965b) reported that more than 80% of peanut flowers were tripped for pollen by a combined bee complex, but they did not specify a particular bee species.

Hammons et al. (1963) reported that the halictids and megachilids were most abundant during early hours of the day, whereas honeybee activity continued throughout the day. Girardeau and Leuck (1967) reported that 6 to 11% increase in yield could be obtained by saturation pollination by honeybees in peanut plantings

8.3 POLLINATION AND FERTILIZATION

Normally, the stigma protrudes above the anthers and is receptive before anthesis. As the stigma is enclosed in the keel and surrounded by the staminal column, self-pollination is most common, with a high frequency of cleistogamy (Murty et al. 1980). Rarely, the stigma and anthers are exerted from the keel, and as the stigma is receptive, some amount of outcrossing may occur. The percentage of outcrossing varied between less than 1% and 3.9%, influenced by factors such as the season, genotype, and location. Various bee species were identified as carriers of pollen (Culp et al. 1968; Gibbons and Tattersfield 1969; Hammons 1963, 1964; Leuck and Hammons 1969; Srinivasalu and Chandrasekharan 1958). Fertilization occurs within approximately 6 hr after pollination, and the flower may wither within 5–6 hr.

8.3.1 The Peg

Generally, fertilization is completed before midday. After fertilization, the flower droops, the corolla closes, the calyx tube bends, and the flower withers. Within a week, the ovary, located at the bottom of the calyx tube, undergoes rapid growth triggered by the activation of the intercalary meristem situated beneath it. The initially green ovary gradually changes color, turning purplish starting from the tip and progressing downward. As the ovary develops further, it penetrates through the floral components, unveiling a lengthening peg. The peg or carpophore is a stalk-like structure that carries the fertilized ovules at its tip. Its growth is positively geotropic until it enters the soil to some depth (up to 5–7 cm). The tip then becomes diageotropic or almost horizontal. Only then does it begin to develop into a fruit. The peg usually withers away if it fails to contact and penetrate the soil after it extends to about 15 cm. In humid conditions, however, some var. *fastigiata* cultivars can form aerial pods. These are underdeveloped, green, and small. True breeding aerial podding types have also been reported (Prasad 1985), and the physiology of such aerial podding, producing normal viable seeds, needs further investigation.

The unique nature of the peg led to several misconceptions about its homology, and it has been described as a gynophore, a carpophore, or an apetalous flower (Brennan 1969; Jacobs 1947). According to Smith (1950), the term "peg" refers to the young fruit in its stalk-like phase of growth. Patel and Narayana (1935) reported that the small plastids that develop in the epidermal walls of the peg after fertilization are responsible for its positive geotropic movement. Jacobs (1947, 1951a, 1951b) observed that the elongation of the peg stops due to the presence of auxin in its distal portion, which moves through the peg tissue and intercalary meristem in a strictly basipetal direction.

Sometimes, the peg or carpophore has been incorrectly named gynophore (Smartt 1976a). Bharathi and Murty (1980) proposed that the peg, located at the base of the ovary, is a highly reduced gynophore and can be identified as a meristem. However, Ramanatha Rao (1988) argues that considering the elongation of the intercalary meristem beneath the ovary, which results in the formation of articulated pods in most wild *Arachis* species, the peg should be classified as a carpophore. Various colors of pegs, including light purple, dark purple, and green (without pigmentation), have been documented. Additionally, a loose correlation has been observed between the green color of pegs and the white color of the testa (Varisai Muhammad et al. 1973c).

Though the development of pigment in the peg is cultivar-dependent, it is greatly influenced by exposure to sunlight, and grades of pink or purple can occur on the same plant. Pegs can be broadly classified as pigmented or green, but the portion in the soil is generally white (colorless). The pegs, in their early stages, exhibit tiny white hairs, creating a soft and fuzzy appearance (Seshadri 1962). As the pegs mature, these hairs are shed. The depth at which the pods are formed depends on how deeply the pegs have penetrated the soil.

Though significant cultivar differences have been reported (Seshadri et al. 1955), variation within the same cultivar for the size of the peg is also very great. The length of the peg was reported to be closely related to the differences in branching pattern (Ono 1979). The length of the peg in wild Arachis species exhibits significant variation and can reach up to 1 m in certain species. The peg's thickness ranges from approximately 1 mm to 2 mm.

An assessment of the thickness of the pegs in the different types showed that the cultivars of sp. *Fastigiata* had thicker pegs than those of sp. *Hypogaea* (Seshadri 1962).

8.3.2 The Fruit

The fruit, known as a pod, is a modified lomentiform indehiscent carpel. Significant variations in the proportion of pegs that penetrate the soil and develop into pods have been observed among different cultivars (Seshadri 1962; Singh et al. 1981). Typically, the mature pod contains a maximum of four seeds.

Occasionally, five or even six seeds per pod have been recorded. Single-seeded pods may be produced when all the ovules except the proximal abort (Smartt 1976a). The pod size may range up to 8.0 cm × 2.7 cm. The fruit is composed of valves that are structurally capable of splitting open, but they do not actually open on their own. Under pressure, pods split along the longitudinal suture. This is the normal line of dehiscence, and the mechanical tissue of the mesocarp is interrupted along this line (Richter 1899).

8.3.3 Pod Development

The ovule undergoes gradual growth until it reaches the flowering stage, and this growth continues to a certain extent after fertilization. The embryo, dormant during peg elongation, begins to develop three to four days after commencement of pod development (Murty and Rao 1979; Murty et al. 1980). The basal ovule develops first. The peg becomes diageotropic and almost horizontal when it reaches its maximum depth in the soil and ceases to grow, and pod development starts. The enlargement of the pod proceeds from base to apex. The pod expands rapidly underground by the growth of a large parenchymatous tissue (endocarp) lying between the ovules and shell layers (Schenk 1961).

The pod reaches its maximum size approximately three weeks after the peg penetrates the soil. As the ovules grow, the endocarp diminishes and eventually disappears completely when the seeds reach maturity. Throughout this period, the inner surface of the shell darkens progressively, indicating a higher tannin content, and ultimately turns a very dark brown upon maturation (Gregory et al. 1973). From fertilization to full maturity, it typically takes around 60 days (Patel et al. 1936). The size of both the mature pod and seeds is influenced by factors such as genotype, soil conditions, and the cultivation method and timing (Shibuya 1935). The normal depth for pod formation is 4–7 cm below the soil surface, with an optimal soil temperature ranging from 31 to 33°C in that zone (Ono 1979). Deviations from these temperature ranges can extend the duration of pod development.

8.3.4 Fertilization

Fertilization in groundnut is porogamous. After pollination, one of the synergids degenerates. The pollen tube enters the embryo sac through this degenerating synergid soon after pollination (Murty and Rao 1979). Triple fusion precedes syngamy (Figure 3.11). Fertilization takes place 12–18 hr after anthesis.

8.3.5 Groundnut Pod and Seed

Pods containing seeds are produced below ground. They vary in shape, size, and texture and may contain up to five seeds per pod. Only after pods attain full size that the seeds inside start to develop. A seed consists of two large cotyledons (does not contain endosperm), a stem axis, and leaf primordial (plumule), hypocotyl and primary root (radicle), and seed coat. Cotyledons comprise about 96% of the seed weight and are the major storage tissue for the developing seedlings.

8.4 CROSS-COMPATIBILITIES AMONG SECTIONS OF ARACHIS

Wild arachis species, a reservoir of several useful genes (diseases and insect pests resistance, including genes for yield-related traits and high oil content), need to be harnessed for improvement of cultivated groundnut.

8.4.1 Breeding Behavior Including Cross-Pollination

Groundnut is regarded a highly self-pollinated crop with less than 1% cross-pollination. However, cross-pollination can reach as high as 10% at locations and in seasons where bee activity is high. In general, natural cross-pollination in groundnut is almost absent. However, at locations where bee activity is intense, depending on the variety and season, natural cross-pollination can occur. Having a sufficient gap between different varieties in seed production fields is crucial to minimize the risk of pollen contamination and accidental mixtures. In India, where natural cross-pollination is minimal, it is recommended to maintain a minimum isolation distance of 3 m between varieties for all categories of certified seeds. The specific isolation distance should be determined based on factors such as location, season, and the level of natural cross-pollination observed for each particular variety.

8.4.2 Contract Seed Production

As local seed traders are the major source of groundnut seed, they could promote quality seed production through contract farming of improved varieties with selected farmers. However, both farmers and traders need to be educated in the technical aspects of groundnut seed production, processing, and storage.

REFERENCES

Bear, J. E. and Bailey, W. K. 1973. Earliness of flower opening and potential for pod development in peanut, *Arachis hypogaea* L. *Journal of the American Peanut Research and Education Association* S, 26–31.

Bharathi, M. and Murty, U. R. 1980. The botany of fruit stalk in groundnut *Arachis hypogaea. Current Science* 49, 948.

Bolhuis, G. G., Frinking, H. D., Leeuwaugh, I. et al. 1965. Observations on the opening of flowers, dehiscence of anthers and growth of pollen tubes in *Arachis hypogaea. Netherlands Journal of Agricultural Science* 7, 138–140.

Brennan, L. R. 1969. The peanut gynophore. *The Biologist* 51, 71–82.

Carley, D. H. and Fletcher, S. M. 1995. An overview of world peanut markets. In: Pattee, H. E. and H. T. Stalker (eds.) *Advances in Peanut Science*. Am. Peanut Res. Edu. Society, Inc., Stillwater, OK, USA, pp. 554–577.

CTA. 2001. Technical centre for agriculture and rural cooperation (CTA). 2001. Groundnuts in a nutshell. Spore 91. CTA, Wageningen, The Netherlands.

Culp, T. W., Bailey, W. K., and Hammons, R. O. 1968. Natural hybridization of peanuts, *Arachis hypogaea* L., in Virginia. *Crop Science* 8, 108–111.

Diwan, V. V. and Salvi, S. R. 1965. Some interesting behavioural features of Apis dorsata Fab. *Indian Bee J.* 27(1), 52.

FAO. 2003. Statistical databases. Available at www.FAO.ORG

Florkowski, W. J. 1994. Groundnut production and trade. In: Smartt, J. (ed.) *The Groundnut Crop-A Scientific Basis for Improvement*, pp. 1–22. London: Chapman & Hal.

Gibbons, R. W. and Tattersfield, J. R. 1969. Out-crossing trials with groundnuts (*Arachis hypogaea* L.). *Rhodesia Zambia Malawi Journal of Agricultural Research* 7, 71–75.

Girardeau, J. H. and Leuck, D. B. 1967. Effect of mechanical and bee tripping on yield of the Peanut. *Journal of Econonmic Entomology* 60, 1454–1455.

Grcgory, W. C., Gregory, M. P. Krapovickas, A. et al. 1973. Structure and genetic resources of peanuts. In: *Peanuts—Culture and Uses*. American Peanut Rcsearch and Education Association, Stillwater, OK, pp. 47–133.

Hammons, R. O. 1963. Artificial cross pollination of the peanut with bee-collected pollen. *Crop Science* 3, 562–563.

Hammons, R. O. 1964. Krinkle: A dominant leaf marker in the peanut *Arachis hypogaea* L. *Crop Science* 4, 22–24.

Hammons, R. O., Krombein, K. V., and Leuck, D. B. 1963. Some bees (Apoidea) associated with peanut flowering. *Journal of Econonmic Entomology* 56, 905.

Hammons, R. O. and Leuck, D. B. 1966. Natural cross-pollination of the peanut, *Arachis hypogaea* L., in the presence of bees and thrips. *Agronomy Journal* 58, 396.

Hassan, M. A. and Srivastava, D. P. 1966. Floral biology and pod development of peanut studied in India. *Journal of Indian Botanical Society* 45, 92–102.

Heide, F. F. R. 1923. Biological observations of Arachis hypogaea L. [Dutch East Indies] Alg. *Proefsta. v. Landb. Meded.* 14, 119.

Isleib, T. G., Wynne, J. C., and Nigam, S. N. 1994. Groundnut breeding. In: Smartt, J. (ed.) *The Groundnut Crop: A Scientific Basis for Improvement*. Chapman & Hall, London, pp. 552–623.

Jacobs, W. P. 1947. The development of the gynophore of the peanut plant, *Arachis hypogaea* L. 1: The distribution of mitoses, the region of greatest elongation and the maintenance of vascular continuity in the intercalary meristem. *American Journal of Botany* 34, 361–370.

Jacobs, W. P. 1951a. Auxin relationships in an intercalary meristem: Further studies on the gynophore of *Arachis hypogaea* L. *American Journal of Botany* 38, 307–310.

Jacobs, W. P. 1951b. The growth of peanut plants at various diurnal and nocturnal temperatures. *Science* 114, 205–206.

Kerner, A. 1897. *The Natural History of Plants* (trans. Oliver, F. W.), Vol. 6. Blackie and Son, London, p. 336.

Krapovickas, A. and Gregory, W. C. 1994. Taxonomia del genero *Arachis* (*Leguminosae*). *Bonplandia*, 8, 1–187.

Leuck, D. B. and Hammons, R. O. 1965a. Further evaluation of the role of bees in natural crosspollination of the peanut, *Arachis hypogaea* L. *Agronomy Journal* 57, 94.

Leuck, D. B. and Hammons, R. O. 1967. Pollen-collecting activities of bees among peanut flowers. *Journal of Econonmic Entomology* 58, 1028–1030.

Leuck, D. B. and Hammons, R. O. 1969. Occurrence of atypical flowers and some associated bees: Apoidea in the peanut. *Arachis hypogaea* L. *Agronomy Journal* 61, 958–960.

Lim, E. S. and Gumpil, J. S. 1984. The flowering: Pollination and hybridization of groundnuts (*Arachis hypogaea* L.). *Pertanika* 7, 61–66.

Lim, E. S. and Hamdan, O. 1984. The reproductive characters of four varieties of groundnuts (*Arachis hypogaea* L.). *Pertanika* 7, 25–31.

Mohammad, A. A. and Khanna, K. L. 1932. Studies on germination and growth in groundnut. *Agriculture and Live-Stock in India* 3, 91–115.

Murty, U. R. and Rao, N. G. P. 1979. A rapid method for the study of fertilization in groundnut, *Arachis hypogaea* L. *Current Science* 48, 492–493.

Murty, U. R., Rao, N. G. P., Kirti, P. B., and Bharati, M. 1980. Fertilization in groundnut. *Arachis hypogaea* L. *Ofeagineux* 36, 73–76.

Nigam, S. N., Ramanatha, R. V., and Gibbons, R. W. 1983. Utilization of natural hybrids in the improvement of groundnuts (*Arachis hypogaea*). *Expl. Agric.* 19, 355–359.

Ono, Y. 1979. Flowering and fruiting of peanut plants. *Japanese Agriculture Research Quarterly* 3, 226–229.

Patel, J. S., John, C. M., and Seshadri, C. R. 1936. The inheritance of characters in the groundnut *Arachis hypogaea*. *Proceedings of the Indian Academy of Sciences* 3, 214–233.

Patel, J. S. and Narayana, G. V. 1935. A rare instance of polyembryony in *Arachis hypogaea*. *Current Science* 4, 32–33.

Prasad, M. V. R. 1985. Aerial podding in groundnut (*Arachis hypogaea* L.). *Indian Journal of Genetics and Plant Breeding* 45, 89–91.

Rao, R. V. 1988. Botany. In: Reddy, P. S. (ed.) *Groundnut*. Indian Council of Agricultural Research, New Delhi, pp. 24–64.

Reed, E. L. 1924. Anatomy, embryology and ecology of *Arachis hypogaea*. *Bot. Gaz.* 78, 289–310.

Richter, E. G. 1899. Beitrage zur Biologie von *Arachis hypogaea*. In: Schricber, A. (ed.) *Inaug. Diss. Kgl. Bot. Gart. Breslau*. Breslau, p. 39.

Sastri, D. E. and Moss, J. P. 1982. Effects of growth regulators on incompatible crosses in the genus *Arachis* L. *Journal of Experimental Botany* 33, 1293–1301.

Schenk, R. V. 1961. Development of the peanut fruit. *Georgia Agricultural Experiment Station Technical Bulletin* 22, 5–53.

Seshadri, C. R. 1962. *Groundnut*. The Indian Central Oilseeds Committee, Hyderabad, India, 274 p.

Seshadri, E. R., Bhavanisankara Rao, M., and Srinivasalu, N. 1955. Studies on the gynophores in the groundnut. Proceedings of the 6th Scientific Workers Conference, Department of Agriculture, Tamil Nadu, Department of Agriculture, Coimbatore, Tamil Nadu.

Shibuya, T. 1935. Morphological and physiological studies on the fructification of peanut (*Arachis hypogaea* L.). *Memoirs of the Faculty of Science and Agriculture, Taihoku Imperial University* 17, 1–120.

Shibuya, T., Sato, H., and Suzuki, M. 1955. Morphological and physiological studies on subterranean fructuation of legume crops. *Crop Science Society (Japan) Proc.* 24, 16–19.

Singh, H., Yadav, A. K., Yadava, T. P., and Chabra, M. L. 1981. Genetic variability and heritability for morpho-physiological attributes in groundnut. *Indian Journal of Agricultural Sciences* 52, 432–434.

Smartt, J. 1976a. Comparative evolution of pulse crops. *Euphytica* 25, 139–143.

Smith, B. W. 1950. *Arachis hypogaea* L. Aerial flower and subterranean fruit. *American Journal of Botany* 37, 802–815.

Smith, B. W. 1954. *Arachis hypogaea*, reproductive efficiency. *American Journal of Botany* 41, 607–616.

Srinivasalu, N. and Chandrasekharan, N. R. 1958. A note on natural crossing in groundnut, *Arachis hypogaea* Linn. *Science and Culture* 23, 650.

Varisai Muhammad, S., Ramanathan, T., and Ramachandran, M. 1973c. Variation in morphological characters of *Arachis hypogaea* L. *Madras Agricultural Journal* 60, 1373–1379.

9 Soybean

9.1 INTRODUCTION

Soybean (*Glycine max* L. (Merr.) Fabaceae) is cultivated worldwide in temperate, subtropical, and tropical regions from ~50° N to ~40° S (Leff et al. 2004). In 2021, global production reached 363,863,000 tons, with Brazil being the largest producer (37%), followed by the United States (33%), Argentina (12%), and China (5%), respectively (USDA Foreign Agricultural Services 2022). Soybean (*Glycine ma*x L. Merrill) is the major oilseed crop in the world, accounting for nearly 50% of total oilseed acreage as well as production (Hermann, 1962). The Food and Agricultural Organization has categorized soybean as oilseed rather than a pulse. It is the main source of protein and oil in the world. This crop provides complete protein, containing eight amino acids essential for human health (Asadi and Faraji 2009). This has been considered as a miracle crop due to its extraordinary nutritional qualities and source of dietary protein in East and Southeast Asia.

North and Central China have been reported as the origin of soybean (Hymowitz 1970; Cuzin 1976). Soybean first emerged as a domestic crop around eleventh century BC in the eastern half of northern China and migrated to southern China, Korea, Japan, and Southeast Asia between the eleventh and third century BC (Hymowitz 1970).

Among oilseed crops, soybean is a cheap source of proteins for masses, especially in developing nations. It has an extraordinary growth, which enables it to fulfill the meal demand and rising consumption of livestock products. Nearly 50% oilseed output of the world is met by soybeans Joshi 2001). Generally, oil crops are best suited for tropical environments and perform well in 0–2,740 altitudes above mean sea level (amsl).

In the year 2014, soybean cultivation reached 108.83 lakh hectares, recording production of 90 lakh metric tons and productivity of 827 kg/ha in India. In Madhya Pradesh during 2014, the area reached about 55.46 lakh hectares, production of 51.60 lakh metric tons, and productivity of 930 kg/ha (SOPA 2014). It contributes around 47% and 25% of the total oilseeds and edible oil pool, respectively, of the country. In addition to a substantial share in edible oil, the crop earns annually around 13,000 crores through the export of de-oiled cake.

In India, soybean (*G. max*) has been the number one oilseed crop in terms of both area and production since 2005. The crop has shown tremendous growth during the last four decades from an area of 30,000 ha and production of 14,000 ton in 1970 to 9.95 million hectares with total production of 12.57 million ton in 2011. Soybeans occupied 42% of India's total oilseeds and 25% of edible oil production. The crop now ranks fourth in terms of global soybean area sown and fifth in terms of soybean production, after USA, Brazil, Argentina, and China.

Soybean is mainly grown as a rainfed crop during the rainy season in several states of India, such as Madhya Pradesh, Maharashtra, Rajasthan, Karnataka, Andhra Pradesh, Chhattisgarh, Nagaland, and Gujarat. The cultivation of the crop has helped in improving the socio-economic conditions of large numbers of small and marginal farmers in the rainfed agro-ecosystems.

The crop has potential of mitigating rampant protein energy malnutrition as well as becoming an ideal food of the country on account of a number of nutraceutical and functional compounds. Shanmugasundaram and Yan (2004) estimated that the total biological yield of vegetable soybean can be as high as 40 t/ha, consisting of 10 t of marketable pods and more than 30 t of residue to enrich the soil or feed animals. Besides several other factors, the productivity of vegetable soybean is limited by insect pests and diseases (Talekar and Chen 1983; Talekar 1987; Yeh et al. 1991; Chen 2003; Cho and Lee 2003; Lai et al. 2004). Several insect species cause considerable damage, beginning from emergence to harvest (Talekar 1987). The crop losses may be as high as 100% (Talekar and Chen 1983). Yield losses due to individual disease/insect/weed species range from 20 to 100%

DOI: 10.1201/9781032656724-10

(Yeh et al. 1991). Indiscriminate use of pesticides seriously eliminates beneficial insects, including pollinators, from soybean ecosystem. However, implementation of integrated pest management has resulted in 30–35% additional yield.

Soybean is grown as an annual crop mostly for its seeds, which are produced in pods containing one to four seeds each. Soybean is predominantly cultivated as a yearly plant primarily valued for its seed production, with pods typically containing one to four seeds. While soybean has traditionally been regarded as predominantly self-pollinating (Free 1993; Roubik 1995), there has been a growing acknowledgment of the involvement of animal pollinators in this crop (Chacoff et al. 2010; Giannini et al. 2015; Melathopoulos et al. 2015).

Soybean flowers have several characters which attract pollinators. They produce nectar and have volatiles and nectar guides (Erickson and Garment 1979; Robacker et al. 1983; Gill and O'Neal 2015; Levenson et al. 2022; Milfont et al. 2013; Monasterolo et al. 2015). In spite of claims of full self-fertilization, plants have been reported to produce up to 6.3% outcrossed seed (Ray et al. 2003), while outcrossing rates as high as 20% were found in wild soybean (Fujita et al. 1997). Many earlier studies have confirmed the benefit of insect pollination in soybean flowers (Garibaldi et al. 2021).

9.2 FLORAL BIOLOGY

Soybeans have typical papilionaceous flowers, with a tubular calix of five unequal sepals. Five petals consisting of (a) one posterior banner petal, (b) two lateral wing petals, and (c) two anterior keel petals are in contact but not fused with each other. When a bud in the axil of a trifoliolate leaf develops into an inflorescence, the stalk of that inflorescence remains stem-like, having the typical stem anatomy, with epidermis, cortex, endodermis, vascular tissue, and considerable secondary growth from a vascular cambium (Dzikowski 1937). The ten stamens collectively called the androecium occur in two groups, in which the filaments of nine of the stamens are fused and elevated as a single structure, whereas the posterior remains separate. The single pistil is unicarpellate, having one to four campylotropous ovules alternating along the posterior suture (Carlson and Lersten 1987). The style is about half the length of the ovary and curves backward toward the free posterior stamen, terminating in a capitate stigma (Figure 9.1).

Soybean flowers are often fertilized by the time the flower opens, which may occur a full day before the flower fully opens (Dzikowski 1936). Some cultivars are entirely cleistogamous, where flower buds do not open and the fertilization takes place with its own pollen. In some cultivars, the opening of flowers depend upon appropriate environmental conditions. All the soybean flowers do not actually develop into pods. Over 80% of the flowers may abscise, producing no yield.

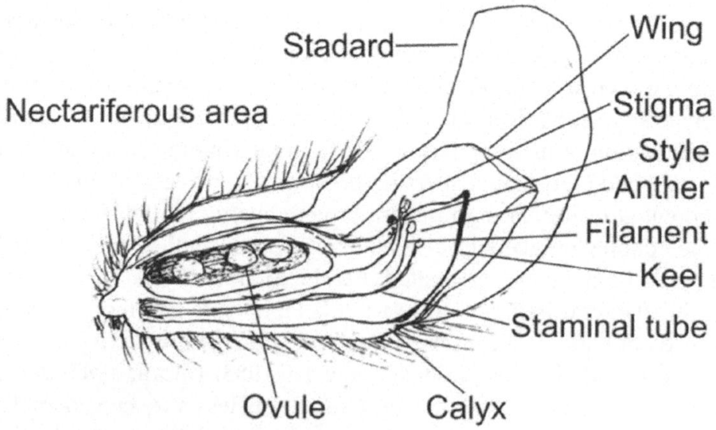

FIGURE 9.1 Diagram of the flower structure of soybean.

However, failure of fertilization is not the cause of floral abscission, since almost all abscised flowers are already fertilized and contain pro-embryos that have undergone two or three cell divisions (Abernathy et al. 1977).

Sowing date affects significantly on the different stages of growth as one of the important farming issues, and it is one of the important factors in determining harvest of maximum cultivar yield in one region. Proper planting dates can help in the optimal utilization of climate factors, such as temperature, humidity, day length, and also anthesis time adaptation with proper temperature (Hashemi 2001). Different varieties of soybean are very sensitive to deviations in environmental conditions where the crop is being grown. Therefore, it is necessary to study the genotype–environment interaction to identify the stable varieties in different environments (Calvino et al. 2003).

Erickson (1976) stated that bees readily forage on soybeans to gather both nectar and pollen, but highlighted that little is known about the level of preference bees exhibit for soybean pollen or its nutritional quality. Bees gather large quantities of soybean nectar in many areas of the United States and under certain conditions in preference to nectar from other sources.

Soybean nectar is abundant and of exceptional quality, with the amount produced depending on the specific soybean variety and environmental factors, particularly soil conditions that promote higher rates of nectar secretion (Erickson 1976). Juliano (1976) and Moreti et al. (1998) emphasized the significance of insect pollination, specifically by bees, for soybean plants. Generally, plants effectively compete for the attention of bees by offering nectar with sugar concentrations exceeding 25%. Soybean flowers produce nectar containing 30–50% dissolved sugar solids, slightly higher than other plants (Erickson 1975a, 1975b), which is sufficient to attract nectar-collecting insects.

9.3 POLLINATION IN SOYBEANS

Soybean flowers can help produce significant amounts of honey (Hambleton 1936; Milum 1940; Johnson 1944; Pellett 1947; Davis 1952; Jaycox 1970a; Pellet 1976). The honey produced is dark in color, sharp in aroma (Clemson 1985; Crane et al. (1984).

In spite of being autogamous, cleistogamous, and self-pollinating plants, soybeans have been reported to benefit from insect pollination (Robacker et al. 1983; Free 1993). Natural cross-pollination in soybeans is very low, from 0.03% (Caviness 1966) to 3.62% (Beard and Knowles 1971). However, Gordienko (1960) found 28% and 44% for two varieties of soybeans caged with honeybees. Several studies (Koelling et al. 1981; Cutler 1934; Beard and Knowles 1971; Abrams et al. 1978; Sadanaga and Grindeland 1981) reported that honeybee pollination of male sterile soybeans in enclosed plots resulted in a significant 477% increase in seed production compared to plots without bees. Discrimination by honeybees between varieties relative to floral color and/or other characteristics has largely been ignored by plant breeders, which needs to be looked into (Brim and Young 1971). Erickson (1975b) reported that floral characteristics of different varieties such as flower size, color, abundance, cleistogamy, aroma, nectar production, and blossom sequence impact the attractiveness of honeybees. Therefore, discrimination of varieties by foraging honeybees has a significant effect on floral fidelity. Soybean flowers are not always attractive to honeybees, leading to the hypothesis that environmental conditions during growth and flowering of plants affect the development of flower characteristics (Robacker et al. 1983). In general, plants that grow in higher temperatures produce more nectar and are more attractive than those grown in cooler climates, up to a maximum of 29°C. Erickson et al. (1978) reported that it is important to consider that the attraction of honeybees was not the same for all varieties. Jaycox (1970b) found that the number of bees varied from 680 to 810/ha, depending on the soybean cultivars (Pedersen 1962), or 20% of the population observed on white clovers (Weaver 1965). Another factor that may influence the attractiveness of soybeans to bees is the distribution of flowers on the plants. The flowers of soybeans are not grouped into large heads but are located at the leaf nodes of the plants. They are usually beneath the foliage on indeterminate cultivars. Heinrich and Raven (1972) reported that inflorescences allow a bee or another large pollinator to fulfill its energy requirements by accessing a cluster of flowers that would not be suitable for individual foraging. The amount of energy expended in walking between flowers can be 100 times less than an equivalent period of flight. Even though honeybees walk between the flowers, the density of flowers per plant is very low. Therefore, more flying would be required to visit a given number of flowers on soybeans than on crops such as alfalfa or clover, justifying the highest number of bees visiting these crops, as contrasted with soybeans. It is important to consider that the soybean flower stigma becomes receptive one or two days before anthesis, while the anthers release the pollen before the flower opens, which are essential conditions for auto-pollination (Fehr 1980; Delaplane and Mayer 2000).

9.4 POLLINATORS

Soybeans are considered to be a self-pollinating legume, which implies that insects are not required to pollinate the crop. However, a wide variety of insects visits soybean flowers for nectar and pollen. Gill and O'Neal (2015) found that the most abundant species collected included *Agapostemon virescens* F. and *Lasioglossum (Dialictus)* species (Hymenoptera: Halictidae), *Melissodes bimaculata* Lepeletier (Hymenoptera: Apidae), and *Toxomerus marginatus* Say (Diptera: Syrphidae). They found soybean pollen alone or intermixed with pollen grains from other plant species on 29 and 38% of the bees examined in 2011 and 2012, respectively. The results suggest a diverse community of pollinators—composed of mostly native, solitary bees—visits soybean fields and forages on their flowers.

Ahrent and Caviness 1994 studied 12 soybean cultivars where both honeybees and indigenous insect populations were present; cross-pollination varied from as low as 0.09% to as high as 1.63% based on a two-year average. Therefore, they concluded that insects, including honeybees, are not required to pollinate soybeans, and the presence of insects would not be expected to significantly improve yields.

Pollinators are attracted to different flowers, depending upon the availability of resources, such as nectar and pollen, its amount, and nutritional quality. There are few studies regarding the diversity and seasonal abundance of pollinators visiting soybean flowers around the world. Soybean flowers attract a huge spectrum of insects, including honeybees and other pollinating insects. Not only the honeybees but also other pollinating insects are found on soybeans. Monasterolo et al. (2015) reported that exposed flowers produced significantly higher yields than in the enclosures. Rust et al. (1980) collected 29 species of wild bees, which included *Melissodes bimaculata* (Lepeletier) and *Halictus confusus* Smith, *Megachile rotundata* (F.), *Megachile mendica* Cresson, and *Dialictus testaceus* (Robertson) as the most abundant pollen carriers. They also collected *Bombus impatiens* Cresson, *Bombus vagans* Smith, *Ceratina calcarata* Robertson, *Melissodes bimaculata* (Lepeletier), *Megachile rotundata* (F.), *Megachile mendica* Cresson, Agapostemon virescens (F.), *Augochlorella striata* (Provancher), *Halictus confusus* (Smith), *Lasioglossum coriaecum* (Smith), Dialictus testaceus (Robertson), *D. tegularis* (Robertson), *D. illinoenis* (Robertson), D. obscurus (Robertson), *D. pilosus* (Smith), *D. imitatus* (Smith), *D. zephyrus* (Smith), *D. versatus* (Robertson), and *D. atlanticus* Mitchell on soyabean flowers. Barella (2009) reported that *Apis mellifera* was the dominant species foraging on soybeans (57%) in Barra do Bugre, MT, while *Meliponini* species represented 29% of the total insects visiting soybean flowers. Soybean blossoms have functional nectaries, and sugar concentrations in soybean nectars are 5–10% higher than those of alfalfa, when growing conditions are favorable (Erickson 1984b). In the Central United States, soybean nectar production and bee visitation occur between 0900 and 1500 h each day. Soybean nectar volume per flower varies significantly in different cultivars, ranging from none to 0.2 μL/flower, with h as exception to some flowers, having as much as 0.5 μL, noting that the honey stomach of worker honeybee holds 35 μL–50 μL (Erickson 1984a). Collection of pollen by honeybees varies considerably depending upon the variety from which collecting pollen (Erickson 1984a, b).

9.5 FRUIT SET/SEED SET

Soybeans are partially dependent on insect pollination (Klein et al. 2003; Gallai et al. 2009; ISSA 1984). Lautenbach et al. (2012) reported benefits of insect pollination on soybean in Brazil, Argentina, India, China, and the USA. Robacker et al. (1982, 1983) have shown higher soybean yields when bees were introduced in the field for pollination purposes. Soybean yields have been shown to be influenced by honeybee visitation (Vila 1988; Vila et al. 1992).

Abrams et al. (1978) reported that the yields of three soybean varieties grown in Indiana experienced an increase of approximately 17.2% when honeybee colonies were present within a radius of up to 32 m. However, the rate of yield increases declined significantly beyond this 32 m radius from the colonies.

In Wisconsin, two cultivars, Corsoy and Hark, yielded 14.8% and 16.4% increases, respectively, in cages with honeybees over those without (Erickson 1975a, 1975c). In Kansas, yield in soybeans cv. Forrest was 20% greater in cages with bees than in cages without bees (Kettle and Taylor 1979). Caged experiments conducted by Erickson et al. (1978) recorded 16% increase in the yields when honeybees were introduced in the cages. Chiari et al. (2005, 2008) evaluated the effect of the honeybee pollination on production and quality of soybean seeds and concluded that seed production was higher in covered areas with honeybee colonies than in covered areas without honeybee colonies. The pod number in covered treatment with honeybees was 61% higher as compared with the covered treatment without honeybees. The study of Chiari et al. (2008) established that the soybean yields in the covered area with honeybees (2,757 kg/ha) and in the uncovered area (2,828 kg/ha) were higher than in the covered area without honeybees (2,000 kg/ha). The number of pods per plant was greater in the covered area with honeybees (38) and in the uncovered area (32) as compared to the covered area without honeybees (21), but no difference was found for the seed weight or its germination. In Ceará, Northeastern Brazil, Milfont et al. (2013) reported that the yield the yield increased 6.3% in open pollination over the caged area.

REFERENCES

Abernathy, R. H., Palmer, R. G., Shibles, R., and Anderson, J. C. 1977. Histological observations on abscising and retained soybean flower. *Can. J. Plant Sci.* 57, 713–716.

Abrams, R. I., Edwards, C. R., and Harris, T. 1978. Yields and cross-pollination of soybeans as affected by honeybees and alfalfa leaf cutting bees. *American Bee Journal* 118, 555–558.

Ahrent, D. K. and Caviness, C. E. 1994. Natural cross-pollination of 12 soybean cultivars in Arkansas. *Crop Science* 34, 376–378.

Asadi, L. and Faraji, L. 2009. *Principles of Applied Agriculture Oilseeds.* Iran's Agricultural Science Publishing, p. 93.

Barella, W. M. 2015. Abelhas polinizadoras na cultura da soja (*Glycine max* L.). Jornada Científica Da Unemat, 2, 2009. Available at www.unemat.br/eventos/jornada2009/5conic.php?content=downloads/prog_conic_paineleoral/painel-05. Accessed 9 January 2015.

Beard, B. H. and Knowles, P. F. 1971. Frequency of cross-pollination of soybeans after seed irradiation. *Crop Science* 11, 489–492.

Calvino, P. A., Sadras, V. O., and Andrade, F. H. 2003. Quantification of environmental and management effects on the yield of late-sown soybean. *Field Crops Res.* 83, 67–77.

Caviness, C. E. 1966. Estimates of natural cross-pollination in Jackson soybeans in Arkansas. *Crop Science* 6, 211–212.

Chacoff, N. P., Morales, C. L., Garibaldi, L. A., Ashworth, L., and Aizen, M. A. 2010. Pollinator dependence of Argentinean agriculture: Current status and temporal analysis. *Am. J. Plant Sci. Biotechnol. Glob. Sci. Books.* 3, 99–105.

Chen, Y. C. 2003. Powdery mildew of vegetable soybean in Taiwan. *Plant Pathol Bull* 12, 209–211.

Chiari, W. C., Toledo, V. de A. A. de, Hoffmann-Campo, C. B., Rúvolo-Takasusuki, M. C. C., Toledo, T. C. S. de O. A. de, and Lopes, T. de S. 2008. Polinização por *Apis mellifera* em soja transgênica [*Glycine max* (L.) Merrill] Roundup ReadyTM cv. BRS 245 RR e convencional cv. BRS 133. *Acta Scientiarum Agronomy* 30, 267–271.

Chiari, W. C., Toledo, V. de A. A. de, Ruvolo-Takasusuki, M. C. C., Oliveira, A. J. B. D., Sakaguti, E. S., Attencia, V. M., Costa, F. M., and Mitsiu, M. H. 2005. Pollination of soybean (*Glycine max* (L.) Merril by honeybees (*Apis mellifera* L.). *Brazilian Archives of Biology and Technology* 48, 31–36.

Cho, E. K. and Lee, S. H. 2003. Cowpea mosaic virus in vegetable soybean in Korea. *Plant Pathol J* 19, 166–170.

Clemson, A. 1985. *Honey and Pollen Flora.* Department of Agriculture New South Wales. Inkata Press, Sydney.

Crane, E., Walker, P., and Day, R. 1984. Directory of Important World Honey Sources. International Bee Research Association London UK.

Cutler, G. H. 1934. A simple method for making soybean hybrids. *Journal of the American Society of Agronomy* 26, 252–254.

Cuzin, V. F. 1976. Soviet greut encyclopaedia (third ed.). *Moscow Publishing Company Soviet Encyclopaedia* 24, 281–282.

Davis, J. H. 1952. Soybeans for honey production. *American Bee Journal* 92, 18–19.

Delaplane, K. S. and Mayer, D. F. 2000. *Crop Pollination by Bees.* CABI Publishing, New York, p. 301.

Dzikowski, B. 1936. Studia nad soja Glycine hispida (Moench) Maxim. Cz. 1. Morfologia. *Mem. Inst. Natl. Pol. Econ. Rurale* 254, 69–100.

Dzikowski, B. 1937. Studia nad soja *Glycine hispida* (Moench) Maxim. Cz. II. Anatomia. *Memories Institute National Polish Economie Rurale* 258, 229–265.

Erickson, E. H. 1975a. Effect of honeybees on yield of three soybean cultivars. *Crop Science* 15, 84–86.

Erickson, E. H. 1975b. Variability of floral characteristics influences honeybee visitation to soybean blossoms. *Crop Science* 15, 767–771.

Erickson, E. H. 1975c. Honeybees and soybeans. *American Bee Journal* 115, 351–353.

Erickson, E. H. 1976. Bee pollination of soybeans. Soybean Seed Research Conference, 6. 1976, Chicago. Report . . . Washington: American Seed Trade Association, pp. 46–49.

Erickson, E. H. 1984a. Soybean floral ecology and insect pollination. *Soybean Genetics Newsletter* 11, 152–162.

Erickson, E. H. 1984b. Soybean pollination and honey production—A research progress report. *American Bee Journal* 124, 115–119.

Erickson, E. H., Berger, G. A., Shannon, J. G., and Robin, J. M. 1978. Honey bee pollination increases soybean yields in the mississippi delta region of arkansas and missouri. *Journal of Economic Entomology* 71, 601–603.

Erickson, E. H. and Garment, M. B. 1979. Soya-bean flowers: Nectary ultrastructure, nectar guides, and orientation on the flower by foraging honeybees. *J. Apic. Res.* 18, 3–11. doi: 10.1080/00218839.1979.11099935

Fehr, W. R. 1980. Soybean. In: Fehr, W. R. and Hadley, H. (ed.) *Hybridization of Crop Plants.* American Society of Agronomy, Madison, pp. 589–599.

Free, J. B. 1993. Insect pollination of crops, 2nd edn. University Press, Cardiff, p. 768.

Fujita, R., Ohara, M., Okazaki, K., and Shimamoto, Y. 1997. The extent of natural cross-pollination in wild soybean (*Glycine soja*). *J. Hered.* 88, 124–128. doi: 10.1093/oxfordjournals.jhered.a023070

Gallai, N., Salles, J. M., Settele, J., and Vaissière, B. E. 2009. Economic valuation of the vulnerability of the world agriculture confronted with pollination decline. *Ecological Economy* 68, 810–821.

Garibaldi, L. A., Schulte, L. A., Nabaes Jodar, D. N., Gomez Carella, D. S., and Kremen, C. 2021. Time to integrate pollinator science into soybean production. *Trends Ecol. Evol.* 36, 573–575. doi: 10.1016/j.tree.2021.03.013

Giannini, T. C., Cordeiro, G. D., Freitas, B. M., Saraiva, A. M., and Imperatriz-Fonseca, V. L. 2015. The dependence of crops for pollinators and the economic value of pollination in brazil. *J. Econ. Entomol.* 108, 849–857. doi: 10.1093/jee/tov093

Gill, K. A. and O'Neal, M. E. 2015. Survey of soybean insect pollinators: Community identification and sampling method analysis. *Environmental Entomology* 44(3), 488–498. doi: 10.1093/ee/nvv001

Gordienko, V. 1960. Sexual hybrids of soya beans obtained by directed bee pollination. In: Mel´Nichenko, A. N. (ed.) *Pollination of Agricultural Plants by Bees.* Izd-vo Minist. Sel´sko Khoz, Moscow, 3, pp. 400–407.

Hambleton, J. I. 1936. Soybean for pollen and nectar. *Bee Culture* 64, 431.

Hashemi, J. M. 2001. Sowing date on the developmental stages and some agronomic and physiological characteristics of five soybean cultivars grown in the second planting. *Crop Sci. J.* 3(4), 49–59.

Heinrich, B. and Raven, P. H. 1972. Energetics and pollination ecology. *Science* 176, 597–602.

Hermann, F. J. 1962. A revision of the genus Glycine and its immediate allies. *US Department of Agriculture Technical Bulletin*, 1268.

Hymowitz, T. 1970. On the domestication of the soybean. *Eco. Bol.* 24, 408–421.

Jaycox, E. R. 1970a. Ecological relationships between honeybees and soybeans. I: Introduction. *American Bee Journal* 110, 306–307.

Jaycox, E. R. 1970b. Ecological relationships between honeybees and soybeans. II: The plant factors. *American Bee Journal* 110, 343–345.

Johnson, A. P. 1944. Honey from soybeans. *American Bee Journal* 84, 306.

Joshi, N. C. 2001. *Weed Control Manual*, 5th edn. Delhi Research Station, New Delhi, p. 538.

Juliano, J. C. 1976. Polinização entomófila da soja. In: *Congresso Brasileiro De Apicultura*, 4. Anais . . ., Curitiba.

Kettle, W. D. and Taylor, O. R. 1979. Ecological interactions of honeybees and soybeans. *Journal of the Kansas Entomological Society* 52, 549.

Klein, A. M., Steffan-Dewenter, I., and Tscharntke, T. 2003. Bee pollination and fruit set of *Coffea arabica* and *C. canephora* (Rubiaceae). *American Journal of Botany* 90, 153–157.

Koelling, P. D., Kenworthy, W. J., and Caron, D. M. 1981. Pollination of male-sterile soybeans in caged plots. *Crop Science* 21, 559–561.

Lai, C. P., Lee, C. L., Chen, P. H., Wu, S. H., Yang, C. C., and Shaw, J. F. 2004. Molecular analyses of the *Arabidopsis* TUBBY-like protein gene family. *Plant Physiol.* 134, 1586–1597.

Lautenbach, S., Seppelt, R., Liebscher, J., and Dormann, C. F. 2012. Spatial and temporal trends of global pollination benefit. *PLoS ONE* 7, 35954.

Leff, B., Ramankutty, N., and Foley, J. A. 2004. Geographic distribution of major crops across the world Glob. *Biogeochem. Cycles*, 18. doi: 10.1029/2003GB002108

Levenson, H. K., Sharp, A. E., and Tarpy, D. R. 2022. Evaluating the impact of increased pollinator habitat on bee visitation and yield metrics in soybean crops. *Agric. Ecosyst. Environ.* 331, Article 107901. doi: 10.1016/j.agee.2022.107901

Melathopoulos, A. P., Cutler, G. C., and Tyedmers, P. 2015. Where is the value in valuing pollination ecosystem services to agriculture. *Ecol. Econ.* 109, 59–70. doi: 10.1016/j.ecolecon.2014.11.007

Milfont, M. de O., Rocha, E. E. M., Lima, A. O. N., and Freitas, B. M. 2013. Higher soybean production using honeybee and wild pollinators, a sustainable alternative to pesticides and autopollination. *Environ. Chem. Lett.* 11, 335–341. doi: 10.1007/s10311-013-0412-8

Milum, V. G. 1940. Bees and soybeans. *American Bee Journal* 80, 22.

Monasterolo, M., Musicante, M. L., Valladares, G. R., and Salvo, A. 2015. Soybean crops may benefit from forest pollinators Agric. *Ecosyst. Environ.* 202, 217–222. doi: 10.1016/j.agee.2015.01.012

Moreti, A. C. de C. C., Silva, E. C. A. da, Alves, M. L. T. M. F., and Silva, R. M. B. da. 1998. Observações sobre a polinização entomófila da cultura da soja (*Glycine max* Merril). *Boletim da Indústria Animal* 55(1), 91–94.

Pedersen, M. W. 1962. An analysis of certain factors associated with pollination and seed production in alfalfa. In 1st Internatl. Symposium on Pollination Proc., Copenhagen, Aug. 1960. Commun. 7, Swedish Seed Growers' Assoc., pp. 64–74.

Pellett, F. C. 1947. *American Honey Plants.* Orange Judd Publ. Co., New York, p. 321.

Pellett, F. C. 1976. *American Honey Plants*, 5th edn. Dadant and Sons, Hamilton, p. 467.

Ray, J. D., Kilen, T. C., Abel, C. A., and Paris, R. L. 2003. Soybean natural cross-pollination rates under field conditions. *Environ. Biosaf. Res.* 2, 133–138. doi: 10.1051/ebr:2003005

Robacker, D. C., Flottum, P. K., and Erickson, E. H. 1982. The role of flower aroma in soybean pollination energetics. Pollination Conference, 10. 1982, Carbondale, IL.

Robacker, D. C., Flottum, P. K., Sammataro, D., and Erickson, E. H. 1983. Effects of climatic and edaphic factors on soybean flowers and on the subsequent attractiveness of the plants to honey bees. *Field Crops Res.* 6, 267–278. doi: 10.1016/0378-4290(83)90067-9

Roubik, D. W. and Food and Agriculture Organization of the United Nations (eds.). 1995. *Pollination of Cultivated Plants in the Tropics, FAO Agricultural Services Bulletin.* Food and Agriculture Organization of the United Nations, Rome.

Rust, R. W., Mason, C. E., and Erickson, E. H. 1980. Wild bees on soybeans, *Glycine max*. Environmental Entomology 9(2), 230–232.

Sadanaga, K. and Grindeiand, L. 1981. Natural cross-pollination in diploid and autotetraploid soybeans. *Crop Science* 21: 503–506.

Shanmugasundaram, S. and Yan, M. R. 2004. Global expansion of high value vegetable soybean. The 7th World Soybean Research Conference, pp. 915–992.

SOPA. 2014. SOPA statistical data. Available at www.sopa.org

Talekar, N. S. 1987. Insects damaging soybean in Asia, pp. 25–45. In S. R. Singh, K. O. Rachie and K. E. Dashiell (eds.), *Soybeans for the tropics: research, production and utilization*. Wiley, New York.

Talekar, N. S. and Chen, B. S. 1983. Seasonality of insect pests of soybean and mungbean in Taiwan. *J Econ Entomol* 76(1), 34–37.

USDA Foreign Agricultural Services. 2022. Soybean explorer [WWW Document]. Available at https://ipad.fas.usda.gov/cropexplorer/cropview/commodityView.aspx?cropid=2222000 accessed 3 March 2022.

Vila, V. P. V. 1988. Efeito das abelhas africanizadas, *Apis mellifera* L., na híbridação e na produtividade da soja, *Glycine max* (L.) Merrill. 58 f. Dissertação (Mestrado em Entomologia)—Universidade Federal de Viçosa, Viçosa.

Vila, V. P. V., Martinho, M. R., Sediyama, T., and Freire, J. A. H. 1992. Effect of africanized bees, *Apis mellifera* L. in the hybridization and productivity of soybeans *Glycine max* (L.) Merrill. In: *International Congress of Apicultural Apimondia*, 32. Apimondia Publishing House, Bucharest, pp. 414–415.

Weaver, N. 1965. Foraging behavior of honeybees on white clover. *Insectes Sociaux* 12, 231–240.

Yeh, C. C., Hartman, G. L., and Talekar, N. S. 1991. Plant protection technology for vegetable soybean. Pp. 85–91. In *Proceedings of the Workshop on vegetable soybean: Research needs for production and quality improvement*, Apr 29–May 2, 1991, Taiwan.

10 Cotton

10.1 INTRODUCTION

Cotton, *Gossypium* spp. (Malvales: Malvaceae), is considered the world's most important fiber crop (Percival et al. 1999; Wendel et al. 1992) and the second most important oilseed crop (Jenkins and Suha 2001). Cotton is currently grown successfully in many countries around the world, and a number of them, such as China, India, USA, Pakistan, Brazil, Uzbekistan, Turkey, and Australia, make significant contributions to the total world production. China ranks number one, followed by India.

10.2 PHENOLOGY OF COTTON

The growth, structure, and development patterns of the cotton plant have been detailed in a number of publications (Brown and Ware 1958; Hector 1963; Mauney 1984; Oosterhuis 1990; Oosterhuis and Jernstedt 1999). Cotton plant physiology has been described by Eaton (1955) and Mauney and Stewart (1986). Plant development in cotton proceeds through a number of stages: (1) germination and seedling establishment, (2) leaf area and canopy development, (3) flowering and boll development, and (4) boll maturation (Oosterhuis 1990).

10.2.1 The Cotton Plant

Upland cotton plants are 0.6–1.5 m high (Clemson 1985), with the shoot system being made up of a main axis stem, leaves, buds, branches, and ultimately, flower buds, flowers, and fruit (bolls) (Figure 10.1–10.6). Agronomically, main stem leaves are predominant in the plant apex and thus serve an important role in light interception and as a primary target for oviposition by certain insects (Oosterhuis and Wullschleger 1988).

10.2.2 Flowering and Bolls

Reproductive growth commences about four to five weeks after planting, with the formation of floral buds in the apical parts of the plant (Beasley 1975) (Figure 10.1). The reproductive stage is followed several weeks later by flower opening (anthesis) and the commencement of fruit (boll) development (Oosterhuis 1990). Floral buds first appear as small green pyramidal structures known as squares. A square is composed of three large bracts forming the epicalyx, which completely enclose the developing flower. The flower is 5–10 cm long and 5 cm wide.

The flower exhibits five noticeable petals, which are joined together for the majority of their extent, forming an obliquely lobed calyx that tightly envelops the petal base (Delaplane and Mayer 2000). Consequently, the corolla may seem to have separate petals (polypetalous), but it is actually composed of fused petals at the base (sympetalous). Moreover, the staminal column is comprised of multiple filaments that are fused both to the corolla and to each other for a significant portion of their length.

Distinct filaments are exerted in pairs along the length of the staminal column, and each bears a two-lobed anther. The staminal column surrounds the style that joins together with two to five lobed stigmas to constitute the female part of the flower.

The ovary is superior and composed of three to five united carpels, each with several ovules. One carpel forms one locule (Smith and Cothren 1999). Cotton ovaries display axile placentation, featuring two rows of ovules in each locule. The number of ovules in each locule ranges from 8 to 15, and

DOI: 10.1201/9781032656724-11

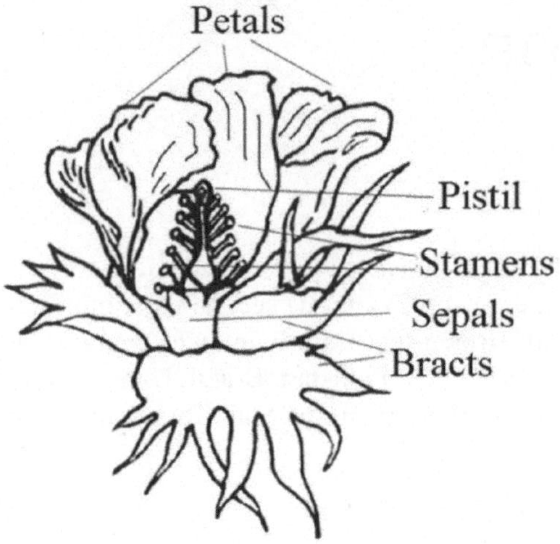

FIGURE 10.1 Diagram showing the different parts of the cotton flower.

this count may differ depending on the variety and growth conditions during flower development. However, it is worth noting that not all these ovules undergo development to become mature seeds. On average, each locule contains approximately eight seeds. These seeds are enveloped by abundant lint-like hairs (Clemson 1985).

Fruits are large rounded capsules called "bolls" which are divided into three to five compartments. Each boll commonly contains 25–35 seeds. The number of pollen grains present on the stigma will determine the probability that the ovary matures into a ripe boll, and also the number of ovules which are fertilized and develop into seeds in the mature boll (Smith and Cothren 1999).

10.2.3 The Flowering Pattern

Cotton has a distinctive and predictable flowering pattern. The first flowers to open are those at the base of the plant, usually at main stem nodes 6 or 7, and at the first partition along a fruiting branch. The order of flowering is thus spirally upward and outward. The cotton plant continues to grow and produce flowers until defoliation or being killed by frost (Smith and Cothren 1999).

Shedding or abscission of squares and young bolls is a natural process in cotton and is further influenced by unfavorable environmental conditions, such as prolonged overcast weather, extreme temperatures, water stress, mechanical damage, certain genetic factors, diseases (Smith and Cothren 1999), and pest insects (Khan and Bauer 2002; Ward 2005). Additionally, insufficient pollination is also considered a potential reason for flower shedding (Meade 1918; McGregor 1976). Moreover, during periods of intense heat, the cotton flower may fail to produce viable pollen, contributing to the shedding phenomenon. When this occurs, the flower normally sheds, unless pollen is brought to it from another flower by pollinators (McGregor 1976).

10.2.4 Floral Biology

The cotton flower is borne individually, and the quantity of flowers produced is influenced by various factors, such as the available plant nutrition, water availability, variety of cotton being grown, plant population density (Percival 1965; McGregor 1976), and prevailing climatic conditions (Sawan et al. 2002). These factors collectively play a role in determining the number of flowers that ultimately develop on the cotton plant. Usually, there are not more than three open flowers per plant per day in *G. hirsutum* (Free 1993). Flowers open in the morning shortly after sunrise (McGregor 1976), then later, anthers of *G. hirsutum* dehisce to produce abundant pollen, and the corollas start to expand at 0630 h until they are fully open at 0800–1000 h (Waller et al. 1981b; Wilson and Stapp 1984). Opening and dehiscence of anthers are also influenced by atmospheric conditions, such as low temperatures and high humidity. The flowers begin to wilt in mid-afternoon or toward sunset and never reopen (Free 1993). Janki et al. (1968) studied the floral biology of five varieties belonging to the three major species of cotton and found that they differed in their commencement and period of flowering.

10.2.5 Nectar and Nectary Glands

Nectaries are plant glands that are specialized for the production of nectar, which is a solution of sugars in water with minor amounts of other substances, such as amino acids, minerals, essential oils, and organic acids (Barker and Barker 1971).

Similar to numerous other plants, cotton secretes nectar, which is a tasteless, odorless, and colorless liquid (Vansell 1944a). This nectar serves as an enticing incentive for attracting insects and is primarily considered a reward for pollinators that visit the flowers. Moreover, the nectar in cotton flowers can also function as an alternative food source for beneficial insects and various arthropod species within cotton crops (Mohan and Kairon 1999). This aspect highlights its significance in promoting biodiversity and potentially aiding in pest management within the cotton ecosystem.

The cotton plant is unusual in that the nectaries are not limited to the flowers. Cotton has four sets of nectaries, one of which is floral, with the other three being extrafloral. All the nectaries outside the flower are usually referred to as extrafloral nectaries (Mound 1962; Butler et al. 1972; McGregor 1976). The floral nectaries are located at the base of the petals of the flower, where there are five small openings through which an insect with a long proboscis can reach the nectar. With regard to the extrafloral nectaries, the circumbracteal nectaries (usually three) are located between the three flower bracts, the subbracteal nectaries (usually three) are located just below the bracts (one nectary below each bract), and a leaf nectary is found on the underside of almost every leaf on the midrib. Occasionally, there are two more leaf nectaries, one on each of the first two main branches of the midrib (McGregor 1976).

The floral nectar secretion of upland cotton has been extensively studied, including the factors that influence it, such as the season and time of day (Butler et al. 1972; Moffett et al. 1976b; Waller et al. 1981b), the variety of cotton (Vansell 1944a; Butler et al. 1972; Waller et al. 1981b), and the characteristics of the soil (Parks 1921). Cotton is considered a slow nectar producer, maintaining a relatively constant secretion rate ranging from 0.5 μL to 2.2 μL per hour (Cruden et al. 1983). On the day of opening, each cotton flower typically secretes a total volume of nectar between 5 and 33 μL, with significant variation observed between different cotton varieties (Vansell 1944a; Butler et al. 1972; Moffett et al. 1976b; Waller et al. 1981b; Loper and Davis 1985).

Quality of cotton floral nectar also varies greatly with variety, with sugar concentration (TDS) levels between 18 and 68% (Vansell 1944a; Butler et al. 1972; Moffett et al. 1976b; Waller and Moffett 1981; Waller et al. 1981b; El-Banby et al. 1985). Free (1993) reported that the different concentrations of floral nectar from different varieties may be explained partly by differences in the amount of evaporation occurring in the flowers. Vansell (1944a) found that the primary sugars in *G. hirsutum* nectar were glucose, fructose, and sucrose, with glucose and fructose being predominant. Butler et al. (1972), Moffett et al. (1975b), and Waller et al. (1981b) subsequently confirmed the high glucose content of both floral and extrafloral nectars. Extrafloral nectary production was investigated by Butler et al. (1972), who reported that the quantity and quality of extrafloral nectars of *G. hirsutum* varied—for example, leaf nectar produced 1.85 ?L (TDS 49.1%)/plant/d, while subbracteal glands produced only 0.6 ?L (TDS 15.3%); however, no circumbraceal nectar was recorded. The proportion of sugars was similar to floral nectar. There was a sharp decline in the availability of leaf nectar between 0800 and 1100 h. Moffett et al. (1976b) showed that floral nectar increased steadily from 0800 to 1700 h. Thus, leaf nectar appeared to diminish in availability as floral nectar increased.

10.3 COTTON POLLINATION

10.3.1 Pollination Ecology

Pollination plays a vital role in sustainable agricultural production (Hoopingarner and Waller 1992) and involves the transfer of pollen onto the stigma. This transfer can occur through close proximity of the anther (self-pollination) or with the help of biotic or abiotic agents (cross-pollination). Cross-pollination leads to greater genetic variability in the offspring compared to self-pollination. As a result, it promotes diversity and potentially enhances the adaptability and resilience of the plant population in changing environmental conditions.

Cotton is commonly regarded as a partially cross-pollinated crop (McGregor 1976; Tanda and Goyal 1979a) but is largely self-fertile and self-pollinating (Free 1993; Delaplane and Mayer 2000).

Flowers of the *Gossypium* spp. typically bloom in the morning, shortly after sunrise (McGregor 1976). In the case of *G. hirsutum*, the anthers dehisce (open up) once the flowers have opened, and they produce a plentiful supply of viable, self-fertile pollen between 0630 and 1000 h (Waller et al. 1981b; Wilson and Stapp 1984).

Each individual upland cotton flower yields an average of 40,000–45,000 pollen grains (Barrow 1981; Loper 1986; Vaissiere 1991b). Freshly dehisced cotton pollen grains have the appearance of creamy powder. They are characterized by being spiny, spherical, and relatively large, measuring approximately 101.5 μm and 122.5 μm with and without spines, respectively (Janki et al. 1968). Additionally, these pollen grains are covered with a sticky substance (viscous material) (McGregor 1976). Despite their abundance, these pollen grains are not easily dispersed to other flowers by wind (Tanda and Goyal 1979c). Instead, they rely on other means of transport, such as pollinators, like insects, to facilitate successful cross-pollination.

The effective initiation of the pollen–pistil interaction depends on the viability of the pollen and the receptivity of the stigma (Shivanna et al. 1997). Pollen *viability* refers to the ability of pollen to deliver functional male gametes to the embryo sac (Shivanna et al. 1991).

Stigma *receptivity* refers to the ability of the stigma to support pollen germination and tube growth upon receiving compatible viable pollen (Shivanna et al. 1997). In general, the stigma becomes receptive by the time the flower opens and the pollen is shed. However, this does not always occur and depends on the pollination mechanism (i.e., self- or cross-pollination) of a particular species (Sedgley 1982).

The receptivity period is a critical time that influences fruit set in many crops that need honeybees for pollination.

Crops with short receptivity periods may encounter difficulties with pollination, especially when unfavorable conditions hinder honeybee flights. In the case of cotton, the maximum secretion of a shiny, sugary liquid from the stigma was observed between 0400 (i.e., six hours prior to the flower opening) and 1100 h. This period was considered the peak of stigma receptivity (Janki et al. 1968).

Cotton pollen grains are shed from the anthers at the two-celled stage, resulting in an immature male gametophyte composed of a tube cell and a generative cell. Once the pollen grains adhere to the stigma, they germinate, and a pollen tube quickly penetrates the stigma, elongating through the style and reaching the ovary. Shortly after pollen tube emergence, the generative nucleus divides to form two sperm cells (Smith and Cothren 1999). This process is crucial for successful fertilization and the subsequent development of seeds.

The peak stigma receptivity in cotton occurs between 1100 and 1600 h on the day of flower opening, but receptivity sharply declines during the afternoon (Janki et al. 1968). Consequently, most pollination in cotton takes place by mid-afternoon.

Once pollen grains are deposited on the stigmatic hairs, they germinate within approximately 30 min (Beasley 1975) and remain viable for about 12 hr (Free 1993). However, the viability of pollen gradually decreases from around 0900 h onwards.

Barrow (1981) reported that pollen germination exceeded 98% within the first 8 hr after flower opening, and Kearney (1923) found that pollen could still fertilize around 86% of flowers at 1700 h on the day of flower opening. However, after 24 hr, pollen viability dropped significantly to 30% (Barrow 1981) or even as low as 11.7% to 18.5% (Janki et al. 1968). This highlights the importance of timely pollination for successful fertilization and seed production in cotton crops.

Viability is temperature-sensitive, with full viability occurring in the range 32–40°C, but with no germination above 42°C (Barrow 1981). Hand-pollination trials have provided some preliminary data about the required number of pollen grains necessary for full seed production in cotton.

According to Ter-Avanesian (1978a), when only two pollen grains were used for pollination, no seeds matured as a result. On the other hand, pollinating with five pollen grains per flower led to a high abscission rate of 80–90% of the flowers compared to fully pollinated flowers, indicating that this level of pollination was inadequate for successful seed development.

Meade (1918) previously reported that bolls failed to set unless at least 25 pollen grains were deposited on the stigma. Even with this minimum number, only one or two seeds matured in each locule, suggesting that for all ovules in a boll to be fertilized, at least 25–55 pollen grains are required to reach the stigma. This highlights the critical role of sufficient pollen deposition for successful fertilization and seed set in cotton.

Ter-Avanesian (1978) found bolls resulting from pollinations made with 20, 100, and 300 pollen grains were heavier and larger in size relative to naturally pollinated control flowers, but with 1,000 grains, there was no significant difference in the average weight of bolls. Vaissiere et al. (1982) also reported that bolls that matured from a stigma with >50 pollen grains produced significantly more (25.6) seeds than bolls produced from stigmas with <50 pollen grains (8). Waller and Mamood (1991) reported that a stigma needed about 100 viable pollen grains to set a full boll (i.e., 25–35 seeds/boll).

In cotton, fertilization typically takes place for the majority of ovules by the end of the first day after flower opening, specifically within 12–24 hr after pollination (Beasley 1975). Temperature plays a significant role as the primary determining factor in this process (Smith and Cothren 1999). Optimal temperatures during this critical period are crucial for successful fertilization, seed development, and subsequent cotton crop production. The post-pollination change in the color of petals, from white to pink after fertilization, begins at the petal tips.

This progresses to encompass the entire corolla, which becomes light red by sunset (Janki et al. 1968; McGregor 1976; Delaplane and Mayer 2000).

The deep-red corolla and staminal column of the cotton flower undergo withering on the second day after blooming and eventually fall off (Delaplane and Mayer 2000). It is worth noting that the color red falls outside the visual range of many insects, including bees (Kevan and Baker 1983).

Interestingly, the color change from white to red in the flowers of some insect-pollinated plants follows a similar pattern. This color transformation is often associated with the cessation of nectar secretion (Gori 1983). The change in color may signal to pollinators that the flower is no longer offering nectar rewards, thus encouraging them to focus on other viable flowers for pollination.

In cotton, the corolla closes up when it turns to red so that floral nectar is not accessible anymore (Moffett et al. 1981; Waller et al. 1981b). Eisikowitch and Loper (1984) observed honeybee foragers ignoring such flowers, indicating that the bees associate change in color and petal condition with lack of reward. They suggested that a change in cotton flower color by itself may not affect bee visitation, unless there is an absence of reward.

After fertilization, the bolls and their developing seeds expand rapidly. Inside the seed, the embryo seedling is fully formed within five weeks after fertilization. The oil and proteins that will provide the energy for germination accumulate mostly during the latter part of seed development (Smith and Cothren 1999). While the embryo takes shape inside the seed, cotton fibers develop from the outermost layer of cells on the seed coat.

Fiber initials (epidermal cells in the ovule) begin their elongation phase on the morning of anthesis (Beasley 1975), and elongation ceases by about 24–28 d post-anthesis, and fibers are mature 50–70 d after anthesis. At maturity, the lint dries and the fibers become twisted in a way that makes them especially suitable for spinning. As bolls dry, they split and flare open, exposing lint and seeds. Before ginning, the lint and seeds are called seed cotton (Flint 1996).

10.3.2 Pollinating Agents

Cotton is commonly regarded as a partially cross-pollinated crop (McGregor 1976). Since pollen grains of cotton are too heavy to be airborne, wind does not play any role in cross-pollination (Tanda and Goyal 1979c; Umbeck et al. 1991), insects are the likely natural agents for any pollen transfer (Figure 10.2–10.6). Most of the research on cotton pollination by insects has not been conducted in recent years, presumably because of high pesticide use. Most of the following references reflect this. Flowers of *Gossypium* spp. produce nectar and pollen which attract many beneficial as well as harmful insects (Butler et al. 1972), so numerous studies have been conducted to identify insect visitors to cotton flowers. The European honeybee, *Apis mellifera* L. (Hymenoptera: Apidae), is considered the most important pollinator of cotton in the United States of America (USA) (Kearney 1923; McGregor 1959; Weaver 1978; Moffett et al. 1976a, 1976c, 1980; Vaissiere et al. 1984; Waller et al. 1985b; Loper and Danka 1991; Free 1993), Egypt (Wafa and Ibrahim 1959), India (Sidhu and Singh 1962; Tanda and Goyal 1979b; Tanda 1983), and Sudan (El-Sarrag et al. 1993). However, other species of honeybees, including *Apis cerana* L., *A. florae* L. (Sidhu and Singh 1961; Tanda and Goyal 1979a), and *A. dorsata* L. (Tanda 1983), have also been reported to pollinate cotton. In USA, native non-*Apis* hymenopteran species such as *Bombus* spp. (Apidae) (Moffett et al. 1980, 1981; Berger et al. 1988), Melissodes spp. (Anthophoridae), and *Agapostemon anglicus*

FIGURE 10.2 *Apis mellifera* foraging on cotton flower.

FIGURE 10.3 *Apis dorsata* foraging on cotton flower.

FIGURE 10.4 *Anthophora* spp. bee foraging on cotton flower.

FIGURE 10.5 *Lithurgus* bee foraging on cotton flower.

FIGURE 10.6 Pollen sticked with the body of *Lithurgus* bee.

Cockerell (Halictidae) (Moffett et al. 1980, 1981), *Xylocopa varipuncta* Patton (Anthophoridae) (Waller et al. 1985b), *Osmia fulviventris* (Panzer) (Megachilidae) (Waller et al. 1978), *Megachile rotundata* (F.) (Megachilidae), *Halictus ligatus* Say (Halictidae), and *Anthophora confusa* Smith (Anthophoridae) (Moffett et al. 1980; Loper and Danka 1991), and stingless bee, *Trigona nigerima* (Apidae) (Loper and Danka 1991), can be important visitors and pollinators of cotton flowers under some circumstances.

The effectiveness of each pollinator species depends on several factors, including their population numbers, their visitation patterns, and the duration they spend on each flower. While other insects may also serve as efficient pollinators, *Apis mellifera* (honeybee) is considered the most significant pollinator for cotton worldwide due to its abundance and ease of handling by humans (Moffett et al. 1978a; Waller et al. 1985a; Tanda and Goyal 1979b; Free 1993).

Moffett et al. (1980) conducted a survey of 13 cotton fields across a wide area of the Texas plains and found that wild bees and wasps were not reliable as pollinators. Their populations exhibited significant fluctuations from year to year and between different fields, making them less consistent in their pollination effectiveness compared to honeybees.

As a result, due to the stable and substantial population of honeybees, along with their manageable nature, they are considered the primary and preferred pollinators for cotton crops, ensuring a more consistent and successful pollination process.

Moffett et al. (1976d) found that high populations of non-*Apis* spp. (wild bees) were observed only in cotton fields near large acreages of native pasture and that, due to modern agricultural practices, populations of wild bees were usually too low and unpredictable to reliably pollinate large acreages for hybrid cottonseed production. They concluded that rental of honeybee colonies appeared to be the best means to ensure adequate bee populations for cotton pollination (Sidhu and Singh 1961; Tanda and Goyal 1979b; Moffett et al. 1980; Vaissiere et al. 1984).

The subfamily Megachilinae comprises five tribes, all of which are well-represented in India. The first tribe, Lithurgini, is represented by the genus *Lithurgus Berthold*, which includes 14 species. These species have a scattered distribution in various parts of the country. Two specific species, *L. atratus* and *L. dentipes*, have drawn significant attention due to their role in pollinating cotton (Malvaceae family) and have been a matter of concern (Abrol 2012).

Additionally, their nesting behavior has also been studied in the past. Researchers such as Horne (1870, 1872), Malyshev (1930), and Lieftinck (1939) have commented on their nests built in hollow sticks. The nesting habits of these species are of interest to understand their life cycle and ecological interactions, especially regarding their role in pollination and their potential impact on cotton crops in India.

In Jammu region, *L. atratus* has been observed since the last three years foraging on cotton, okra, and hollyhock flowers, with the initiation of flower opening around 1000 h (personal observation, Dr. Uma Shankar). Even the *Lithurgus* bees were found mating inside the flowers of cotton and okra and fighting for resources availability.

Tribe Megachilini now consists of only two genera, namely, Megachile, which include the leaf-cutting bees and mason bees. Both genera are found almost everywhere throughout the country (Gupta 2003). Species of genus Megachile (exceeding 130 in number) have attracted many authors with regard to their pollination and nesting behavior.

10.3.3 The Role of Honeybees as Pollinating Agents in Cotton

Cross-pollination by honeybees in commercial production of cotton involves a substantial cost for growers. Since beehives can be moved readily wherever honeybees are needed for pollination, beekeepers in the USA have provided colonies to pollinate over 120 commercial crops, including cotton (McGregor 1976).

Although the introduction of honeybees for cotton pollination is regarded as a supplementary management practice, bee pollination has consistently been reported to increase boll set, boll weight, seed per boll, seed weight, lint per boll, total cotton yield, and uniformity of boll ripening (Kuliev 1958; McGregor et al. 1955; Moffett et al. 1978b; Vaissiere et al. 1984; Waller et al. 1985b; Rhodes 2002).

It also promotes earliness of seed set and decreases boll shedding (McGregor et al. 1955), improves seed germination (Radoev and Bozhirov 1961; El-Sarrag et al. 1993), increases seed oil content (El-Sarrag et al. 1993), and improves lint quality (McGregor et al. 1955; Avetisyan 1958; Kuliev 1958; Kaziev 1960; Tanda and Goyal 1979a; El-Sarrag et al. 1993; Rhodes 2002).

Several authors have reported increased yield (total boll weight) in various studies. These studies demonstrated yield increments of 15–20% (Wafa and Ibrahim 1960b; Sidhu and Singh 1962; Stith 1970; Tanda and Goyal 1979b; Rhodes 2002), 20–25% (Mahadevan and Chandy 1959; McGregor and Todd 1956; McGregor et al. 1955; Radoev and Bozhirov 1961; Tanda and Goyal 1979a), 25–30% (Avetisyan 1958; Kuliev 1958; Kaziev 1960), and even more than 30% (Shishikin 1952; Mahadevan and Chandy 1959; El-Sarrag et al. 1993).

Various methods have been employed to evaluate the efficacy of honeybee pollination. These techniques include comparing fields with and without managed honeybees (Shishikin 1952; Vaissiere et al. 1984; Vaissiere 1991a), assessing yield in relation to bee visitation levels (Rhodes 2002), and conducting experiments with bagged and unbagged flowers (Radoev and Bozhirov 1961). However, the most prevalent approach has been the use of cages, either to exclude bees from flowers or to enclose them with flowers (Shishikin 1952; Avetisyan 1958; McGregor and Todd 1956; Mahadevan and Chandy 1959; Sidhu and Singh 1962; Waller et al. 1985b; Tanda and Goyal 1979a; El-Sarrag et al. 1993; Rhodes 2002).

For instance, McGregor et al. (1955) investigated the impact of bee activity on boll set, seed number, and yield by confining honeybees with Pima and upland cotton plants under plastic screen cages and comparing them with caged plants without bees. They observed a 24.5% higher yield of cottonseed from Pima cotton when caged with bees, while there was no yield increase in upland cotton, although they noted an earlier fruit set.

Honeybee visits to cotton flowers not only increase the amount of pollen deposited on the stigmas of the same plant but may also increase the amount of foreign pollen introduced from other plants (of the same species) to the stigma. As an example, Kearney (1923) documented that bagged cotton flowers often had stigmas that were not entirely covered with pollen. However, this situation was uncommon in open flowers, especially when there was a high level of insect visitation, particularly by honeybees. Increased yield as a result of cross-pollination may be related to such factors as increased numbers of pollen grains over the entire surface of the stigma and increased rate of pollen tube formation in pollen grains from other flowers (Kearney 1923). These factors may result in higher fertilization of ovules, causing higher levels of boll setting and seed production. Thus, the measurable benefits from bee activity probably derive from bees increasing the amount and distribution of pollen on stigmas. This was further confirmed by Vaissiere et al. (1984), who reported that in the absence of bees, the number of pollen grains on stigmas was very low (0–7). In the presence of managed honeybees, this figure was often much higher. However, it was highly variable; for example, a range of between 1 and 324 pollen grains per stigma was recorded from a high visitation rate of 3.19 bees per 100 flowers. Since Vaissiere et al. (1984) collected their samples 24 hr after pollination, they suggested that factors other than bee visitation, such as flower or plant biology or the microclimate around flowers, were important in determining the number of pollen grains transferred onto stigmas within any one day.

The percentage of cotton flowers that develop into mature bolls is generally low. Even under the most favorable conditions, many buds do not reach the blooming stage, and many flowers that open fail to set bolls. According to Smith and Cothren (1999), a cotton plant will normally shed about

60% of its squares and young bolls under typical crop-growing conditions. However, McGregor (1976) showed that honeybee pollination increased boll set in cotton crops, and this finding was subsequently supported by Moffett et al. (1978b), who reported that cotton plants placed in cages with bees set 23% more bolls per flower than plants not caged with bees. In addition, they found that the bolls harvested from bee-pollinated flowers were heavier and contained more seeds per boll than bolls that were bagged or caged to exclude bees.

According to Vaissiere et al. (1984), honeybee pollination had a significant impact on the seed yield of the male sterile parent, primarily affecting boll weight and boll set. They proposed that the number of bolls per plant could be indicative of the level of pollinator visitation throughout the season, while boll weight could be seen as a measure of the efficiency of pollen transfer during those visits. McGregor et al. (1955) discovered that cross-pollination between different cotton varieties resulted in a better yield compared to cross-pollination within the same variety.

Several studies have also shown that honeybees can improve the quantity and quality of cotton through intervarietal pollination of *Gossypium* spp. (Avetisyan 1958; Kuliev 1958; Free 1993; McGregor 1976), *G. arboretum* (Tanda 1983, 1984), and *G. herbaceum* (Tanda and Goyal 1979a, 1979b).

Honeybees have historically been most commonly used in cotton to produce hybrid varieties. Hybrid cottonseed is reported to only be produced when adequate numbers of bees are present to transfer pollen from the normal male fertile (MF) pollen parent to the male sterile (MS) seed parent (Waller 1982; Moffett and Stith 1972a; Vaissiere et al. 1984). There has been only one report on efficacy of honeybees in cotton in Australia. Rhodes (2002) found, in the conventional cotton variety Siokra V-16, that plots with high bee visitation rates produced more harvestable bolls, and that they were heavier and contained more seeds and lint than those with low or no bee visitation. He also reported that the harvested lint from these bolls had improved micronaire.

10.4 MANAGEMENT OF HONEYBEE COLONIES DURING THE POLLINATION SEASON

Using honeybees to pollinate a crop requires a sufficient number of high-quality colonies and ensuring that these colonies can maintain good foraging activity throughout the flowering of the target crop. After moving bee colonies to a site, there are some other important techniques that are likely to assist in maintaining colony quality, such as providing the colonies with shade and easy access to water in hot and arid areas and managing the crop pests without harming bees (Martin and McGregor 1973; Moffett et al. 1974, 1977b; Johansen 1977; Goebel 1984; Jay 1986). There are three important considerations when colonies are imported to ensure they adequately pollinate the target crop.

10.4.1 The Stocking Rate

The stocking rate is defined as the number of imported colonies per unit area of target crop to pollinate. However, stocking rate is the major component of the cost of pollination, and the economics of using honeybees for pollination is a key consideration in assessing profitability of the target crop production. There is currently no recommended stocking rate of honeybee colonies to pollinate cotton, but some recommendations have been made in terms of the bee visitation rate (i.e., the number of bees in 100 flowers). McGregor (1959), using 200 colonies in 16 and 32 ha fields, concluded that ten or more honeybees per 100 flowers were indicative of a good boll set in Arizona. He suggested that the bee–flower ratio should be used as an indicator of the number of colonies required in a cotton crop for effective pollination. This technique excludes any bee working in extrafloral nectaries, which is unlikely to contribute to pollination of the crop. McGregor (1959) proposed that the required number of colonies to achieve the desired rate of flower visitation in cotton fields might differ depending on factors such as the amount of cotton cultivated per hectare and the densities of other nearby flowering plants that could potentially compete for honeybees' attention. The suggestion stems from the fact that honeybees are not attracted to cotton, making the presence of adequate bee colonies crucial for effective pollination in such environments. Thus, ten colonies per hectare may not be enough for a small field surrounded by competing flowering plants, whereas one colony may be sufficient for a more isolated field of several hectares.

In 1974, Moffett reiterated that between one and ten honeybees per 100 cotton flowers were sufficient for adequate pollination. Subsequently, Moffett et al. (1976d) suggested that a pollination rate of 0.5% was enough to achieve satisfactory boll set in Arizona, a conclusion similar to that reached by Waller et al. (1985a).

The stocking rate for honeybee colonies varied in different studies, with values of 17.7 colonies per hectare reported by Moffett et al. (1978a), 7.5–15.5 by Vaissiere et al. (1984), and 12.2 by Waller et al. (1985a). As mentioned earlier, the stocking rate required to achieve desirable bee visitation rates in cotton varied depending on several factors. Vaissiere et al. (1984, 1991a) conducted research with a wide range of stocking rates, from 0 to 35.5 colonies per hectare, and concluded that cotton yield correlated with the number of honeybee colonies provided.

Researchers such as Moffett et al. (1981), Waller et al. (1985a), and Loper and Danka (1991) reported that cotton yield decreased as the distance from the bee colonies increased, although this reduction in yield was less than the decline in bee visitation rate. Numerous other studies have also investigated honeybee stocking rates in cotton, particularly in hybrid cotton production, often evaluating the associated bee visitation rates.

These are summarized in Table 1.3, and they demonstrate that there is no consistent relationship between these two parameters. While there is some consensus of an optimum bee visitation rate of 0.5 bees/100 flowers to maximize pollination in cotton, the practical calculation of the required stocking rate to achieve this is clearly difficult, both for farmers and beekeepers.

10.4.2 Scheduling the Arrival and Removal of Honeybee Colonies in Relation to Flowering of the Target Crop

The time of placement of colonies is very important for optimum pollination. It is suggested that colonies should be introduced at the onset of flowering of the target crop so that they do not become conditioned to the other competing floral sources (Ribbands 1951; Free and Williams 1974). The general recommendation is for beekeepers to bring their colonies when the target crop is at 10–20% flowering (Free et al. 1960; Hoopingarner and Waller 1992). However, timing appears to be crop-dependent, because in some cases, early introduction of colonies may result in bees foraging out of the target crop, while a later time means that a high portion of the crop may not be cross-pollinated (Free et al. 1960; Moeller 1972). In cotton, there is no recommended time for importing hives, or a recommended optimal pollination period to economically increase production. However, yield data might be the best indicator of the proper time for colonies to be introduced rather than bee counts and pollen trap analyses, which are used in most studies of other crops, because these methods only provide indirect measures of pollination efficiency (Moeller 1972).

Bee visitation to cotton flowers is highly variable and influenced by several factors, including location, time of day, and date (Moffett et al. 1980). According to Phillips and Simpson (1989), optimizing honeybee management and pollination timing can enhance the quality of hybrid seed production, suggesting that using bees solely during the first half of the flowering period could be beneficial.

However, regardless of the ideal timing for bee pollination, it seems that bee visitation cannot be completely controlled. As reported by Moffett et al. (1980), even in cotton fields with substantial bee populations, bees may sometimes disregard the flowers and focus on collecting extrafloral nectar. Therefore, achieving a consistent level of 0.5 bees per 100 flowers for reliable pollination can be challenging.

10.4.3 The Spatial Distribution of the Colonies in Relation to the Target Crop Area

Placement of honeybee colonies in or near the crop can be an important aspect of efficient pollination, although requirements vary for different crops (Liven 1961). Todd and McGregor (1960), for example, reported that placing a single apiary on the edge of the field or orchard is an inefficient use of pollinators, and they suggested locating colonies in the center of fields. Although this might be possible in orchards such as almonds, it is impractical in many field crops, such as cotton, because it may interfere with the other necessary agricultural practices.

Numerous studies conducted in various crops have consistently demonstrated that optimal pollination occurs when honeybee colonies are strategically dispersed. For instance, Oldershaw (1970) found that scattering honeybee colonies singly was highly effective. Similarly, Free et al. (1960) observed that forming small groups of colonies, typically consisting of four to five colonies, distributed throughout the target crop yielded favorable results. Alternatively, even situating the colonies along most sides of the field contributed to enhanced pollination outcomes. (Moffett et al. 1974). In cotton, Vaissiere et al. (1984), Moffett et al. (1981), and Waller et al. (1985a) confirmed that pollination efficiency of honeybees was dependent not only on the number of colonies but also on their distribution in the field. Bees usually visit flowers close to their hives when they are available (Ribbands 1953; Free and Williams 1974; Free 1993), and consistent with this, Waller et al. (1985a), Loper and Danka (1991), and Vaissiere et al. (1984) have reported reduced bee foraging activity with increased distance of flowering cotton plants from honeybee colonies. Thus, locating colonies in small batches close to a large cotton crop appears to be important to achieve higher and uniform bee flower visitation.

10.5 HONEYBEE FORAGING BEHAVIOR IN COTTON FLOWERS

Honeybee foraging behavior is one of the important research areas in pollination biology. Honeybees change their foraging behavior in response to the needs of the colony and the availability of food sources in the field (Seeley 1985, 1995). Subsequently, the foraging of individual bees varies with the size of the foraging population and the overall competition for nectar and pollen resources. In cotton plants, most foraging honeybees have been routinely observed collecting nectar (from both floral and extrafloral nectary glands) rather than pollen.

Several authors have recorded diurnal foraging patterns of honeybee in cotton (Wafa and Ibrahim 1959; Moffett et al. 1980, 1981; Waller et al. 1985a; El-Sarrag et al. 1993). They concluded that extrafloral nectaries were preferred by bees in the early morning, and suggested that the foragers divide their time by collecting extrafloral nectar throughout the morning and floral nectar during the afternoon.

Honeybee activity showed variations throughout the day, with the lowest numbers observed in late afternoon. Tanda and Goyal (1979c) reported that insect visitation to *G. arboreum* flowers began as soon as they opened at 0800 h, reaching maximum activity between 1000 and 1100 h, and decreasing after 1300 h. However, their findings appear somewhat inconsistent with Moffett et al.'s (1980) report, which indicated that bee visits to *G. hirsutum* flowers were highest between 1000 and 1600 h, peaking at 1300–1400 h.

Honeybees are primarily attracted to cotton flowers for their nectar rather than pollen. Therefore, understanding the quantity, quality, and availability pattern of nectar in different cotton genotypes is crucial for comprehending bee foraging activity and pollination biology. The foraging behavior of honeybees in cotton can vary based on factors such as the presence and abundance of extrafloral and floral nectaries and the time of day (Eisikowitch and Loper 1984).

Moffett et al. (1975b) reported that cotton genotypes with floral nectar containing higher sugar concentrations usually attract more honeybees to their flowers, and Waller (1972) found that honeybees preferentially visited artificial flowers with high sugar (sucrose) concentration, with the optimum total sugar content being between 30 and 50%. Since the sugar concentration of floral nectar in most commercial cotton plants is lower than this optimum, Moffett et al. (1978a) have suggested that high sugar concentration of nectar be considered in selecting parental lines for hybrid cotton. Apart from the frequency of flower visits, another important aspect of honeybee foraging behavior which influences pollination efficacy is the length of time they spend on individual flowers (handling time). There are several reports on handling time and other infloral behavior of honeybees (*A. mellifera* and *A. cerana*) in cotton (Sidhu and Singh 1961; Moffett et al. 1975a; Tanda and Goyal 1979b; Tanda 1983; Buchmann and Shipman 1990; Loper and Danka 1991), with mean handling times reported from between 3.9 and 12.8 sec/flower. Reports on the seasonality and patterns of bee visitation to cotton flowers are somewhat conflicting. Moffett et al. (1975a, 1976b, 1978b, 1980) reported that honeybee visits to cotton flowers vary greatly within and between flowering seasons, locations, and varieties. Several studies have consistently revealed that honeybee visits to cotton flowers are relatively low during mid-season. Instead, higher percentages of visits tend to occur at the beginning and toward the end of the flowering season (McGregor 1976; Moffett et al. 1979a). Moffett et al. (1975b) further found that none of the 81 cotton genotypes they studied were sufficiently attractive to honeybees to overcome the mid-season drop in bee visitations that occur in southern Arizona. Weaver (1978) documented a significant decline in total bee activity during the latter part of the flowering period in isolated cotton fields, even when no insecticides were applied. In another study by Moffett et al. (1980), higher bee visits to cotton flowers were recorded in the first two weeks compared to the following weeks. However, earlier investigations (Moffett et al. 1975a, 1979a) had suggested an increase in bee visits during late flowering. These contrasting observations could potentially be attributed to the variations in flower numbers at the beginning and end of the season, as flower availability can influence bee activity patterns. Although nectar-gathering honeybees often emerge from a cotton flower liberally dusted with pollen, they generally visit several flowers in quick sequence until they collect a full nectar load. Then, before returning to the hive, they spend between 15 min (Loper 1986) and 20–30 min (Loper and DeGrandi-Hoffman 1994) to scrape the pollen grains from their body with their legs: this implies that the bees actively reject cotton pollen.

10.5.1 Pollen–Gathering Activity: Pollen Is the Male Germplasm of the Plant

It is practically the sole source of proteins, fatty substances, minerals, and vitamins that are necessary for rearing of larvae and the development of newly emerged bees (Haydak 1970). A shortage of pollen and/or the availability of poor-quality pollens which do not contain the necessary nutrients limit the growth of bee colonies (Loper and Berdel 1980). Cotton flowers produce very high amounts of pollen of an acceptable nutrient value.

Stace (1994) conducted a chemical analysis of cotton pollen, revealing a crude protein (CP) level of 19.4%. This value falls just below the 20% CP threshold required by honeybees, as specified by De Groot (1953). However, the cotton pollen exhibited a favorable essential amino acid profile, with only isoleucine being slightly below the level required by honeybees. Although cotton plants offer both nectar and pollen to visiting insects, honeybees rarely collect its pollen grains (Vansell 1944a; McGregor 1959; Moffett et al. 1975a; Loper and Davis 1985; Waller et al. 1985a; Vaissiere 1991a). Tanda and Goyal (1979b) reported that only 12% of honeybees visiting *G. arboreum* gathered pollen. Loper and DeGrandi-Hoffman (1994) reported up to 23.4% of honeybee visitors collected pollen from *G. hirsutum*, but less (7.1%) collected from *G. barbadense*. This difference may be a result of relative flower attractiveness. Moffett and Stith (1972a) and Waller and Moffett (1981) reported that honeybee preference for some hybrid cotton lines was due to pollen avoidance, since flowers of both lines produced similar quantities of nectar. Other authors, however, have reported pollen collection by honeybee foragers under certain situations.

For example, Danka (2005) reported that approximately 80% of pollen gatherers entering hives located near cotton crops had collected cotton pollen. Eisikowitch and Loper (1984) found that honeybees normally bypass cotton to forage on more attractive flowers, but in a no-choice situation, they will collect cotton pollen. Vaissiere (1991a) and Waller et al. (1981b) also agreed that honeybees will collect cotton pollen under stress conditions, with Vaissiere commenting that he observed only 1 field out of 23 where honeybee foragers were gathering pollen; and this was because these colonies were experiencing a severe dearth of pollen. Danka (2005) suggested that high humidity might also stimulate pollen collection. The factors discussed might explain why Loper and Davis (1985) observed that honeybees only collected corbicular loads of cotton pollen during the first week after colonies were moved into the area. Each cotton flower produces an average of 30,000–40,000 pollen grains, representing a weight of 19–26 mg of fresh pollen (Vaissiere 1991b). However, the average size of a honeybee pollen load from honeybees observed foraging in *G. arboreum* was only 10.3 mg (Tanda 1984), whereas the average size of a honeybee pollen load on most crops is 15 mg (Maurizio 1953, cited in Keller et al. 2005).

Cotton pollen grains possess a larger size and a spiky texture, making them susceptible to becoming entangled in the hairs of apoids while they search for floral nectar (Vansell 1944a; Berger et al. 1988). In some cases, the bee's body can become so coated with cotton pollen that it impedes its ability to fly away (Vaissiere 1991a). Honeybees exhibit meticulous grooming behavior to remove cotton pollen from their bodies, which has been interpreted as a way to avoid or reject the pollen (Loper and Davis 1985). Several authors have remarked on the apparent "repellency" of cotton pollen (Tanda and Goyal 1979b; Loper and Davis 1985), prompting multiple studies aimed at understanding honeybees' avoidance behavior toward cotton pollen (Loper 1986; Vaissiere and Vinson 1994). The distinctive characteristics of cotton pollen and its interaction with bees have garnered significant research attention, leading to ongoing studies unravelign the underlying mechanisms and implications of honeybees' avoidance behavior toward cotton pollen.

Vansell (1944a) and Buchmann and Shipman (1990) proposed that the large size of cotton pollen grains and their coverage with beads of viscous material made it difficult for honeybees to pack these grains into their corbiculae (pollen baskets). Additionally, Moffett (1983) suggested that the presence of gossypol, a chemical found in cotton pollen, might act as a toxicant or repellent. However, Loper (1986) conducted a chemical analysis of *G. hirsutum* pollen and found very little or no gossypol content (<0.003%). Consequently, he concluded that gossypol was unlikely to act as a toxicant or repellent to honeybees. Other physical reasons for the relative "repellency" of cotton pollen have also been proposed.

Vaissiere and Vinson (1994) conducted a study on honeybee pollen collection efficiency from different spheroidal pollen types, including cotton and five others. They found that the rate of pollen collection was 0%, 16%, and 71% of landing foragers in okra, cotton, and pumpkin, respectively. In contrast, pollens of corn, pigweed, and sunflower were readily collected by nearly all foragers. This suggests that some physical characteristics of cotton pollen may make it less attractive or more challenging for honeybees to collect compared to other types of pollen.

Pollen traps have also been used to assess pollen preference by bee foragers (Olsen et al. 1979; Dimou and Thrasyvoulou 2007). A number of studies of hive-collected pollen during cotton flowering seasons have confirmed that cotton pollen is not honeybees' first choice when other, more attractive plants are blooming. Loper and Davis (1985) reported that colonies fitted with pollen traps dramatically increased their pollen collection when *Larrea divaricata* Cav. (Zygophyllaceae) commenced to bloom, apparently diverting bees from cotton. Waller et al. (1985a) reported that weekly amounts of pollen collected from eight colonies fitted with pollen traps and adjacent to hybrid cotton fields varied from 403 g to 1,316 g/colony/week; however, only four cotton pollen pellets were found among more than 10,000 examined. In Australia, Rhodes (2002) also reported that no cotton pollen was detected in pellets collected by a pollen trap on a hive in an apiary located next to a flowering cotton crop. However, when Waller et al. (1981b) analyzed pollen from traps of two bee colonies, they recorded 18% and 44% cotton pollen in trapped pollen pellets, and Danka (2005) reported that approximately 26% of pollen pellets were from cotton. Thus, the conclusion from all the field observations and pollen trapping investigations is that while honeybees may collect cotton pollen, they prefer pollen from other floral sources. It should be noted, however, that cotton pellets are small and may not be as likely to be collected in pollen traps of 30% efficacy.

10.5.2 Nectar-Gathering Activity

The majority of honeybee foragers collect nectar of various plant species rather than pollen (McGregor 1976). The density of foraging bees on cotton is primarily influenced by the nectar, which is secreted by both floral and extrafloral nectaries (Butler et al. 1972), and honeybee foragers switch readily between the two types of nectaries, with their proportions depending on the cultivar (Moffett et al. 1975b; Waller et al. 1981b; Eisikowitch and Loper 1984) and the time of day (Moffett et al. 1976b; Waller et al. 1981b). The variability in nectar concentration, discussed in Section 10.2.5, is likely to be related to the floral morphology and microclimate inside the flower. Cotton has cup-shaped flowers, and the nectar accumulates at the base of the calyx, which fits snugly against the petals, reducing the passive concentration of the nectar by exposure to the air. Vaissiere (1991a) suggested it also made the floral nectar accessible only to bees which enter inside the corolla and probe between the petals. Wafa and Ibrahim (1959), reporting hat honeybees may collect floral nectar whilst remaining outside cotton flowers, also described their normal floral visitation behavior as: "the honeybee enters the flower, passes round the anthers and pushes the proboscis in the space between the base of petals to suck nectar."

Reports in the literature of the effect of floral and extrafloral nectaries on honeybee foraging behavior are somewhat confusing (Grout 1955; McGregor 1976; Free 1993) reported that honeybees prefer extrafloral nectaries. However, Moffett et al. (1975b) noted that the absence of extrafloral nectaries did not affect the attractiveness of cotton plants to honeybees.

In Loper's study (1986), it was observed that honeybees did not show any preference in their visitations to flowers of glanded and glandless *G. hirsutum* and *G. barbadense* cotton plants. On the topic of nectar collection, Tanda and Goyal (1979b) found that 80% of the nectar gatherers collected from extrafloral nectaries, while Rhodes (2002) reported a slightly lower value of 70% for nectar collectors from extrafloral nectaries.

In contrast, Eisikowitch and Loper (1984) reported different nectar collection patterns in hybrid cotton lines. They observed that 75 to 100% of nectar-gathering bees collected floral nectar in *G. hirsutum*, whereas only 24% collected floral nectar in *G. barbadense*. These studies indicate that honeybee behavior regarding nectar collection can vary depending on the specific cotton species or lines being examined. Different factors such as the presence of glands and the availability of extrafloral nectaries may influence the preferences of honeybees when it comes to collecting nectar from cotton plants.

Waller and Moffett (1981) suggested that bees divide their time by collecting extrafloral nectar throughout the morning and floral nectar during the afternoon. However, Eisikowitch and Loper (1984) reported that during the afternoon, when secretions from floral nectaries dwindle, bees switch again to extrafloral nectaries, although individual bees may switch between either nectar source even during the same flight.

It seems that honeybee foragers have developed a pattern of learning regarding cotton flowers. They have learned that cotton flowers typically open 2 to 3 hr after sunrise and that these flowers do not secrete any nectar until sometime after they have fully opened (Vansell 1944a).

In the study conducted by Tanda and Goyal (1979b), it was observed that the sugar concentration of cotton nectar collected by honeybees from both floral and extrafloral nectaries was inversely proportional to the relative humidity of the air. This suggests that the sugar content of cotton nectar changes based on the moisture levels in the environment. Honeybees might adjust their foraging behaviors accordingly to adapt to these variations in nectar quality.

Indeed, honeybees prefer sugar solutions above 30% TDS (Waller 1972). Moffett et al. (1976b) found that the attractiveness of cotton was influenced by the sugar concentration rather than the volume of the floral nectar, and thus bees may not visit cotton flowers even when nectar is abundant (Moffett et al. 1980; Delaplane and Mayer 2000). The different nectaries of a cotton plant produce nectar which contains about 20% sugar concentration (Vansell 1944a). Evaporation of water from the nectar is greater in the exposed extrafloral nectaries than the floral nectaries, causing a more rapid increase in sugar concentration in the extrafloral nectaries (Vansell 1944a). Thus, the sugar concentration in the extrafloral nectaries can reach a maximum of 60–82% compared to 20–54% in floral nectaries (Parks 1921; Grout 1955). Furthermore, while the floral nectaries secrete only on the day a flower is open, the extrafloral nectaries secrete for a number of days, and any nectar that remains from a previous day has a longer time to become more concentrated. Consequently, honeybees do not visit the floral nectaries until the extrafloral ones are exhausted. Therefore, increasing bee populations by concentrating honeybee colonies in or near cotton fields may increase bee visitation to floral nectaries, by exhausting extrafloral nectaries earlier (Grout 1955; McGregor 1959).

10.5.3 Cotton as a Honey Plant

Nectar production by the cotton plant has historically been of economic importance to commercial beekeepers in a number of countries, such as the USA (Parks 1921; Butler et al. 1972; Martin and McGregor 1973; Waller 1982), USSR (Kuliev 1958), and Egypt (Wafa and Ibrahim 1959; El-Banby et al. 1985), although there are no recent references of its current status, particularly since the increased use of modern synthetic pesticides. However, this has not been the case in Australia.

Different authors' reports on honey production potential from cotton, based on actual measurements of nectar production from flowers, show a considerable variation. For instance, 1 ha of *G. barbadense* cotton has been estimated to yield approximately 31 (Parks 1921) to 44 (Waller et al. 1981b) kg of honey. However, some studies suggest that the yield can be as low as 15.1 kg per hectare (Butler et al. 1972). Among the 200 nectar-producing plants listed by Crane (1975), cotton falls within the lowest class, with a honey yield ranging from 0 to 25 kg per hectare. Daily honey yields have been estimated to range between 0.12 and 0.76 kg per day per hectare (Butler et al. 1972; Waller et al. 1981b; El-Banby et al. 1985). Comparing different cotton varieties, *G. hirsutum* has been found to yield much less honey than *G. barbadense* (Butler et al. 1972). Similarly, when measuring cotton honey production based on the actual honey collected by honeybee colonies (measured by hive weight differences), there is a noticeable variation in the results.

Average collection of honey per colony was reported to be 43 kg (Grout 1955), 15–20 kg (Avetisyan 1958), presumably over the cotton flowering period, although daily honey collection has been reported to be as low as 0.114 kg (Waller et al. 1981a), 0.589 kg (Butler et al. 1972), to 0.8–2.5 kg (Kuliev 1958). Waller et al. (1981b) confirmed Vansell's (1944b), McGregor and Todd's (1956), and Butler et al.'s (1972) view that *G. barbadense* was a better nectar producer than *G. hirsutum*, by reporting an estimated 44 kg and 28.6 kg honey yield per hectare from *G. barbadense* and *G. hirsutum*, respectively. McGregor and Todd (1956) reported *G. hirsutum* provided good honey crops only on some occasions at certain localities. These findings align with the results reported by Vaissiere (1991a). In one *G. hirsutum* field, a non-significant increase of 3.4 kg in mean hive weight was observed. However, in another field, there was a significant increase of 21.5 kg in mean hive weight, indicating a notable impact on honeybee colonies' weight due to the environment and availability of cotton as a pollen source. The preceding data should be interpreted with caution, because the reported figures of cotton honey collection are based on changes in hive weight, with no attempt to determine whether the nectar had come from cotton or from other nearby, non-cotton plants.

10.5.4 Attractiveness of Flowering Cotton Plants

The quantity of nectar and the concentration and type of constitution sugars in floral nectars determine their attractiveness to bees; however, less is known about what attracts bees to particular pollens (Hopkins et al. 1969). Competition from abundant and more attractive non-cotton nectar and pollen sources within the flight range is probably responsible for the reported low honeybee visits to cotton flowers (Moffett et al. 1975a; Butler et al. 1972). Vansell (1944a) reported that nectar collection from cotton is influenced adversely by the presence of other plants with richer nectars, particularly alfalfa, *Medicago sativa* L. (Leguminosae), although it is clear that other crops, non-crop plants, and weeds may also be more attractive to bees than cotton.

Various studies examining hive-collected pollen during the cotton flowering season reveal that significant amounts of pollen were collected from various sources, including creosote bush (Loper and Davis 1985), pearl millet (*Pennisetum americanum* L.; Poaceae), ragweed (*Ambrosia* sp.; Asteraceae) (Waller et al. 1985a), smartweed (*Polygonum pensylvanicum* L.; Polygonaceae) (Waller et al. 1985a; Vaissiere 1991a), sorghum (*Sorghum bicolour* (L.) Moench; Poaceae) (Waller et al. 1985a; Rhodes 2002), sunflower (*Helianthus annuus* L.; Asteraceae) (Waller et al. 1985a; El-Sarrag et al. 1993), and pigweed (*Amaranthus palmeri* S. Wats; Amaranthaceae) (Vaissiere 1991a) when these plants were in bloom. The presence of these alternative blooming plants seemed to attract bees away from the cotton fields.

Researchers such as Grout (1955), Moffett et al. (1975a), and Waller (1982) found that as midsummer approached, weeds started to appear in cotton fields, leading to increased pollen availability and enhanced brood rearing in honeybee colonies.

Furthermore, flowering trees, such as tamarisk trees (*Tamarix aphylla* L.; Tamaricaceae), were found to be highly attractive to bees, drawing large numbers of honeybees away from the cotton flowers. The competition from these alternative sources likely played a significant role in causing a mid-season decline in bee visits to cotton flowers (Moffett et al. 1972a, 1975a, 1979a).

10.6 TECHNIQUES USED TO INCREASE ACTIVITY OF HONEYBEES, AND THEIR POLLINATION EFFICACY

Honeybee pollen collection from the target crop is the most important concern for both growers and beekeepers. Previous researchers have reported some techniques used to encourage pollen-gathering activity of honeybees during the pollination season of various crops. Honeybee colonies can be stimulated to collect pollen through genetic selection (Hellmich et al. 1985; Danka 2005), by manipulating the amounts of brood and pollen (Chambers 1985; Fewell and Winston 1992; Eckert 1942; Eckert et al. 1994; Dreller et al. 1999; Thorp 2000; Danka 2005), by feeding pollen supplements (additional pollen) or substitutes (Purdie and Doull 1964; Doull 1980a, 1980b; Goodwin et al. 1994), by removal of pollen from the hive (Hirschfelder 1951; Free 1967; Camazine 1990; Tsirakolglou et al. 1997), by trapping pollen to prevent its entry into hives (Lindauer 1953; Rashad and Parker 1958; Liven and Loper 1984; Chambers 1985), by substituting honey for pollen stores (Free 1967; Barker and Barker 1971), by providing sugar syrup (Goodwin 1986, 1997), and by using brood pheromone in the hive (Pankiw et al. 1998, 2004), or applying bee attractants to flowering target crops (Burgett and Fisher 1979; Mayer et al. 1989; Currie et al. 1992; Naumann et al. 1994; Ambrose et al. 1995; Neira et al. 1997).

10.6.1 The Use of Pollen Traps to Increase Bee Activity

The use of pollen traps by beekeepers has greatly increased over a number of years, with the increased use of pollen for bee feeding and for human consumption (Abreu 1992; Block et al. 1994; Bonvehi and Jorda 1997; Diaz-Losada et al. 1998; Kroyer and Hegedus 2001).

Many beekeepers have added pollen production as an income-producing supplement to honey production and crop pollination. Pollen traps may be used to determine the suitability of a permanent location for an apiary (Nabors 1997; Dimou and Thrasyvoulou 2006) or to monitor pollen availability near apiaries so that colonies can be provided with pollen substitutes or pollen supplements when required (Cook 1985; Dimou and Thrasyvoulou 2006). They may also be used to control varroa mite, *Varroa jacobsoni* Oudemans (Acari: Varroidae) (Hart and Nabors 1999). Pollen traps vary in size and design, although they have almost the same mode of action. The traps are normally placed under the hive, and they force bees to enter a series of screens that scrape the pollen off the hind legs of returning field bees, with the pollen pellets being caught in a collection tray underneath.

The efficiency of pollen traps can vary based on their design, including factors like wire size, spacing between wires, and in certain cases, the number and size of round holes in a perforated plate (Loper et al. 1984). Depending on these design aspects, some traps may collect more pollen than others. For instance, Bienkowska and Pohorecka (1996) found that the optimal diameter for holes in the pollen scraping insert was between 4.77 mm and 4.79 mm, suggesting that this specific design allowed for efficient pollen collection. Considering the diverse trap designs available and their impact on pollen collection, beekeepers should carefully select a design that suits their specific needs and objectives in order to effectively manage and utilize pollen resources in their apiaries.

However, the percentage of the pollen loads that are trapped depends on the size of the loads, which can vary considerably (Synge 1947), as well as by adverse weather conditions which cause bees to return to hives before full loads have been collected. The density of honeybee flight activity and pollen collection has been correlated with the amount of brood rearing (Todd and Reed 1970; Al-Tikrity et al. 1972) and the amount of pollen stored (Barker 1971; Eckert et al. 1994; Vaissiere and Vinson 1994) in the colony. Also, egg laying by the queen is correlated with the amount of pollen collected (Cale 1968). According to Lindauer (1955), in colonies with large quantities of brood and a shortage of pollen, more of the foragers will gather pollen. Lindauer (1953) also found a larger percentage of pollen gatherers in colonies fitted with pollen traps. This suggests that pollen traps could be used to stimulate pollen collection by honeybee foragers.

Currently, scientists use pollen traps to collect information relating to colony food intake, brood rearing activity, bee foraging behavior, pollination activity, field pollen flow, pollen preference by bees, and population dynamics (Hirschfelder 1951; Allen and Jeffree 1956; McLellan 1976; Goodwin and Perry 1992; Nabors 1997; Dimou and Thrasyvoulou 2007). The literature on pollen traps frequently reports their stimulating effect on pollen collection. Webster et al. (1985) reported that bee foraging activity (number of incoming bees per minute) was higher in trapped colonies. An increased number of foragers in hives with traps has also been reported by Rashad and Parker (1958) (that 53% efficient pollen traps increased foraging activity by 80%), Hirschfelder (1951), Lindauer (1953), Rybakov (1961), Moriya (1966) and Chambers (1980).

While Liven and Loper (1984) reported that traps did not affect the number of bees leaving a colony, they noted a higher proportion of pollen-gathering bees in colonies with traps. Thorp (1979) and Webster et al. (1985) found both a greater number of foraging bees in almond orchards from colonies with pollen traps and higher percentages of foragers carrying pollen. However, the amount of stored pollen in hives with traps can be adversely affected (Hirschfelder 1951; McLellan 1974; Goodman 1974). The literature is inconsistent about the effects of pollen traps on colony behavior and development. Waller et al. (1981a) reported a decrease in brood rearing and bee population after AOC traps were used for two months. Similarly, Ibrahim and Selim (1974) reported a decrease in sealed brood area, averaging 39.5%, when using pollen traps for the greater part of the active season. This detrimental effect of trapping pollen on brood production and adult bee populations has also been reported by Eckert (1942), Rashad and Parker (1958), Loper et al. (1984), and Duff and Furgala (1986). However, a number of other investigators have reported little or no effect on brood rearing resulting from the presence of traps (Hirschfelder 1951; Rybakov 1961; Lavie and Fresnaye 1963; Lavie 1967; Goodman 1974; McLellan 1974; Cook 1985). This may be because of the different periods of time the traps were used, as well as differences in trap efficiency.

The use of pollen traps can have various effects on bee activities and behavior. For instance, the installation of pollen traps has been found to negatively impact honey production in colonies. Studies by Hirschfelder (1951), Rashad and Parker (1958), Moriya (1966), Lavie (1967), and Duff and Furgala (1986) have reported decreased honey production in trapped colonies, with reductions ranging from 13% to as much as 41%. Conversely, Rybakov (1961) reported a 13% increase in honey production from trapped colonies, showing some variability in the results.

Moreover, the presence of pollen traps can adversely affect the honey moisture content in colonies (Duff and Furgala 1986). Long-term use of pollen traps may also lead to negative consequences for colonies, such as an increased tendency to swarm (Goodman 1974) and reduced overwintering ability (McLellan 1974). These findings highlight the importance of carefully considering the potential impacts of pollen traps on bee colonies before implementing them in beekeeping practices.

10.6.2 Provision of Supplementary Food and Its Effect on Bee Activity

Pollen is the sole source of protein, lipids, minerals, and vitamins for honeybee colonies. Shortage of pollen results in decreased brood rearing, decreased life span in adult workers, and poor honey production (Haydak 1945, 1970; Maurizio 1950; Doull 1980b). Since pollen is often not present in adequate quantities in the field, beekeepers commonly feed bees with pollen substitutes (commonly containing soybean flour, skim milk, and sugar), supplements containing pollen with other materials, or commercially trapped pollen. These are used to stimulate brood rearing, build up colonies to exploit honey flows, produce colonies suitable for pollination, as well as for queen rearing and package-bee production. Several commercial bee food mixtures have been developed, such as Sojapyl (Purdie and Doull 1964), Krawaite (Forster 1966), Bee Feast and Pollenex (Herbert and Shimanuki 1980), WheastTM (Chalmers 1980), and Feed-Bee® and Bee Pro® (Saffari et al. 2004). Consumption rates of pollen supplements or substitutes are closely related to the amount of brood being reared in the hive (Palmer-Jones 1947; Forster 1966, 1968a, 1968b; Standifer et al. 1971, 1973; Herbert and Shimanuki 1980; Waller et al. 1981a), although Waller et al. (1981a) reported that provision of supplements did not significantly affect the brood pattern. Doull (1973) presented a detailed analysis of the correlations between feeding of pollen supplements and brood rearing, pollen gathering, and honey production. He reported that there were positive correlations between the rate of brood rearing and the consumption of pollen supplements by colonies. Increasing the amount of brood in the colony also leads to an increase in pollen collection (Filmer 1932; Free 1967; Todd and Reed 1970; Al-Tikrity et al. 1972; Hellmich and Rothenbuhler 1986; Eckert et al. 1994), and to a temporary increase in the amount of pollen stored in the hive (Hellmich and Rothenbuhler 1986). Many studies have examined the effect of providing colonies with pollen or a pollen substitute on their subsequent field pollen collection and nectar collection activity (Seeley 1985).

Feeding bees with pure pollen, as compared to pollen supplements or substitutes, further reduces their pollen collection (Free and Williams 1971; Moeller 1972). Honey-gathering activity may be affected positively by provision of supplementary food or pollen.

Purdie and Doull (1964) conducted a study where they found that supplementary feeding during a blue gum (*Eucalyptus globulus* Labill) honey flow in South Australia resulted in a significant increase of 35.8 kg of honey per colony. Similarly, in another study, Doull (1980b) reported a substantial 38% increase in honey production per colony due to supplementary feeding during the same honey flow period. These findings suggest that providing supplementary food to honeybee colonies during specific flowering events can have a positive impact on honey production.

Also, Nabors (2000) reported an increase in honey production after spring feeding with a pollen substitute, while Shoreit and Hussein (1993) reported increase in honey production following provision of soybean flour. In contrast, Goodwin et al. (1994) found that pollen substitutes had no effect on honey production in colonies used for kiwifruit pollination, nor for colonies managed for honey production.

10.6.3 Use of Pollination-Enhancement Chemicals on Bee Activity and Pollination

The importance of insect pollination in many crops and the key role of honeybees as pollinators have been discussed previously by McGregor (1976), Free (1993), and Delaplane and Mayer (2000). However, some crops do not receive adequate pollination even when honeybees are abundant due to factors such as unattractive flowers, poor weather during the pollination period, and/or competition from more attractive plant species nearby (McGregor 1976; Jay 1986; Free 1993). A number of bee-attracting chemicals have therefore been used to increase bee visitation to flowers of the target crop, with a view to increasing its pollination.

The first reported use of a bee attractant was with high-concentration sugar syrups. However, Free (1965) reported that spraying apple trees with the syrups to increase bee visits in an apple orchard was counter-productive because bee activity was diverted to collecting sugar that had deposited on the leaves and petals of the apple tree rather than seeking nectar in the flowers. Free (1968) therefore concluded that the major obstacle in pollination management of bees was the problem of attracting bees away from more attractive competing flowers or nectar sources. To overcome the failure of sugar product sprays to attract bees to flowers of target crops, an alternative strategy in pollination management of bees was to use sprays that relied on olfactory attractants.

In Waller's study (1970), citral, geraniol, and anise oil were applied in alfalfa, but there was no observed increase in floral bee visitation or yield. Similarly, Mayer and Johansen (1982) used geraniol and anise oil in apple (*Malus domestica* Borkh.), pear (*Pyrus communis* L.), and dandelion (*Taraxacum officinale* Weber), but these treatments did not result in a higher bee visitation or improved yield. On the other hand, Woyke (1981) reported an increase in bee visits to onion (*Allium cepa* L.) when using these products. The effectiveness of citral, geraniol, and anise oil in enhancing bee visitation and yield seems to vary depending on the specific plant species and context of the study.

Although a controversial area, the use of honeybee attractants continues to be of great interest to growers and beekeepers for improving crop pollination. As a consequence, a number of commercial products are in the marketplace. While the actual composition of some of these products may be not published for commercial reasons, they can, nevertheless, be placed into two broad categories: (1) those based on bee foods, such as sugars and proteins, and (2) those based on pheromones, the volatile chemicals honeybees use for communication.

Numerous studies have been conducted to evaluate the effectiveness of these commercial bee attractants. Studies reported results of non-pheromone attractants used, although they are generally disappointing. For example, Rajotte and Fell (1982) reported that BeeLure (Custom Chemical, USA) did not increase the attractiveness of apple blossoms to bees, and Mayer and Johansen (1982) reported that Pollenaid-D® (Crop King Chemicals, USA) was ineffective in apples, dandelion, and pears. BeeLine® (Helena Chemical, USA) has frequently been reported to be unsuccessful in a range of crops, including red clover, *Trifolium pratense* L. (Fabaceae) (Burgett and Fisher 1979); apples (Mayer and Johansen 1982; Margalith et al. 1984); pears (Mayer and Johansen 1982); carrot, *Daucus carota* L subsp. sativus (Apiaceae) (Belletti and Zani 1981; Jay 1986); cucumber, *Cucumis sativus* L. (Cucurbitaceae) (Rapp 1981; Margalith et al. 1984; Schultheis et al. 1994); watermelon, *Citrullus lanatus* (Thunb.) (Cucurbitaceae) (Schultheis et al. 1994; Ambrose et al. 1995); and sunflower (Singh and Sinha 1997). The discovery of a number of honeybee pheromones and the ability to synthesize them have enabled their evaluation as attractants to aid in pollination. Some products contain only pheromone, while others contain pheromones and other compounds. Results with pheromone-based bee attractants have been variable.

Bee-Scent® (Yellowstone Intl., USA), for example, has been reported to increase honeybee foraging and fruit set in apple, cherry, pear, and plum, *Pyrus domestica* L. (Rosaceae) (Mayer et al. 1989), to increase the number of foraging honeybees on the day of application, but without a corresponding yield increase, in watermelons (Loper and Roselle 1991) and to improve yield of watermelon (with increased seed number) but without increasing bee activity (Elmstrom and Maynard 1990). However, no significant increases in bee activity or yield have been reported by other researchers in strawberry, Fragaria × ananassa Duch. ex Rozier (Rosaceae), (Butts 1991), cucumber, or watermelon (Schultheis et al. 1994; Ambrose et al. 1995). Bee-HereTM (Troy Biosciences, Inc., USA) was reported to increase bee visitation and yield in sweet orange, Citrus sinensis L. Osbeck (Rutaceae) (Malerbo-Souza et al. 2004), and raspberry, Rubus idaeus L cv Meeker (Rosaceae) (Neira et al. 1997), but showed no increased bee activity or fruit set when tested on red clover (McGourty 1992) and kiwi fruit (Tsirakolglou et al. 1997). Queen mandibular pheromone (QMP) contains a mixture of pheromones extracted from mated queen honeybees. Since its synthesis and commercial release, it has been used to manipulate beehives such as for swarm prevention (Winston et al. 1989, 1993), and as an attractant for pollination (Winston and Slessor 1993), although in this latter use, results are sometimes inconclusive. It has shown promise as its commercial product, Fruit Boost® (PheroTech, Canada), in increasing the effectiveness of bee pollination and yield in pears (Currie et al. 1992); cranberry, *Vaccinium macrocarpon* Ait. (Ericaceae); and blueberry, *Vaccinium corymbosum* L. (Ericaceae) (Currie et al. 1992).

On the other hand, there are reports of it increasing bee activity without any yield increase, in apples (Currie et al. 1992) and cranberry (Mackenzie and Averill 1992). Fruit Boost® has also been reported by some researchers to have no effect on bee visits, fruit size, or fruit set in trials in apricots, *Prunus armeniaca* L. (Rosaceae) (McLaren et al. 1992); pear; and sweet cherry (Naumann et al. 1994), or kiwifruit (Howpage 1999). In cotton plants, inadequate cross-pollination may occur because of low numbers of bee visits associated with the unattractiveness of the cotton flower to honeybees. While the use of chemical attractants to improve honeybee pollination has been widely reported in other crops, no investigations have been conducted in cotton. If chemical attractant(s) are shown to be effective in increasing bee activity on cotton flowers, the resulting cross-pollination may increase yield and lint quality.

10.7 THE ROLE OF BEETLES IN POLLINATION

Cotton is commonly regarded as a partially cross-pollinated crop (McGregor 1976), and insects are the natural agents for pollen transfer. Surveys of cotton-visiting insects have been reported from a number of countries (Wafa and Ibrahim 1959; Moffett et al. 1976, 1978b, 1980; Tanda and Goyal 1979c; Tanda 1983; El-Sarrag et al. 1993), with pest species, pollinators, and biological control agents being the most common. Of the beneficial species, those in the order Hymenoptera were generally predominant, with Lepidoptera and beetles (Coleoptera), although relatively rare, being the next most common. Moffett (1983) observed collops, *Collops vittatus* (Say) (Melyridae), and sap beetles, *Conotelus mexicanus* Murray (Nitidulidae), present in cotton flowers in Arizona and suggested that they may contribute to the pollination of flowers. In Australia, a dried fruit or pollen beetle, *Carpophilus aterrimus* Macleay (Nitidulidae), is commonly found in cotton flowers (Rhodes 2002; Llewellyn et al. 2007), although its role and contribution to pollination in this crop are not known.

In the last decade, investigations conducted in a number of countries have suggested that beetles may pollinate a wide range of plant species, although the specificity of the plant–pollinator interaction may vary (Endress 1994). In most beetle pollination, scents from flowers (Gottsberger 1990; Gibernau et al. 1999; Garcia-Robledo et al. 2004) or visual cues (Young 1986; Steiner 1998a; Van Kleunen et al. 2007) act as primary attractants. A number of species of nitidulid, sap, or pollen beetles are found in flowers, although the majority are saprophagous and mycetophagous, feeding on dried and fresh fruits, decaying fruit, fermenting plant juices, peanuts, honey, fungi, rotting vegetable matter, and plant sap (Hinton 1945; Okumura and Savage 1974). One species, the small hive beetle, *Aethina tumida* Murray, is a parasite of honeybee colonies, and another, *Brachypeplus auritus* Murray, feeds on the wax and honey of *Trigona* spp. (Habeck 2002). However, other nitidulid beetles, such as *Carpophilus hemipterus* L., C. mutilates Erichson, *Uroporus humeralis* F., and *Haptoncus luteolus* Erichson, are effective pollinators of atemoya, *Annona squamosa* L. × *A. cherimola* Mill. (Annonaceae) (Gazit et al. 1982). Recent research in USA (Nagel et al. 1989; Nadel and Pena 1994), Australia (George et al. 1989; Blanche and Cunningham 2005), Colombia (Garcia-Robledo et al. 2004), and Japan (Kono and Tobe 2007) has confirmed the role of some nitidulids in pollination and has expanded the number of species involved to include *Carpophilus chalybeus* Murray, *C. freemani* Dobson, *C. marginellus* Motschulsky, *C. fumatus* Boheman, *C. maculatus* Murray, *C. oiosellus* Motschulsky, *C. dimidiatus* F, *Lobiopa insularis* Castelnau, *Colopterus posticus* Erichson, *Colopterus truncates* Randall, *Aethina (Circopes) australis* Kirejtshuk, *Brachypeplus instriatus* Kirejtshuk, and *Macrostola costulata* Reitter.

10.8 HAZARDOUS EFFECTS OF INSECTICIDES ON COTTON POLLINATION

The use of insecticides to control crop pests is not always accompanied by a significant increase in yield. One reason is their lethal effects on beneficial arthropods, such as pollinators (Atallah et al. 1998), or biological control agents, such as predators (Salman et al. 1983). Historically, cotton is one of the most insecticide-intense cropping systems (Waller 1982), accounting for about 40% of the US crop insecticides used in 1976 (Eichers 1981), and 25% of the world's insecticides in the mid-1990s (Anon 1995). Since then, with the development of transgenic cotton varieties, these figures are likely to have decreased significantly. These pesticide applications in cotton have been regarded as a major cause of honeybee (Moffett et al. 1978b; Weaver 1978; Waller 1982; Robertson and Rhodes 1992) and native pollinator (Moffett et al. 1978b; Weaver 1978) losses in the field. The seriousness of bee mortality to beekeepers lies not only in the direct loss of bees but also in the loss of opportunities for bee foraging sites, as well as pesticide contamination of honey, pollen, and nectar (Wafa et al. 1968a, 1968b; Hanny et al. 1983; Cabras et al. 1994; Kubik et al. 2000).

Honeybee foragers are often killed after they contact insecticide-treated flowers or foliage while searching for nectar or pollen (Estesen et al. 1992). Moffett et al. (1979b) reported that all colonies in his investigations lost weight during the cotton spray season due to heavy bee mortality. There have been a number of reports detailing the loss of honeybees in cotton as a result of pesticide use. Moffett et al. (1978b, 1979a) reported that when honeybees visited cotton flowers, they set seed adequately, but after insecticide application, the bee visitation rate dropped by 90%, with a corresponding decline in seed set. Weaver (1978) also reported a cessation of bee activity in regularly insecticide-treated cotton crops. The safe time interval after pesticide application for honeybees varies with the type of product used and environmental conditions. Stoner et al. (1981), for example, observed that by the second day post-treatment with methyl parathion, acephate, folex, and chlordimeform, foraging had returned to normal, whereas Moffett et al. (1978b) reported that bee visitations did not return to their normal level until at least one week after application of methyl parathion, toxaphene, and chlordimeform.

The most common insecticides used in cotton over the past several decades have been synthetic pyrethroids, many of which are also highly toxic to honeybees. For these reasons, despite honeybees being used as pollinators in many crops, most beekeepers avoid providing bees for cotton in USA (Bennett 1993) and Australia (Stace1994). Attempts have been made to minimize the hazardous effects of insecticide sprays applied to cotton on honeybees. Timing of pesticide application is critical. Atallah et al. (1998) suggested that insecticides should be applied early in the cotton season before the commencement of bee foraging activity. Eisikowitch and Loper (1984) also suggested that since peak honeybee activity in cotton was between 1000 and 1500 h, insecticide sprays should be delayed until after these times. It should be noted that this would only prevent direct contact of bees with sprays but would not prevent their contact with active pesticide residues on plants. Locating apiaries away from direct pesticide exposure is also important in reducing bee losses. Colonies located in a treated field are likely to sustain more losses than colonies situated beyond the treated area, at the edge of or outside the field (Anderson and Atkins 1968).

In their study, Moffett and Stith (1972b, c) discovered that bee mortality caused by parathion decreased as the distance from a sprayed cotton field increased. They observed that honeybee colonies located at the edge of the field, within an apiary, experienced less damage from the pesticide compared to similar colonies situated in the middle of the field. This finding suggests that the proximity to the sprayed area influenced the extent of bee mortality due to parathion exposure.

Another method is the removal of honeybee colonies at times when pesticides are to be applied. Moffett and Stith (1972a) suggested that since honeybee activity resulted in good cotton boll set early in the season, bee colonies could be removed from the cotton fields during the mid-flowering season to enable insecticide use. Stoner et al. (1981) recommended a more practical way of achieving this: by placing hives permanently on mounted trailers and moving apiaries out of farms when insecticides were to be applied. However, many beekeepers with large apiaries do not find it economically practical or feasible to move their colonies to pesticide-free areas kilometers away (Anderson and Atkins 1968). It is not always easy to move colonies away from an area before spraying, so confining bees to their colonies for varying lengths of time during pesticide application (Jaycox 1963), feeding the bees pollen and sugar syrup, shading the hives, equipping hives with Miller bottom boards and Wardecker waterers, and covering colonies with burlap are other practical alternatives (El-Khashab 1967; Moffett et al. 1977a, 1979b, 1981).

With the commercial release of Bollgard® cotton in Australia, it was predicted that pesticide use would be reduced by up to 75% over conventional varieties (Anon 2004), improving the environmental sustainability and increasing profitability of the cotton industry by up to A$ 1.7 billion per year (Anon 2004). In Australia, for example, overall pesticide use has been reduced by 70% in the areas where it was planted (Doyle et al. 2005), and in the USA, Bollgard® cotton has reduced the annual frequency of insecticide applications to cotton from 14.1 during 1995 (Williams 1996) to 5.5 in 2001 (Williams 2002). However, in Australia, populations of secondary pests such as green mirid, *Creontiades dilutus* (Stål) (Hemiptera: Miridae); cotton aphid, *Aphis gossypii* Glover (Hemiptera: Aphididae); and thrips, *Thrips tabaci* Lind. (Thysanoptera: Thripidae) have increased in importance, requiring insecticidal control (Whitehouse et al. 2005; Knight et al. 2007).

REFERENCES

Abreu, M. 1992. Food use of pollen in relation to human nutrition. *Alimentaria* 235: 45–46.

Abrol, D. P. 2012. *Pollination Biology: Biodiversity Conservation and Agricultural Production.* Dordrecht (The Netherlands) and New York: Springer. 792 p.; ill.; index. ISBN: 978-94-007-1941-5. 2012.

Allen, M. D. and Jeffree, E. P. 1956. The influence of stored pollen and of colony size on the brood rearing of honeybees. *Annals of Applied Biology* 44, 649–656.

Al-Tikrity, W. S., Benton, A. W., Hillman, R. C., and Clarke, W. W. Jr. 1972. The relationship between the amount of unsealed brood in honeybee colonies and their pollen collection. *Journal of Apicultural Research* 11, 9–12.

Ambrose, J. T., Schulteis, J. R., Bambara, S. B., and Mangum, W. 1995. An evaluation of selected commercial bee attractants in the pollination of cucumbers and watermelons. *American Bee Journal* 135, 267–272.

Anderson, L. D. and Atkins, E. L. 1968. Pesticide usage in relation to beekeeping. *Annual Review of Entomology* 13, 213–238.

Anon. 1995. *Cotton: The Crop and Its Agrochemicals Market.* Allen Woodburn Associates Ltd./Managing Resources Ltd. American Crop Protection Association, California USA.

Anon. 2004. Bollgard II – the new generation in GM cotton. http://www.csiro.au/index.asp?type=feq&id=Bollgard&stylesheet=divisionFaq

Atallah, M. A., Zeutoun, Z. A., and Hassan, A. R. 1998. The relative toxicity of some synthetic pyrethriod and carbamate insecticides to honeybees (*Apis mellifera* L.) in the field. Proceedings of 14th International Conference on Apiculture in Tropical Climates, Cairo, Egypt. November 6–10, International Bee Research Association, London, UK, pp. 218–223.

Avetisyan, G. 1958. Bees: The cotton farmers' helper. *Indian Bee Journal* 20, 28–29.

Barker, H. G. and Barker, I. 1971. Nectar constitution and pollinator-plant coevolution. In: Gilbert, L. E. and Raven, P. A. (eds.) *Animal and Plant Coevolution*. University of Texas, Austin, TX, pp. 100–140.

Beasley, C. A. 1975. Developmental morphology of cotton flowers and seed as seen with the scanning electron microscope. *American Journal of Botany* 62, 584–592.

Belletti, A. and Zani, A. 1981. A bee attractant for carrots grown for seed. *Sementi Elette* 27, 23–27.

Bennett, G. J. 1993. Pollination is big business in Arizona. *Bee Culture* 121, 646–649.

Berger, L. A., Vaissiere, B. E., Moffett, J. O., and Merritt, S. 1988. Bombus spp. (Hymenoptera: Apidae) as pollinators of male-sterile upland cotton on the Texas high plains. *Environmental Entomology* 17, 789–794.

Bienkowska, M. and Pohorecka, K. 1996. Results of pollen trapping dependent on the size of holes in the insert scraping pollen pellets [Polish] Abstract (Eng.). *Pszzelnicze Zeszyty Naukowe* 40, 95–101.

Blanche, R. and Cunningham, S. 2005. Rain forest provides beetles for Atemoya crops. *Journal of Economic Entomology* 98, 1193–1201.

Block, G., Sinha, R., and Gridley, G. 1994. Collection of dietary-supplement data and implication for analysis. *American Journal of Clinical Nutrition* 59, 234–239.

Bonvehi, J. S. and Jorda, R. E. 1997. Nutrient composition and microbiological quality of honeybee-collected pollen in Spain. *Journal of Agricultural and Food Chemistry* 45, 725–732.

Brown, N. B. and Ware, J. O. 1958. *Cotton*. McGraw-Hill, New York.

Buchmann, S. L. and Shipman, C. W. 1990. Pollen harvesting rates for *Apis mellifera* L. on *Gossypium* (Malvaceae) flowers. *Journal of Kansas Entomological Society* 63, 92–100.

Burgett, M. and Fisher, G. C. 1979. An evaluation of Beeline as a pollinator attractant on red clover. *American Bee Journal* 119, 356–357.

Butler, G. D., Loper, G. M., McGregor, S. E., Webster, J. L., and Margolis, H. 1972. Amounts and kinds of sugars in the nectar of cotton (*Gossypium* spp.) and the time of their secretion. *Agronomy Journal* 64, 364–368.

Butts, K. M. 1991. Bee attractant: Improving strawberry quality? *Citrus and Vegetable Magazine* 55, 16–17.

Cabras, P., Martini, M. G., Floris, I., and Spanedda, L. 1994. Residues of cymiazole in honey of honey bees. *Journal of Apiculture Research* 33, 83–86.

Cale, G. H. 1968. Pollen gathering relationship to honey collection and egg-laying in honeybees. *Journal Apicultural Research* 108, 8–9.

Camazine, S. 1990. From individual to colony-level behavior in honeybees: The regulation of pollen foraging. In: Veeresh, G. K., Mallik, B., and Viraktamath, C. A. (eds.) *Social Insects and the Environment, Proc 11 Int. Congr. IUSSI.* Oxford & IBH Publ. Co., New Delhi, pp. 566–567.

Chalmers, W. T. 1980. Fish meals as pollen-protein substitutes for honeybees. *Bee World* 61, 89–96.

Chambers, S. R. 1980. Management of honeybee colonies on pollen trap. *Australian Beekeeper* 81, 208–209.

Clemson, A. 1985. *Honey and Pollen Flora*. Inkata Press, Department of Agriculture of N.S.W., Australia.

Cook, V. A. 1985. Trapping pollen from the bee. *Grower* 103(9), 2.

Crane, E. 1975. *Honey: A Comprehensive Survey.* International Bee Research Association, Heinemann, London.

Cruden, R. W., Hermann, S. M., and Peterson, S. 1983. Patterns of nectar production and plant animal coevolution. In: Bentley, B. and Elias, T. (eds.) *The Biology of Nectarines.* Colombia University Press, New York, pp. 126–152.

Currie, R. W., Winston, M. L., and Slessor, K. N. 1992. Effect of synthetic queen mandibular pheromone sprays on pollination of fruit crops by honey bee (Hymenoptera: Apidae). *Journal of Economic Entomology* 85, 1293–1299.

Danka, R. G. 2005. High levels of cotton pollen collection observed for honey bees (Hymenoptera: Apidae) in South-Central Louisiana. *Journal of Entomological Science* 40, 316–326.

De Groot, A. P. 1953. Protein and amino acid requirements of the honey bee (*Apis mellifera* L.). *Physiologia comparata et D'Oecologia* 3, 197–285.

Delaplane, K. and Mayer, D. 2000. *Crop Pollination by Bees*. University Press, Cambridge, UK.

Diaz-Losada, E., Ricciardelli, G. D., Pilar, S. A. A., and Otero, M. 1998. The possible use of honeybee pollen loads in characterizing vegetation. *Grana* 37, 155–163.

Dimou, M. and Thrasyvoulou, A. 2006. Efficient use of pollen traps to determine the pollen flora used by honey bee. *Journal of Apicultural Research* 45, 42–46.

Dimou, M. and Thrasyvoulou, A. 2007. A comparison of three methods for assessing the relative abundance of pollen resources collected by honey bee colonies. *Journal of Apicultural Research* 46, 144–148.

Doull, K. M. 1973. Relationships between consumption of a pollen supplement, honey production and broodrearing in colonies of honeybees, *Apis mellifera* L. I. *Apidologie* 11(4), 361–365.

Doull, K. M. 1980a. Relationships between pollen, broodrearing and consumption of pollen supplement by honeybees. *Apidologie* 4(4), 285–293.

Doull, K. M. 1980b. Relationships between consumption of a pollen supplement, honey production and brood rearing in colonies of honeybees, *Apis mellifera* L. II. *Apidologie* 11(4), 367–374.

Doyle, B., Reeve, I., and Coleman, M. 2005. A survey of cotton growers' and consultants' experience with Bollgard in the 2004–2005 season. Cotton Research, Institute for Rural Futures, University of New England.

Dreller, C., Page, R., and Fondrk, K. M. 1999. Regulation of pollen foraging in honeybee colonies: Effects of young brood, stored pollen, and empty space. *Behavioral Ecology and Sociobiology* 45, 227–233.

Duff, S. R. and Furgala, B. 1986. Pollen trapping honey bee colonies in Minnesota, Part II: Effect on foraging activity, honey production, honey moisture content, and nitrogen content of adult workers. *American Bee Journal* 126, 755–758.

Eaton, F. M. 1955. The physiology of the cotton plant. *Annual Review of Plant Physiology* 6, 299–328.

Eckert, C. D., Winston, M. L., and Ydehnberg, R. C. 1994. The relationship between population size, amount of brood, and individual foraging behavior in the honey bee, *Apis mellifera* L. *Oecologia* 97, 248–255.

Eckert, J. E. 1942. The pollen required by a colony of honeybee. *Journal of Economic Entomology* 35, 309–311.

Eichers, T. R. 1981. Use of pesticides by farmers. In: Pimentel, D (ed.) *Handbook of Pest Management in Agriculture*, Vol. 2. CRC Press, Boca Raton, Flo.

Eisikowitch, D. and Loper, G. M. 1984. Some aspects of flower biology and bee activity on hybrid cotton in Arizona, USA. *Journal of Apiculture Research* 23, 243–248.

El-Banby, M. A., El-Sherif, M. E., and Shatla, M. A. 1985. Productivity and chemical analysis of nectar and its mechanisms of flow in cotton flowers, *Gossypium barbadense*. Proceedings, 15th International Apiculture Congress of Apimondia, Japan.

El-Khashab, A. M. 1967. The practical methods for protecting honeybees from insecticides. MSc thesis. Faculty of Agriculture, Cairo University, Cairo, Egypt.

Elmstrom, G. W. and Maynard, D. N. 1990. Attraction honey bees to watermelon with bee attractant. *Proceedings of the Florida State Horticultural Society* 103, 130–133.

El-Sarrag, M. S., Ahmed, H. M., and Siddig, M. A. 1993. Insect pollinators of certain crops in the Sudan and the effect of pollination on seed yield and quality. *Journal of King Saud University, Agriculture Sciences* 5, 253–262.

Endress, P. K. 1994. *Diversity and Evolutionary Biology of Tropical Flowers*. Cambridge University Press, Cambridge.

Estesen, B. J., Buck, N. A., Waller, G. D., Taylor, K. S., and Mamood, A. 1992. Residual life and toxicity to honey bees (Apis mellifera L.) of selected insecticides applied to cotton in Arizona. *Journal of Economic Entomology* 85, 700–709.

Fewell, J. H. and Winston, M. L. 1992. Colony state and regulation of pollen foraging in the honey bees (*Apis mellifera* L.). *Behavioral Ecology and Sociobiology* 30, 387–393.

Filmer, R. S. 1932. Brood area and colony size as factors in activity of pollinating units. *Journal of Economic Entomology* 25, 336–343.

Forster, I. W. 1966. Pollen supplement for honey bee colonies. *New Zealand Beekeeper* 28, 14–21.

Forster, I. W. 1968a. Pollen supplements for honey bee colonies New Zealand Beekeeper, 30 (1968), pp. 2–8.

Forster, I. W. 1968b. Pollen supplements for honey bee colonies New Zealand Beekeeper, 30 (1968), pp. 16–17.

Free, J. B. 1965. Attempts to increase pollination by spraying crops with sugar syrup. *Journal of Apicultural Research* 4, 61–64.

Free, J. B. 1967. Factors determining the collection of pollen by honeybee foragers. *Anim. Behav.* 15, 134–144.

Free, J. B. 1968. The conditions under which foraging honey bees expose their Nasonov gland. *Journal of Apicultural Research* 7, 139–145.

Free, J. B. 1993. *Insect Pollination of Crops*, 2nd edn. Academic Press, London, UK.

Free, J. B., Nancy, W., and Jay, S. C. 1960. The effect on foraging behaviour after moving honeybee colonies to crops before or after flowering has began. *Journal of Economic Entomology* 53, 464–466.

Free, J. B. and Williams, I. H. 1971. The effect of giving pollen supplement to honeybee colonies on the amount of pollen collected. *Journal of Apicultural Research* 10, 87–90.

Free, J. B. and Williams, I. H. 1974. Influence of location of honeybee colonies on their choice of pollen sources. *Journal of Applied Ecology* 11, 925–935.

Garcia-Robledo, C., Kattan, G., Murica, C., and Quintero-Marin, P. 2004. Beetle pollination and fruit predation of *Xanthosoma daguense* (Araceae) in an Andean cloud forest in Colombia. *Journal of Tropical Ecology* 20, 459–469.

Gazit, S., Galon, I., and Podoler, H. 1982. The role of nitidulid beetles in natural pollination of annona in Israel. *Journal of the American Society for Horticultural Science* 107, 849–852.

George, A. P., Nissen, R. J., Ironside, D. A., and Anderson, P. 1989. Effects of nitidulid beetles in natural pollination of fruit set of Annona spp. hybrids. *Journal of the American Society for Horticultural Science* 39, 289–299.

Gibernau, M., Barabe, D., Cerdan, P., and Dejean, A. 1999. Beetle pollination of *Philodendron solimoesense* (Araceae) in French Guiana. *International Journal of Plant Science* 160, 1135–1143.

Goebel, R. 1984. Honey bees for pollination. *The Australian Beekeeper* 85, 166–174.

Goodman, R. D. 1974. The rate of brood rearing and effect of pollen trapping on honey bee colonies. *The Australian Beekeeper* 76, 39–41.

Goodwin, R. M. 1986. Increased kiwifruit pollen collection after feeding sugar syrup to honey bees within their hive. *New Zealand Journal of Crop and Horticultural Science* 14, 57–61.

Goodwin, R. M. 1997. Feeding sugar syrup to honey bee colonies to improve pollination: A review. *Bee World* 78, 56–62.

Goodwin, R. M., Houten, A. T., and Perry, J. H. 1994. Effect of feeding pollen substitute to honey bee colonies used for kiwifruit pollination and honey production. *New Zealand Journal of Crop and Horticultural Science* 22, 459–462.

Goodwin, R. M. and Perry, I. H. 1992. Use of pollen traps to investigate the foraging behavior of honey bee colonies in kiwifruit orchards. *New Zealand Journal of Crop and Horticultural Science* 20, 23–26.

Gori, D. F. 1983. Post-pollination phenomena and adaptive floral changes. In: Jones CE, Little RJ (eds.). *Handbook of experimental pollination biology*. New York: Van Nostrand Reinhold, pp. 31–45.

Gottsberger, G. 1990. Flowers and beetles in the South American Tropics. *Botanica Acta* 103, 360–365.

Grout, R. A. 1955. Honey bees make hybrid cotton possible. *American Bee Journal* 95, 10–11.

Gupta, R. K. 2003. An annotated catalogue of the bee species of the Indian region. Available at http://geocities.com/beesind2

Habeck, D. H. 2002. Nitidulidae Latreille 1802. In: Arnett, R. H., Thomas, M. C., Skelley, P. E., and Frank, J. H. (eds.) *American Beetles*, Volume 2. CRC Press, Boca Raton, FL, pp. 311–315

Hanny, B., Loper, G., and Harvey, J. 1983. Chemical detection of Penncap-M capsules in pollen, and methyl parathion residues in honeybees (*Apis mellifera* L.) and bee products from colonies near Arizona cotton fields treated with Penncap-M. *American Bee Journal* 123, 526–529.

Hart, T. and Nabors, R. 1999. Pollen traps as a method of varroa control. *American Bee Journal* 139(5), 366–367.

Haydak, M. 1945. Value of pollen substitutes for brood rearing of honeybees. *Journal of Economic Entomology* 38, 484–487.

Haydak, M. 1970. Honey bee nutrition. *Annual Review of Entomology* 15, 143–156.
Hector, J. M. 1963. Introduction to the botany of field crops. II: Non-cereals. *Central News Agency, Johannesburg*, 814–928.
Hellmich, R. L., Kulincevic, J. M., and Rothenbuhler, W. C. 1985. Selection for high and low pollen-hoarding honey bees. *Journal of Heredity* 76, 155–158.
Hellmich, R. L. and Rothenbuhler, W. C. 1986. Relationship between different amounts of brood and the collection and the use of pollen by the honey bees (*Apis mellifera*). *Apidologie* 17, 13–20.
Herbert, E. W. and Shimanuki, H. 1980. An evaluation of seven potential pollen substitutes for honey bees. *Journal of Apicultural Research* 19, 349–350.
Hinton, H. E. 1945. *A Monograph of the Beetles Associated with Stored Products, Vol. 1.* British Museum, London, 5 pp.
Hirschfelder, H. 1951. Quantitative untersuchungen sum pollenentragen der bienevülker [Quantitative investigations on pollen collection by bee colonies]. *Z. Bieneforsch* 1, 67–77.
Hoopingarner, R. and Waller, G. 1992. Crop pollination. In: Graham, J. (ed.) *The Hive and the Honey Bee.* Dadant & Sons Pub., Hamilton, III, pp. 1043–1082.
Hopkins, C. Y., Jevans, A. W., and Boch, R. 1969. Occurrence of octadeca-trans-2, cis-9, cis-12-trienoic acid in pollen attractive to the honey bee. *Canadian Journal of Biochemistry* 47, 433–436.
Horne, C. 1870. Notes on the habits of some hymenopterous insects from the North-west Provinces of India: With an appendix, containing descriptions of some new species of Apidae and Vespidae collected by Mr. Horne: By Frederick Smith, of the British Museum. *Trans. Zool. Soc. Lond.* 7, 161–196.
Horne, C. 1872. Notes on the habits of some hymenopterous insects from the northwest provinces of India. *Trans. Zool. Soc. Lond.* 7, 161–196.
Howpage, D. 1999. Pollination biology of kiwifruit: Influence of honey bees, Apis mellifera L, pollen parents and pistil structure. PhD thesis, University of Western Sydney, Australia.
Ibrahim, S. H. and Selim, H. A. 1974. Effect of pollen traps on honeybee colonies. *Agricultural Research Review* 52, 109–114.
Janki, R., Singh, P., Prakesh, R., and Haque, M. F. 1968. Floral biology of cotton (*Gossypium* spp.). *Indian Agricultural Science* 38, 528–535.
Jay, S. C. 1986. Spatial management of honey bees on crops. *Annual Review of Entomology* 18, 387–391.
Jaycox, E. R. 1963. Confinement of honeybee colonies to avoid pesticide losses. *Journal of Apicultural Research* 2(1), 43–49.
Jenkins, J. N. and Saha, S. 2001. *Genetic improvement of cotton: emerging technologies.* USDA.
Johansen, C. A. 1977. Pesticides and pollinators. *Annual Review of Entomology* 22, 177–192.
Kaziev, T. 1960. The influence of bee pollination on cotton quality. *Pchelovodstvo* 37, 33–35.
Kearney, T. H. 1923. Self fertilization and cross fertilization in Pima cotton. Bulletin U.S. Department of Agriculture, No. 1134.
Keller, I., Fluri, P., and Imdorf, A. 2005. Pollen nutrition and colony development in honey bees: Part I. *Bee World* 86, 3–10.
Kevan, P. G. and Baker, H. G. 1983. Insects as flower visitors and pollinators. *Annual Review of Entomology* 28, 407–453.
Khan, M. and Baure, R. 2002. Damage assessment, monitoring and action thresholds of stinkbug pests in cotton. In: Proceedings of the 11th Australian Cotton Conference, Brisbane, Queensland. Australian Cotton Growers Research Association, pp. 395–400.
Knight, K. M. M., Brier, B. H., Lucy, M. J., and Kopittke, R. A. 2007. Impact of mirid (*Creontiades* spp.) (Hemiptera: Miridae) pest management on *Helicoverpa* spp. (Lepidoptera: Noctuidae) outbreaks: The case for conserving natural enemies. *Pest Management Science* 63, 447–452.
Kono, M. and Tobe, H. 2007. Is *Cycas revolute* (Cycadaceae) wind- or insect-pollinated? *American Journal of Botany* 94, 847–855.
Kroyer, G. and Hegedus, N. 2001. Evaluation of bioactive properties of pollen extracts as functional dietary food supplements. *Innovative Food Science & Emerging Technologies* 2, 171–174.
Kubik, M., Nowacki, J., Pidek, A., Warakomska, Z., Michalczuk, L., Goszczyński, W., and Dwużnik, B. 2000. Residues of captan (contact) and difenoconazola (systemic) fungicides in bee products from an apple orchard. *Apidologie* 31, 531–541.
Kuliev, A. M. 1958. The use of honeybees to increase cotton yield. 17th International Beekeeping Congress (Summary in Bee World Apicultural Abstracts). Apimondia, Bucharest, 39: 378/58, pp. 502–509.

Lavie, P. 1967. Effect of using pollen traps on honey yield from honeybee colonies [Influnce de l'utilisation du piege a pollen sur le rendement en mile des colonies d'abeilles] [Summary in English]. *Annals Abeille* 10, 83–95.

Lavie, P. and Fresnaye, J. 1963. Experimental study with a pollen trap at the top of the hive [Étude éxperimentale de la trappe á pollen en position superiéure]. *Annals Abeille* 6, 277–301.

Lieftinck, M. A. 1939. Uit het leven van *Lithurgus atratus*, een Indisch houtbijtje. *De Tropische Naturae* 28, 193–201.

Lindauer, M. 1953. Division of labour in the honeybee colony. *Bee World* 34, 85–90.

Lindauer, M. 1955. The water and temperature regulation of the honeybee colony. *Bee World* 36, 62–72.

Liven, M. D. 1961. Distribution of foragers from honey bee colonies placed in middle of a large field of alfalfa. *Journal of Economic Entomology* 54, 431–433.

Liven, M. D. and Loper, G. M. 1984. Factors affecting pollen trap efficiency. *American Bee Journal* 124, 721–723.

Llewellyn, D., Tyson, C., Constable, G., Duggan, B., Beale, S., and Stell, P. 2007. Contamination of regulated genetically modified cotton in the field. *Agriculture, Ecosystem & Environment* 121, 419–429.

Loper, G. M. 1986. Cotton pollen: Honey bee avoidance and absence of gossypol. *Journal of Economic Entomology* 79, 103–106.

Loper, G. M. and Berdel, R. L. 1980. A nutritional bioassay of honeybee brood-rearing potential. *Apidologie* 11, 181–189.

Loper, G. M. and Danka, R. G. 1991. Pollination tests with Africanized honeybees in southern Mexico. *American Bee Journal* 131, 191–193.

Loper, G. M. and Davis, D. D. 1985. Disparity of cotton pollen dispersal by honey bees visiting upland and Pima pollen parents. *Crop Science* 25, 585–589.

Loper, G. M. and DeGrandi-Hoffman, G. 1994. Does in hive pollen transfer by honey bees contribute to cross pollination and seed set in hybrid cotton. *Apidologie* 25, 94–102.

Loper, G. M., Levin, M. D., and Waller, G. D. 1984. Pollen trap efficiency, brood rearing, and trapping precautions. *American Bee Journal* 124, 291.

Loper, G. M. and Roselle, R. M. 1991. Experimental use of BeeScent to influence honey bee visitation and yield of watermelon. *American Bee Journal* 131, 777.

Lusas, E. W. and Jividen, G. M. 1987. Glandless cottonseed: A review of the first twenty-five years of processing and utilization. *Journal of the American Oil Chemists' Society* 64, 839–854.

Mackenzie, K. E. and Averill, A. L. 1992. A new honey bee attractant, the queen mandibular pheromone. *Cranberries* 56, 13–14.

Mahadevan, V. and Chandy, K. C. 1959. Preliminary studies on the increase in cotton yield due to honeybee pollination. *Madras Agricultural Journal* 46, 23–25.

Malerbo-Souza, D. T., Nogueira-Couto, R. H., and Couto, L. A. 2004. Honey bee attractants and pollination in sweet orange, *Citrus sinensis* (L.) Osbeck, var. Pera-Rio. *Journal of Venomous Animals and Toxins Including Tropical Diseases* 10, 144–153.

Malyshev, S. I. 1930. Nistgewohnheiten der Steinbienen *Lithurgus* Latr. *Zeitschrift für Morphologie und Ökologie der Tiere* 19, 116–134.

Margalith, R., Lensky, Y., and Rabinowitch, H. D. 1984. An evaluation of Beeline as a honey bee attractant to cucumber and its effect on hybrid seed production. *Journal of Apicultural Research* 23, 50–54.

Martin, E. C. and McGregor, S. E. 1973. Changing trends in insect pollination of commercial crops. *Annual Review of Entomology* 18, 207–226.

Mauney, J. R. 1984. Anatomy and morphology of cultivated cottons. In: Kohel, R. J., Lewis, C. F. (eds.) *Cotton*. Agronomy Monograph No 24. American Society of Agronomy, Madison, WI, pp. 25–79.

Mauney, J. R. and Stewart, J. M. 1986. Cotton physiology. Cotton Foundation National Cotton Council, Memphis, TN.

Maurizio, A. 1950. The influence of pollen feeding and brood rearing on the length of life and physiological condition of the honeybee. *Bee World* 31, 9–12.

Maurizio, A. 1953. Weitere untersuchungen an pollenhoschen. *Beihefte zur Schweizerizerischen Bienen-Zeitung* 20, 486–556.

Mayer, D. F. and Johansen, C. A. 1982. Field evaluation of chemical pollinator attractants on tree fruits. *American Bee Journal* 122, 287–289.

Mayer, D. F., Britt, R. L. and Lunden, J. D. 1989. An evaluation of Bee scent as a honey bee attractant. *Good Fruit Grower*, 40, 40.

McGourty, M. A. 1992. A field evaluation of 'Bee-Here' TM pheromone-based forage attractant for honey bees *Apis mellifera* L. applied to red clover, *Trifolium pratense* L. Master of Agriculture Research Paper, Department of Entomology, Oregon State University, Oregon.

McGregor, E. S. 1959. Cotton-flower visitation and pollen distribution by honeybees. *Science* 129, 97–98.

McGregor, E. S. 1976. Insect pollination of cultivated crop plants. USDA Agriculture Handbook No. 496, pp. 171–189.

McGregor, E. S., Rhyne, C., Worly, S., Jr., and Todd, F. E. 1955. The role of honey bees in cotton pollination. *Agronomy Journal* 47, 23–25.

McGregor, E. S. and Todd, F. E. 1956. Cotton is benefited by honey bees. *Prog. Agric. Ariz.* 7, 12.

McLaren, G. F., Fraser, J. A., and Grant, J. E. 1992. Pollination of apricots. *Orchardist of New Zealand* 65, 22–23.

McLellan, A. R. 1974. Some effects of pollen traps on colonies of honeybee. *Journal of Apicultural Research* 13, 143–148.

McLellan, A. R. 1976. Factors effecting pollen harvesting by the honeybee. *Journal of Applied Ecology* 13, 801–811.

Meade, R. M. 1918. Beekeeping may increase cotton crop. *Journal of Heredity* 9, 282–288.

Moeller, F. E. 1972. Honey bee collection of corn pollen reduced by feeding pollen in the hive. *American Bee Journal* 772, 210–212.

Moffett, J. O. 1983. Hybrid cotton. In: Jones, C. E. and Little, R. J. (eds.) *Handbook of Experimental Pollination Biology*. Van Nostrand Reinhold, New York, pp. 508–514.

Moffett, J. O., Cobb, H. B., and Rummel, R. D. 1980. Bees of potential value as pollinators in the production of hybrid cottonseed on the High Plains of Texas. In: Brown, J. M. (ed.) *Proceedings of the Beltwide Cotton Production Research Conferences Dallas, Texas*. National Cotton Council of America, Memphis, TN, pp. 268–270.

Moffett, J. O., McGregor, S. E., and Stith, L. S. 1974. Pollination practices with hybrid cotton production. In: Brown, J. M. (ed.) *Proceedings of the Beltwide Cotton Production Research Conferences Dallas, Texas*. National Cotton Council of America, Memphis, TN, p. 82.

Moffett, J. O. and Stith, L. S. 1972a. Honeybees as pollinators of hybrid cotton. *Environmental Entomology* 1, 368–370.

Moffett, J. O. and Stith, L. S. 1972b. Pollination by honey bees of male-sterile cotton in cages. *Crop Science* 12, 476–478.

Moffett, J. O., Stith, L. S. 1972c. Bees losses from parathion decreased as distance from sprayed field increased. *American Bee Journal* 112, 174–175.

Moffett, J. O., Stith, L. S., Burkhart, C. C., and Shipman, C. W. 1975a. Honey bee visits cotton flowers. *Environmental Entomology* 4, 203–206.

Moffett, J. O., Stith, L. S., Burkhart, C. C., and Shipman, C. W. 1975b. Influence of cotton genotypes on floral visits of honey bees. *Crop Science* 15, 782–784.

Moffett, J. O., Stith, L. S., Burkhart, C. C., and Shipman, C. W. 1976a. Insect visitors to cotton flowers. *Arizona Academic Science* 11, 47–48.

Moffett, J. O., Stith, L. S., Burkhart, C. C., and Shipman, C. W. 1976b. Nectar secretion in cotton flowers and its relation to visits by honey bees. *American Bee Journal* 116, 32, 34, 36.

Moffett, J. O., Stith, L. S., Burkhart, C. C., and Shipman, C. W. 1976c. Fluctuation of wild bee and wasp visits to cotton flowers. *Arizona Academic Science* 11, 64–68.

Moffett, J. O., Stith, L. S., and Shipman, C. W. 1976d. Influence of distance from pollen plant on seed produced by male-sterile cotton. *Crop Science* 16, 765–766.

Moffett, J. O., Stith, L. S., and Shipman, C. W. 1978a. Producing hybrid cotton seed on a field scale using honey bee as pollinators. In: Brown, J. M. (ed.) *Proceedings of the Beltwide Cotton Production Research Conferences*. National Cotton Council of America, Memphis, TN, pp. 77–80.

Moffett, J. O., Stith, L. S., and Shipman, C. W. 1978b. Effect of honey bee visits on boll set, seed produced, and yield of Pima cotton. In: Brown, J. M. (ed.) *Proceedings of the Beltwide Cotton Production Research Conferences Dallas, Texas*. National Cotton Council of America, Memphis, TN, pp. 80–82.

Moffett, J. O., Stith, L. S., and Shipman, C. W. 1979a. Honey bees and the production of hybrid cotton seed. *American Bee Journal* 119, 492, 493, 542.

Moffett, J. O., Stoner, A., and Wardecker, A. L. 1977a. The wardecker waterer. *American Bee Journal* 117, 364, 365, 378.

Moffett, J. O., Stoner, A., and Wardecker, A. L. 1977b. Methods of reducing losses of honey bees caused by spraying cotton with insecticides. *Journal of Economic Entomology* 70, 737–741.

Moffett, J. O., Waller, G. D., and Rummel, R. D. 1981. 1980 study on insect pollinators and hybrid cotton production in the High Plains of Texas. In: Brown, J. M. (ed.) *Proceedings of the Beltwide Cotton Production Research Conferences Dallas, Texas*. National Cotton Council of America, Memphis, TN, p. 88.

Moffett, J. O., Wilson, W. T., Stoner, A., and Wardecker, A. 1979b. Effect of fifteen combinations of four management methods on losses of honey bees caused by spraying insecticides on cotton. *Journal of Economic Entomology* 72, 453–455.

Mohan, P. and Kairon, M. S. 1999. Nectar glands and honey bee pollination in cotton (*Gossypium*). *Advances in Plant Sciences* 12, 625–626.

Moriya, K. 1966. Effects of pollen trap on numbers of pollen foragers in honeybee colony. *Japanese Journal of Ecology* 16, 105–109.

Mound, L. A. 1962. Extra-floral nectaries of cotton and their secretion. *Empire Cotton Grower Review* 39, 254–261.

Nabors, R. A. 1997. Trapping pollen collection of honeybee *Apis mellifera* L. to determine pollen flow periods. *American Bee Journal* 137, 215–216.

Nabors, R. A. 2000. The effects of spring feeding pollen substitutes to colonies of *Apis mellifera*. *American Bee Journal* 140, 322–323.

Nadel, H. and Pena, J. E. 1994. Identity, behaviour, and efficacy of nitidulid beetles (Coleptera: Nitidulidae) pollinating commercial *Annona* species in Florida. *Environmental Entomology* 23, 878–886.

Nagel, J., Peña, J. E., and Habeck, D. 1989. Insect pollination of atemoya in Florida. *Florida Entomology* 72, 208–211.

Naumann, K., Winston, M. L., Slessor, K. N., and Smirle, M. J. 1994. Synthetic honey bee (Hymenoptera) queen mandibular gland pheromone applications affect pear and sweet cherry pollination. *Journal of Economic Entomology* 87, 1595–1599.

Neira, M. C., Lobos, N. S., Riveros, M. G., Carrillo, R. L., Pessot, R. Z., and Mundaca, N. B. 1997. Insect fauna associated with flowers o of raspberries (*Rubus idaeus* L., cv Meeker) and evaluation of pollination activity of *Apis mellifera* L. under the influence of a pheromonal attractant. *Revista Chilena de Entomologia* 24, 37–44.

Okumura, G. T. and Savage, I. E. 1974. Nitidulid beetles most commonly found attacking dried fruits in California. *Natl. Pest Control Operators News* 34, 2–7.

Oldershaw, D. 1970. The pollination of highbush blueberries. *Report of Pollination Conference University of Arkansas, 9th MP* 127, 171–176.

Olsen, L. G., Hoopingarner, R., and Martin, E. C. 1979. Pollen preference of honeybees sited on four cultivated crops. *Journal of Apicultural Research* 18, 196–200.

Oosterhuis, D. M. 1990. Growth and development of the cotton plant. In: Miley, W. N. and Oosterhuis, D. M. (eds.) *Nitrogen Nutrition in Cotton: Practical Issue, Proceedings, Southern Branch Workshop for Practicing Agronomist*. American Society of Agronomy, Madison, WI, pp. 1–24.

Oosterhuis, D. M. and Jernstedt, J. 1999. Morphology and anatomy of the cotton plant. In: Smith, C. W. and Cothren, T. J. (eds.) *Cotton: Origin, History, Technology, and Production*. Wiley, New York, NY, pp. 175–206.

Oosterhuis, D. M. and Wullschleger, M. J. 1988. Cotton main-stem leaves in relation to vegetative development and yield. Proceedings Beltwide Cotton Physiology Research Conference, New Orleans, pp. 82–84.

Palmer Jones, T. 1947. Use of pollen supplements in New Zealand. *New Zealand journal of agriculture* 75, 147–150.

Pankiw, T. 2004. Brood pheromone regulates foraging activity of honey bees (Hymenoptera: Apidae). *Journal of Economic Entomology* 97, 748–751.

Pankiw, T., Page, R. E. Jr., and Fondrk, M. K. 1998. Brood pheromone stimulates pollen foraging in honey bees (*Apis mellifera*). *Behavioral Ecology and Sociobiology* 44, 193–198.

Pankiw, T., Roman, R., Sagili, R.R. and Zhu-Salzman, K. 2004. Pheromone-modulated behavioral suites influence colony growth in the honey bee (*Apis mellifera*) *Naturwissenschaften* 91, 575–578.

Parks, H. B. 1921. The cotton plant as a source of nectar. *American Bee Journal* 10, 391–393.

Percival, A. E., Wendel, J. F., and Stewart, J. M. 1999. Taxonomy and germplasm resources. In: Smith, C. W. and Cothren, T. J. (eds.) *Cotton; Origin, History, Technology, and Production*. Wiley, New York, NY, pp. 33–63.

Percival, M. S. 1965. *Floral Biology*. Pergamon Press, New York (reprinted 1979).

Phillips, S. A., Jr. and Simpson, J. L. 1989. Hybrid cotton pollination in relation to accumulated degree days. *Agronomy Journal* 81, 975–980.

Purdie, J. D. and Doull, K. M. 1964. Supplementary protein feeding of honeybees on a pollen deficient honey flow. *The Australian Beekeeper* 66, 76–80.

Radoev, L. and Bozhirov, M. 1961. Studies on bee pollination and honey productivity of cotton. Proceedings 19th International Congress of Apiculture, Liblice, Czechoslovakia, p. 99.

Rajotte, E. G. and Fell, R. D. 1982. A commercial bee attractant inactive in enhancing apple pollination. *HortScience* 17, 230–231.

Rapp, R. 1981. The effect of pollination by honeybees (*Apis mellifera* L.) on cucumber cv Delila seed production. MSc Thesis, Hebrew University of Jerusalem.

Rashad, S. and Parker, R. L. 1958. Major pollen sources in the Manhattan Kansas area and the influences of weather factor upon pollen collection by honeybees. *Proceedings 10th International Congress of Entomology* 10, 1037–1046.

Rhodes, J. 2002. Cotton pollination by honey bees. *Australian Journal of Experimental Agriculture* 42(4), 513–518.

Ribbands, C. R. 1951. The flight range of honeybees. *Journal of Animal Ecology* 20, 302–310.

Ribbands, C. R. 1953. *The Behavior and Social Life of Honeybees.* Bee Research Association, University of Minnesota, Minneapolis, Minnesota, 352 pp.

Robertson, L. N. and Rhodes, J. W. 1992. Honey bee (Apis mellifera L.) deaths near sprayed cotton and observations on bee foraging behaviour in flowering cotton (Hymenoptera: Apidae). *Journal of the Australian Entomological Society* 31, 243–246.

Rybakov, M. N. 1961. Pollen traps and the activity of bees [In Russian]. *Pchelovodstvo* 38, 15–16 *(From Apicultural Abstracts* 429/64 (1964), 86. 62–69.

Saffari, A., Kevan, P. G., and Atkinson, J. 2004. A promising pollen substitute for honey bees. *American Bee Journal* 144, 230–231.

Salman, A. G. and Abd-El-Raof, T. K. 1983. Effect of certain pesticides used against cotton pests on three predaceous insects and honey bee workers. *Summary* (*En*). *Bulletin of the Entomological Society of Egypt*. 11, 155–162.

Sawan, Z. M., Hanna, L. I., Gad, E. L., Karim, G. H. A., and McCuisition, W. L. 2002. Relationships between climatic factors and flower and boll production in Egyptian cotton. *Journal of Arid Environments* 52, 499–516.

Schultheis, J. R., Ambrose, J. L., Bambara, S. B., and Managum, W. A. 1994. Selective bee attractants did not improve cucumber and watermelon yield. *HortScience* 29, 255–258.

Sedgley, M. 1982. Anatomy of the unpollinated and pollinated watermelon stigma. *Journal of Cell Science* 54, 341–355.

Seeley, T. D. 1985. *Honeybee Ecology: A Study of Adaptation in Social Life.* Princeton University Press, New Jersey.

Seeley, T. D. 1995. *The Wisdom of the Hive.* Harvard University Press, Cambridge, MA.

Shishikin, E. A. 1952. Moskva Gosudarstvennoe Izdatel'stvo Sel' skokhoz. [In Russian]: 95–103. (English summary: Effect of pollination by honeybees in increasing the productivity of cotton). *Bee World Apicultural Abstracts* 36, 44/55.

Shivanna, K. R., Cresti, M., and Ciampolini, F. 1997. Pollen development and pollen pistil interaction. In: Shivanna, K. R. and Sawhney, V. K. (eds.) *Pollen Biotechnology for Crop Production and Improvements.* New Delhi, India: University of Cambridge Press, pp. 15–39.

Shivanna, K. R., Linskens, H. E., and Cresti, M. 1991. Pollen viability and pollen vigor. *Theoretical Applied Genetics* 81, 38–42.

Shoreit, M. N. and Hussein, M. H. 1993. Field tests with some protein supplements for feeding bees at Assiut Governorate. *Egyptian Journal of Applied Science* 8, 366–375.

Sidhu, A. S. and Singh, S. 1961. Studies on agents of cross pollinating of cotton. *Indian Cotton Growing Review* 15, 341–353.

Sidhu, A. S. and Singh, S. 1962. Role of honey bees in cotton production. *Indian Cotton Growing Review* 16, 18–23.

Singh, P. B. and Sinha, S. N. 1997. Effect of Bee-Q on honey bee visit and the seed yield of the hybrid sunflower. *Seed Research* 24, 151–153.

Smith, C. W. and Cothren, T. J. 1999. *Cotton; Origin, History, Technology, and Production.* John Wiley & Sons, Inc, New York, USA.

Stace, P. S. 1994. Pollen quality: Yellow vine or cat's head (*Tribulus terrestris*) and cotton (*Gossypium hirsutum*). *Bee Briefs* 10, 6–7.

Standifer, L. N., Owens, C. D., Hydak, M. H., Millsn, J. P., and Levin, M. D. 1973. Supplementary feeding of honeybee colonies in Arizona. *Journal of Apicultural Research* 12, 298–301.

Standifer, L. N., Waller, G. D., Hydak, M. H., Millsn, J. P., and Levin, M. D. 1971. Simulative feeding of honeybee colonies in Arizona. *Journal of Apicultural Research* 10, 27–34.

Steiner, K. E. 1998a. Beetle pollination of peacock moraeas (Iridaceae) in South Africa. *Plant Systematics and Evolution* 209, 47–655.

Stith, L. S. 1970. A bee in hybrid cotton production. *Rep. Pollination Conf. Univ. Ark., 9th MP* 127, 28–32.

Stoner, A., Moffett, J. O., Wilson, W. T., and Standifer, L. N. 1981. *Apis mellifera*: Effect of management of one- and two-queen colonies on bee losses by spraying insecticides on cotton. *Journal of the Georgia Entomological Society* 16, 323–351.

Synge, A. D. 1947. Pollen collection by honey bees (*Apis mellifera*). *Journal of Animal Ecology* 16, 122–138.

Tanda, A. S. 1983. Assessing the role of honey bees in a field of Asiatic cotton (*Gossypium arboreum* L.). *American Bee Journal* 123, 593–594.

Tanda, A. S. 1984. Bee pollination increase yield of 2 interplanted varieties of Asiatic cotton (*Gossypium arboreum* L.). *American Bee Journal* 124, 539–540.

Tanda, A. S. and Goyal, N. P. 1979a. Preliminary observation on the effect of intervarietal bee pollination of desi cotton (*Gossypium arboreum* Linn.). *Indian Journal of Entomology* 41, 281–282.

Tanda, A. S. and Goyal, N. P. 1979b. Insect pollination in Asiatic cotton (*G. arboreum*). *Journal of Apicultural Research* 18, 64–72.

Tanda, A. S. and Goyal, N. P. 1979c. Pollen dispersal by insects in desi cotton (*G. arboretum* L.). *Seed & Farms* 5, 56–59.

Ter-Avanesian, D. 1978. Significance of pollen amount for fertilization. *Bulletin of the Torrey Botanical Club* 105, 2–8.

Thorp, R. W. 1979. Honeybee foraging behavior in California almond orchards. In: Caron, D. M. (ed.) *Proceedings of Fourth International Symposium on Pollination*. Maryland Agric. Exp. Sta. Special Misc. Publ. No. 1, pp. 148–152. Cornell University, USA.

Thorp, R. W. 2000. The collection of pollen by honeybees. *Plant Systematics and Evolution* 222, 211–223.

Todd, E. F. and McGregor, S. E. 1960. The use of honey bees in the production of crops. *Annual Review of Entomology* 5, 265–278.

Todd, F. E. and Reed, C. B. 1970. Brood measurement as a valid index to the value of honey bees as pollinators. *Journal of Economic Entomology* 63, 148–149.

Tsirakolglou, V., Thrasyvoulou, A., and Hatjina, F. 1997. Techniques to increase the attractiveness of kiwi flowers to honey bees. *Acta Horticulturae* 444, 439–445.

Umbeck, P. F., Barton, K. A., Nordheim, E. V., McCarty, J. C., Parrot, W. L., and Jenkins, J. N. 1991. Degree of pollen dispersal by insects from a field test of genetically engineered cotton. *Journal of Economic Entomology* 84, 1943–1950.

Vaissiere, B. E. 1991a. Honey bee stocking rate, pollinator visitation, and pollination effectiveness in upland cotton grown for hybrid seed production. The Sixth International Symposium on Pollination; Ruijter, A. de. International Society for Horticultural Science (ISHS), Wageningen, Netherlands, pp. 359–363.

Vaissiere, B. E. 1991b. Honey bees *Apis mellifera* L. (Hymenoptera: Apidae), as pollinators of upland cotton. *G. hirsutum* L. (Malvaceae) for hybrid seed production. PhD Dissertation. College station, Texas, USA: Texas A&M University.

Vaissiere, B. E., Loper, G. M., and Rummel, D. R. 1982. Pollen transfer in hybrid cotton on the high plains of Texas. In: Brown, J. M. (ed.) *Proceedings of the Beltwide Cotton Production Research Conferences*. National Cotton Council of America, Memphis, TN, pp. 91–92.

Vaissiere, B. E., Moffett, J. O., and Loper, G. M. 1984. Honey bees as pollinators of hybrid cotton seed production on the Texas high plains. *Agronomy Journal* 76, 1005–1010.

Vaissiere, B. E. and Vinson, S. B. 1994. Pollen morphology and its effect on pollen collection by honey bee, *Apis mellifera* L., with special reference to upland cotton, *G. hirsutum* L. *Grana* 33, 128–138.

Van Deynze, A. E., Sundstrom, F. J., and Bradfrod, K. J. 2005. Pollen mediated gene flow in California cotton depends on pollinator activity. *Crop Science* 45, 1565–1570.

Van Kleunen, M., Nänni, I., Donaldson, J. S., and Manning, J. C. 2007. The role of beetle marks and flower colour on visitation by monkey beetles (hopliini) in the greater cape floral region, South Africa. *Ann Bot.* 100(7), 1483–1489.

Vansell, G. H. 1944b. Some western nectars and their corresponding honeys. *Journal of Economic Entomology* 37, 530–533.

Wafa, A. K., Ali, A. M., and Bishara, R. H. 1968a. Contamination of nectar with insecticides in cotton fields. *Bulletin of the Entomological Society of Egypt, Economic Series* 2, 241–245.

Wafa, A. K., Ali, A. M., and Bishara, R. H. 1968b. Contamination of pollen grains of cotton plants with certain insecticides. *Bulletin of the Entomological Society of Egypt, Economic Series* 2, 227–321.

Wafa, A. K. and Ibrahim, S. H. 1959. Pollinators of chief sources of nectar and pollen grain plants in Egypt. *Bulletin of the Entomological Society of Egypt* 43, 133–154.

Wafa, A. K. and Ibrahim, S. H. 1960b. The effect of the honeybee as a pollinating agent on the yield of clover and cotton. Bulletin of Faculty of Agriculture Cairo University. No. 206.

Waller, G. D. 1970. Attracting honey bees to alfalfa with citral, geraniol, and anise. *Journal of Apicultural Research* 9, 9–12.

Waller, G. D. 1972. Evaluating responses of honey bees to sugar solutions using an artificial flower feeder. *Annals of the Entomological Society of America* 65, 857–861.

Waller, G. D. 1982. Hybrid cotton pollination. *American Bee Journal* 122, 555–560.

Waller, G. D., Caron, D. M., and Loper, G. M. 1981a. Pollen patties maintain brood rearing when pollen is trapped from the honey bee colonies. *American Bee Journal* 121(2), 101–103, 105.

Waller, G. D. and Mamood, A. N. 1991. Upland and Pima cotton as a pollen donor for malesterile upland seed parent. *Crop Science* 31, 265–266.

Waller, G. D. and Moffett, J. O. 1981. Pollination of hybrid cotton on the Texas High Plains: Nectar production and honey bee visits. In: Brown, J. M. (ed.) Proc. Beltwide Cotton Prod. Res. Conf., New Orleans, Louisiana. National Cotton Council of America, Memphis, TN.

Waller, G. D., Moffett, J. O., Loper, G. M., and Martin, J. H. 1985a. An evaluation of honey bee foraging activity and pollination efficacy for male-sterile cotton. *Crop Science* 25, 211–214.

Waller, G. D., Torchio, P. F., and Stith, L. S. 1978. A comparison of A*pis mellifera* and *Osmia fulviventris* as pollinators of caged cotton in Arizona. Proceedings of the Fourth International Symposium on Pollination, Maryland, pp. 299–305.

Waller, G. D., Vaissiere, B. E., Moffett, J. O., and Martin, J. H. 1985b. Comparison of carpenter bees (*Xylocopa varipuncta*) and honey bees (A*pis mellifera* L.) as pollinators of male-sterile cotton in cages. *Journal of Economic Entomology* 78, 558–561.

Waller, G. D., Wilson, F. D., and Martin, J. H. 1981b. Influence of phenotype, and time-of-day on nectar production in cotton. *Crop Science* 21, 211–214.

Ward, A. L. 2005. Development of a treatment threshold for sucking insects in determinate Bollgard II transgenic cotton grown in winter production areas. *Australian Journal of Entomology* 44, 310–315.

Weaver, J. B. Jr. 1978. Observations on bee activity in several genotypes of cotton. In: Brown, J. M. (ed.) Proceedings of the Beltwide Cotton Production Research Conferences. National Cotton Council of America, Memphis, TN, pp. 76–77.

Webster, T. C., Thorp, R. W., Briggs, D., Skinner, J., and Parisian, T. 1985. Effect of pollen traps on honeybee foraging and brood rearing during almond and prune pollination. *Environmental Entomology* 14, 683–686.

Wendel, J. F., Burbaker, C. L., and Percival, E. 1992. Genetic diversity in *Gossypium hirsutum* and the origin of upland cotton. *American Journal of Botany* 79, 1291–1310.

Whitehouse, M. E., Wilson, L. J., and Fitt, G. P. 2005. A comparison of arthropod communities in transgenic *Bt* and conventional cotton in Australia. *Environmental Entomology* 34, 1224–1241.

Williams, A. 1996. Opportunities for chiral agrochemicals. *Pest Management Science* 46(1), 3–7.

Williams, M. R. 2002. Cotton insect losses. In: Nashville, T. N., Hardee, D. D., Ottens, R. S., Burris, G., and Ottea, J. A. (eds.) *Proceedings of the Beltwide Cotton Conferences.* National Cotton Council of America, Memphis, TN, pp. 1217–1273.

Wilson, F. D. and Stapp, B. R. 1984. Crossing success in cotton in Arizona as affected by irrigation, number of flowers pollinated, and time of emasculation. *Agronomy Journal* 76, 457–460.

Winston, M. L. and Slessor, K. N. 1993. Application of queen honeybee mandibular pheromone for beekeeping and crop pollination. *Bee world* 74, 11–128.

Winston, M. L., Slessor, K. N., Rubink, W. L., and Villa, J. D. 1993. Enhanced pheromone lures to attract honey bee swarms. *American Bee Journal* 133, 58–60.

Winston, M. L., Slessor, K. N., Williams, L. G., Naumann, H. A., Higo, M. H., Kaminski, L. A. 1989. The influence of queen mandibular pheromones on worker attraction to swarm cluster and inhibition of queen rearing in the honey bee (*Apis mellifera* L.). *Insect Sociaux* 36, 15–27.

Woyke, H. W. 1981. Some aspects of the role of the honeybee in onion seed production in Poland. *Acta Horticulturae* 111, 91–98.

Young, H. J. 1986. Beetle pollination of *Dieffenbachia longispatha* (Araceae). *American Journal of Botany* 73, 931–944.

11 Castor

11.1 INTRODUCTION

Castor (*Ricinus communis* L.) is an important non-edible oilseed crop grown in arid and semi-arid areas of the world. It is grown as an ornamental plant in landscapes which require minimal care. It is native to Africa and grown extensively in various parts of world, including Asia. Cultivation of castor dates back to 7000 BC, and the seeds have been found in ancient tombs. Castor is mostly grown in tropical regions, and the trees may attain the height 40 ft tall. The oil was used for lamps by ancient Egyptians and also for medicinal purposes.

11.2 AREA AND DISTRIBUTION

India exports 4.3 lakh tons of castor oil and its derivatives per annum as against the world production 12.5 lakh tons of castor seed 5.5 lakh tons of castor oil. India, China, and Brazil are the major castor-producing countries, accounting for 90% of the world's production. The topmost country in the list is India, with around 83% of the share in production, followed by China with 9% of share in 2012–13. India is the largest producer of castor after Brazil. A postal stamp represents the significance of castor in Brazil. In India, Andhra Pradesh accounts for nearly 65% of the total castor area in India (Mittal et al. 1991).

In India, it is cultivated in an area of 14.53 lakh hectares with production and productivity of 22.95 lakh tons and 1578.66 kg ha^{-1}, respectively. Almost all the Indian states such as Gujarat, Andhra Pradesh, Tamil Nadu, Rajasthan, and Karnataka are important castor-producing states. The highest production occurs in Gujarat 76%, followed by Rajasthan 17%, and Andhra Pradesh 5%.

11.3 USES OF CASTOR

The oil produced from castor beans is used in paints, lubricants, soaps, hydraulic brake fluids, polymers, and perfumery products. Besides derivatives from castor oil, they are used in many industries, such as cosmetics, shampoo, soap, hand lotion, laxative, fuel for lamps, and as a high-speed lubricant (Capasso et al. 1994; Copley et al. 2005; Morris et al. 2011). It is also being cultivated for biodiesel (Melo et al. 2008; Sailaja et al. 2008; Vanaja et al. 2008). One of the most important features of this crop is that it is not used as food and can be grown on marginal lands. Castor plants are 6–8 ft in height, depending on the variety. As ornamentals, castor beans are sometimes planted as an annual hedge or screen.

11.4 CASTOR AS POLLEN SOURCE

Castor (*R. communis*) crop is a very good source of pollen for bees. Its average yield has been reported to vary from 1,307.1 to 904.7 and 567.7 kg seeds/hectare in India, China, and Brazil, respectively. This crop has a great potential to become the most important vegetable crop for industrial oil production throughout the world (Roetheli et al. 1990).

11.5 FLORAL BIOLOGY AND POLLINATION REQUIREMENTS

Castor is both self- and cross-pollinated by wind, varying from 5 to 36%, depending on the weather conditions. Pollen sheds readily between 26 and 29°C, with a relative humidity of 60%. Purseglove (1968) thought that although castor is mostly wind-pollinated, insects may play some part, as extrafloral glands on leaves below the flowers produce much nectar (Koptur 1992). However, Alex

DOI: 10.1201/9781032656724-12

(1957a, b) reported that although honeybees visit staminate flowers of castor and collect large amounts of pollen, the tests showed that they are of no value as pollinators.

The flower of castor is greenish-yellow in color and contains three to five united sepals and no petals. The male flower has numerous branched stamens, and the female flower has a superior three-celled ovary terminated by three styles (Figure 11.1). Both sexes occur in the same panicle; the female flowers in the upper part have mostly set before the male flowers in the lower part have opened (Purseglove 1968). Inflorescence of castor bean is available throughout the year in dense terminal clusters, with female flowers just above the male flowers.

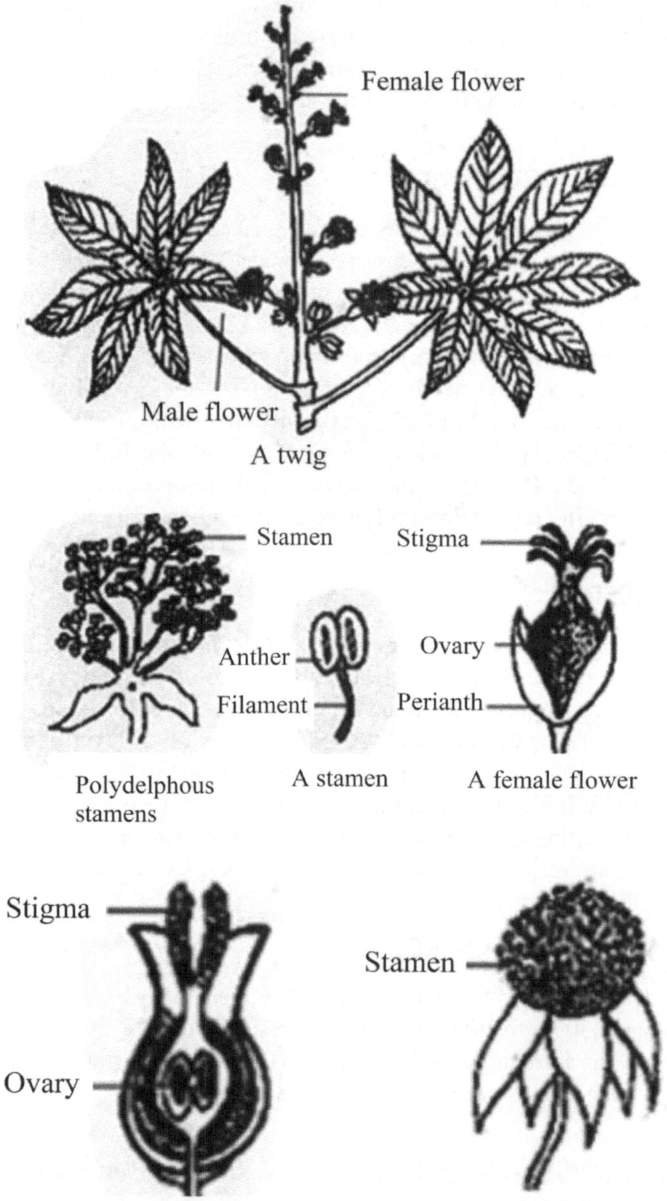

FIGURE 11.1 *Ricinus cuminus* branch and floral structure: (a) twig, (b) floral structure, (c) LS female flower, (d) male flower.

Commercial production of castor has largely been limited due to the presence of toxins in castor seed, lack of proper market, and efficient technologies to produce and process the crop (Brigham 1993). Development of improved production and genetic technologies will help ensure the economic production and potential of this crop. The majority of flowering plants are dependent upon insects or other animals for pollination, wherein bees are the most important pollinating insects (Westerkamp 1996). However, several plant species contain secondary compounds in nectar and pollen that could be toxic to pollinators, including bees (Adler 2000; Adler and Irwin 2005; Praz et al. 2008; Mesquita et al. 2010). Juniora et al. (2011) reported that the pollen of *R. communis* is toxic to honeybees (*Apis mellifera*) under laboratory conditions.

Castor is a very good source of nectar, and if a good number of plants are available, bees can make a good harvest (Figure 11.2a–11.2f). The plants have extrafloral nectaries located at the base of flowers. Bees have also been observed to collect nectar from young seed pods.

Purseglove (1968) thought that although castor is mostly wind-pollinated, insects may play some part, as extrafloral glands on leaves below the flowers produce much nectar. However, Alex (1957a, b) reported that although honeybees visit staminate flowers of castor and obtain large amounts of pollen, tests showed they are of no value as pollinators. Stafford (1976) reported that castor is largely cross-pollinated.

FIGURE 11.2A *Apis cerana* collecting pollen.

FIGURE 11.2B *Apis dorsata* collecting pollen.

FIGURE 11.2C *Apis mellifera* foraging on male flowers.

FIGURE 11.2D *Apis cerana indica* foraging on female flowers.

FIGURE 11.2E Pollen load on *Apis cerana* bee.

FIGURE 11.2F Pollen load on *Apis mellifera* bee.

11.6 POLLINATION REQUIREMENTS OF CASTOR BEAN AND POLLINATION EFFICIENCY OF *APIS MELLIFERA* L.

Rizzardo et al. (2012) revealed *R. communis* bears a mixed breeding system favoring selfing by geitonogamy, although the wind promotes mostly outcrossing. They also found that the honeybee (*Apis mellifera* L.) foraging on castor bean can both transfer pollen from male to female flowers within the same raceme and boost the release of airborne pollen by male flowers.

The treatments hand geitonogamy and open with the introduction of honeybee colonies presented the highest fruit set, over 90%, and did not differ ($p > 0.05$) from each other (Table 11.1). However, they differed significantly ($p < 0.05$) to all the other treatments. Both situations increase geitonogamy rates, raising significantly fruit set and seed yield. The area in which colonies of *A. mellifera* were introduced produced 389.35 kg seeds/hectare, while that without honeybee introduction, only 298.10 kg/ha. Even low, the yield in the area with honeybee hives was 90.35 kg seeds/hectare (30.61%), significantly ($p < 0.05$) greater than in the area with no hives.

The results from the hand geitonogamy treatment producing the highest fruit set indicates that the castor bean is primarily autogamous, and because the flowers are unisexual, geitonogamy is the only mean for selfing. Other authors have reported on the autogamy of castor bean (Foster et al. 2010). The castor bean floral arrangement with female flowers situated above male flowers in the racemes favors pollen dispersion for cross-pollination (Ramprasad and Bandopadhyay 2010).

The introduction of honeybee colonies to castor bean plantations produced 17% increment in fruit set, reaching a figure equivalent to that of hand geitonogamy and maximizing fruit set in this crop. It also produced augments of 30.61% in seed yield per hectare. This result also confirms the occurrence of pollination deficit, as defined by Vaissière et al. (2011), in the area studied. It also indicates that *A. mellifera* bees can promote the transference of pollen grains from male to female flowers of the same or adjacent racemes when foraging for pollen and extrafloral nectar in *R. communis*.

Development of improved production and technological interventions will help ensure rural people across the world fetch the economic potential of this crop for better livelihood.

TABLE 11.1
Fruit Set of Castor Bean *(Ricinus communis)* Flowers in Response to Seven Different Pollination Treatments

Treatments	Number of Flowers	Number of Fruit Set	Fruit Set (%)
Hand geitonogamy pollination	75	74[a]	99
Open pollination with the introduction of *Apis mellifera* colonies	47	44[a]	94
Hand cross-pollination	82	66[b]	80
Open pollination without the introduction of *Apis mellifera* colonies	60	48[b]	80
Bagged with muslin bag	327	259[b]	79
Bagged with paper bag	124	19[c]	15
Emasculated and bagged with paper bag	23	0	0

Note: Values followed by the same letters do not differ at $p > 0.05$.

REFERENCES

Adler, L. S. 2000. The ecological significance of toxic nectar. *Oikos* 91, 409–420.

Adler, L. S. and Irwin, R. E. 2005. Ecological costs and benefits of defenses in nectar. *Ecology* 86, 2968–2978.

Alex, A. H. 1957a. Honey bees aid pollination of cucumbers and cantaloupes. Texas Agricultural Experimental Station Progress Report, U.S.A., 1936, 4p.

Alex, A. H. 1957b. Pollination of some crops by honey bees. Texas Agricultural Experimental Station Progress Report, U.S.A., No. 1960.

Brigham, R. D. 1993. Castor: Return of an old crop. In: Janick, J. and Simon, J. E. (eds.) *New Crops*. Wiley, New York, pp. 380–383.

Capasso, F., Mascolo, N., Izzo, A. A., and Gaginella, T. S. 1994. Dissociation of castor oil-induced diarrhoea and intestinal mucosal injury in rat: Effect of NG-nitro-L-arginine methyl ester. *Br. J. Pharmacol.* 113, 1127–1130.

Copley, M. S., Bland, H. A., Rose, P., Horton, M., and Evershed, R. P. 2005. Gas chromatographic, mass spectrometric and stable carbon isotopic investigations of organic residues of plant oils and animal fats employed as illuminants in archaeological lamps from Egypt. *Analyst* 130, 860–871.

Foster, J. T., Allan, G. J., Chan, A. P., Rabinowicz, P. D., Ravel, J., Jackson, P. J., and Keim, P. 2010. Single nucleotide polymorphisms for assessing genetic diversity in castor bean (*Ricinus communis*). *BMC Plant Biol.* 10, 13.

Juniora, E. M. A., Fernandesa, I. M. S., Santosa, C. S., Mesquitaa, L. X., Pereiraa, R. A., Maracajáb, P. B., and Soto-Blancoa, B. 2011. Toxicity of castor bean (Ricinus communis) pollen to honeybees. *Agriculture, Ecosystems and Environment* 141, 221–223.

Koptur, S. 1992. Extrafloral nectary-mediated interactions between insects and plants. In: Bernays, E. (ed.) *Insect-Plant Interactions*. CRC Press, Boca Rotan, pp. 81–129.

Melo, W. C., Santos, A. S., Santa Anna, L. M. M., and Pereira, Jr., N. 2008. Acid and enzymatic hydrolysis of the residue from Castor Bean (*Ricinus communis* L.) oil extraction for ethanol production: Detoxification and biodiesel process integration. *J. Brazil Chem. Soc.* 19, 418–425.

Mesquita, L. X., Maracajá, P. B., Sakamoto, S. M., and Soto-Blanco, B. 2010. Toxic evaluation in honey bees (Apis mellifera) of pollen from selected plants from the semi-arid region of Brazil. *J. Apicult. Res.* 49, 265–269.

Mittal, J. P., Dhawan, K. C., and Thyagraj, C. R. 1991. Energy scenario of castor crop under dryland agriculture of Andhra Pradesh. *Energy Convers. Mgmt.* 32(5), 425–430.

Morris, J. B., Wang, M. L., and Morse, S. A. 2011. Ricinus. In: Kole, C. (ed.) *Wild Crop Relatives: Genomic and Breeding Resources—Oilseeds*. Springer, Berlin, Heidelberg, Germany, pp. 251–260.

Praz, C. J., Müller, A., and Dorn, S. 2008. Specialized bees fail to develop on non-host pollen: Do plants chemically protect their pollen? *Ecology* 89, 795–804.

Purseglove, 1968. J. W. Purseglove. Manihot. Tropical Crops. Dicotyledons, Wiley, New York, NY (1968), pp. 171–180.

Ramprasad, R. and Bandopadhyay, R. 2010. Future of Ricinus communis after completion of the draft genome sequence. *Curr. Sci. India* 99, 1316–1318.

Rizzardo, R. A. G., Milfont, M. O., Da Silva, E. M. S., and Freitas, B. M. 2012. *Apis mellifera* pollination improves agronomic productivity of anemophilous castor bean (*Ricinus communis*). *Annals of the Brazilian Academy of Sciences* 84(4), 1137–1145.

Roetheli, J. C., Glaser, L. K., and Brigham, R. D. 1990. Castor: Assessing the feasibility of U.S. production. Workshop Proceed. Growing Industrial Material Series. USDA-CSRS and Texas A&M University, Plainview, TX.

Sailaja, M., Tarakeswari, M., and Sujatha, M. 2008. Stable genetic transformation of castor (*Ricinus communis* L.) via particle gun-mediated gene transfer using embryo axes from mature seeds. *Plant Cell Rep.* 27, 1509–1519.

Vaissière, B. E., Freitas, B. M., and Gemmill-Herren, B. 2011. *Protocol to Detect and Assess Pollination Deficits in Crops: A Handbook for Its Use*. FAO, Rome, p. 81.

Vanaja, M., Jyothi, M., Ratnakumar, P., Vagheera, P., Raghuram Reddy, P., Jyothi Laksmi, N., Yadav, S. K., Maheshwari, M., and Venkateswarlu, B. 2008. Growth and yield responses of castor bean (*Ricinus communis* L.) to two enhanced CO2 levels. *Plant Soil. Environ.* 54, 38–46.

Westerkamp, C. 1996. Pollen in bee—flower relations: Some considerations on melittophily. *Bot. Acta* 109, 325–332.

12 Oil Palm

12.1 INTRODUCTION

The palm family has around 2,450 species which occur in tropical and subtropical regions of the world. They are most commonly distributed in South America and the Malaysian region, with just 65 species found in Africa. *Elaeisguineensis* Jacq., which is a high-yielding perennial oil palm crop, is extensively cultivated for commercial purposes in Malaysia and Indonesia (Purseglove 1975; Hartley 1988). This plantation crop has been introduced to southern parts of India about more than two decades ago (Rethinam 1992). At present, only 3.70 lakh hectares are under oil palm cultivation in India (Figure 12.1). The area under oil palm cultivation is proposed to be increased to 10 lakh hectares by 2025–26 and to 16.7 lakh hectares by 2029–30. Particular focus is given on the northeast and the Andaman and Nicobar Islands. This will help not only in achieving self-sufficiency in palm oil production but also in generating income to the farmers. India is the world's largest importer and second-largest consumer of palm oil. In the last 20 years, the consumption of palm oil in India has risen by around 230%, making India the leading global importer of palm oil. More than 90% of the imported palm oil originates from Indonesia, Malaysia, and Thailand. India produces approximately 10 million tons domestically and imports an additional 15 million tons from other nations.

12.2 FLORAL BIOLOGY AND POLLINATION EFFICIENCY

Pollination plays a crucial role in the production of 75% of global food crop species (Food and Agriculture Organization 2023). Pollination is therefore economically important for agriculture and global food security. However, pollinator populations are declining globally at an alarming rate due to multiple factors, such as habitat loss, pesticide exposure, disease, climate change, and changes in land use practices. Pollinator decline is expected to have significant impacts, resulting in decreased crop yields, lower-quality fruits and seeds, and potential cascading effects on ecosystem functioning and biodiversity.

The oil palm ecosystem is likely to be seriously affected due to a decline of pollinators, as the crop requires cross-pollination to produce fruit, and any reduction thereof can affect the yield of the crop. It is important to develop strategies to maintain the pollinator populations and ensure effective crop pollination in this important agricultural system.

12.2.1 Floral Biology

12.2.1.1 Plant

The oil palm is an erect monoecious plant up to 30 ft in height, with a trunk of 12 in or more. It produces clusters of nuts, each of which has two locules about 1 to 1.5 in long. The nuts are of three types, depending upon shell thickness: *dura* (3 to 8 mm), *tenera* (up to 3 mm), and *pisefera* (with no shell). The plant possesses a compact cluster of feather-like leaves measuring between 10 and 15 ft in length.

12.2.1.2 Inflorescence

Elaeis guineensis, a plant belonging to the Arecaceae family (Palmae), is characterized as a monoecious species. It produces inflorescences that contain both male and female flowers, and these inflorescences are borne alternately in cycles of varying durations (Tandon et al. 2001). The inflorescences of this plant are composed of compound spikes and become visible on young palms at the age of 27 months. It is possible to distinguish the sex of the inflorescence approximately 14 months after its initiation (Figure 12.2).

DOI: 10.1201/9781032656724-13

FIGURE 12.1 (a) Palm oil plantation; (b) fruits on palm oil tree.

FIGURE 12.2 Palm: (a) male inflorescence, (b) female inflorescence.

FIGURE 12.3 Fertilized female flowers turn into cluster of fruits.

The pollen viability and stigma receptivity remain around five to six days and 3 to 4 days in the field, respectively (Soh 2012). A male flower on a spikelet produces 20 to 25 g pollen per inflorescence within two to four days. The female inflorescence consists of around 200 spikelets arranged in a spiral pattern, with 10 to 30 flowers present in each spikelet (Figure 12.3).

The flowers open at different times of the day for about a week, and the stigma remains receptive within 36 to 48 h. The inflorescences are tightly enclosed with a prophyll and peduncular bract for about six weeks before the flowers mature (Adam et al. 2005).

12.2.2 Floral Rewards

Palm flowers produce rewards for flower visitors, which may be nutritional, such as pollen grains, food tissues, and nectar and/or ovipositional sites. The majority of insect visitors to palm inflorescences are attracted to the nutritional rewards. This concept is applicable to beetles, certain pollen-feeding flies, as well as flies and bees that seek nectar (Uhl and Moore 1977). Honeybees collect pollen grains and transport to their hives to feed the growing brood. They are attracted to inflorescences by visual cues and are able to detect minor morphological differences between male and female flowers. In the case of flowers that exhibit sexual dimorphism, certain bees exclusively visit the male flowers to gather pollen, resulting in an antagonistic interaction with the palm. For instance, *Licuala peltata*, a type of palm, often produces abundant nectar to attract bees and compete with other plants that rely on bee pollination (Barfod et al. 2003). Additionally, palm inflorescences offer sites for insect visitors to lay their eggs, providing a reward for their presence. Beetles, for example, deposit their eggs in fleshy tissues, ensuring that the hatched larvae have a ready source of food. In the Phytelepheae tribe, male inflorescences have a higher temperature during anthesis (10–20°C above ambient temperature) and are densely perforated by insect boreholes.

Hoppe (2005) made an observation regarding a specific type of weevil that displayed a tendency to congregate in large numbers beneath the sizeable leathery bracts of the *Nypa fruticans*, a type of mangrove palm. This weevil was identified as a potential pollinator for the plant. Silberbauer-Gottsberger et al. (2001) proposed that the role of the protective bracts in palms and Cyclanthaceae resembles that of the pollen chamber found in cantharophilous members of the Annonaceae family, which is formed by the thick fleshy petals, as well as the "kettle" found in Araceae, formed by the spathe. Numerous studies have indicated that floral scent plays a significant role in attracting pollinators, and these scents often consist of one or a few dominant compounds (Knudsen et al. 2001).

12.2.2.1 Pollination Requirements

Separate male and female inflorescences require pollen vectors for effective pollination. The female inflorescences are found on the axillary branches and form fruit bunches. The bunches have the potential to contain several hundred individual fruits, while the male flowers are found on spike-like inflorescences that emerge from the leaf axils, in contrast.

The male inflorescence comprises a central axis that bears many branches, each containing numerous small flowers (Hartley 2003). The male flowers produce pollen, which is then transferred to the female flowers by the wind (anemophilous pollination) and various insects, such as weevils (entomophilous pollination) (Hardy et al. 2020). Evidently, cross-pollination is required, as both inflorescences of the same plant are seldom simultaneously receptive (Turner and Gillbanks 1974). However, a series of observations conducted by Syed (1979) in Cameroon and Malaysia has confirmed the role of insects in oil palm pollination.

Although male and female flowers are in the same individual tree, the male and female flowers usually bloom at different times. The role of pollinating insects is very important, as the structure of the oil palm flower is very unique. The crowns on the female petals serve as a covering, requiring pollinating agents capable of passing through them. Moreover, both male and female oil palm flowers release a noticeable and potent scent which acts as an allure for insects. Male flowers are arranged in spikelet and are located above female flowers, while female flowers are arranged in bunches. Male flowers are very few in one spikelet, and the anthesis starts from the bottom to the top. Even though the anthesis period is shorter than the female flower, it gradually alternates from the bottom up, while female flowers have a longer receptive period of three to five days. At the onset of anthesis in the male inflorescences, pollinating insects begin to arrive in small numbers, reaching maximum at full anthesis, and begin to decline thereafter (Syed et al. 1982).

12.2.2.2 Insect Pollination in Palms

Wood (1874) reported that the hidden virtues of the weevils have yet to be discovered; they are so deeply hidden that they are quite unknown. His optimism has been recognized by the recent knowledge that among their hidden virtues can be counted their importance in the pollination of palms. Flower pollination can occur through two main mechanisms: biotic, which involves pollination carried out by animals, and abiotic, which involves non-living factors, such as wind or water, as the pollinating agents. The success of the angiosperms owes a great deal to entomophily (insect pollination) (Pellmyr et al. 1991), and its development in some of the palms has led to great diversity of pollination mechanisms.

Palms exhibit several features commonly associated with wind pollination (Bertin 1989), including reduced attractive floral components, minimal production of nectar, faint scent, and the generation of copious amounts of pollen grains. Additionally, the pollen of some palms may possess adhesive properties, indicating potential insect pollination.

A formerly widely accepted view was that anemophily was of greater importance in palms than entomophily. The excellent review by Henderson (1986) leads to the conclusion that pollen transfer by insects, especially beetles, predominates in the palm family (Corner 1966). Over the past few decades, there has been a notable change in perspective regarding the role of insects in pollinating palms. Specifically, the shift has occurred from having a certain opinion to acknowledging a state of limited understanding (Schmid 1970a, 1970b; Essig 1971, 1973).

Uhl and Moore (1977) studied the floral structure and pollination in Thrinax, Phoenix, Nypa, Ptychosperma, Bactris, and Asterogyne. They observed that the last four species were primarily entomophilous. The initial two species exhibited anemophily, yet they still possessed certain characteristics that appeared to be connected to entomophily. During the study, only approximately 5%

of genera had their floral biology thoroughly documented. Several of the mentioned traits could also serve as defenses against herbivores.

Pollinating insects and flowers engage in a mutualistic association, which is facilitated by factors such as the scent and color of the flower, as well as the nutritional value of nectar and pollen (Harborne 1988). The flowers of palms have odors which are important where visual stimuli are lacking, and many insects are very sensitive to low concentrations of volatile chemicals (Olesen and Balslev 1990). The flowers may be of delicate hues, especially in contrast to many brightly colored tropical blooms. Beetle pollinators often prefer dull, cream, or greenish flowers and, having a poor color sense, are often more reliant on other stimuli to attract them to inflorescences.

The nutritive value of the tissues of the reproductive structures is also important, which may be low in both the chemical and the physical defenses associated with older tissues. The presence of nectar and pollen is also important for herbivores, providing a significant amount of energy, amino acids, and lipids (Harborne 1988). Pollen provides balanced foodstuff for beetles (Bertin 1989).

TABLE 12.1
Insects Associated with Palm Pollination

Palm Subfamily and Tribe	Palm Genus	Insects associated with Palm Flowers and Floral Characteristics Suggestive of Entomophily
Coryphoideae		
Corypheae	*Chelyocarpus*	Odor probably attractive to beetles.
	Cryosophila	*Derelominus* sp.; *Mystrops* sp. Strong lilac smell. Inflorescence temperature elevation.
	Itaya	*Trigona* sp. bees dominate. Beetles and flies also important.
	Thrinax	Beetles and thrips on *Thrinax excelsa*. Pink scented flowers.
	Trachycarpus *Rhapidophyllum*	*Trachycarpus fortunei* possesses non-sepal nectaries. Visited by *Notolomus* sp. (Derelomini). Flowers may be brightly colored, with dull bracts. Musky odor, especially from male flowers.
	Chamaerops.	Two pollen-eating *Derelomus* species recorded. *Chamaerops humilis* sometimes secretes nectar.
	Rhapis	*Derelomusueoni* breeds in *Rhapis excelsa.*
	Livistona	Nectaries present.
	Johannesteijsmannia	Different species have flowers, with odors ranging from sweet to sour. Nitidulid and staphylinid beetles, flies, thrips, ants, and termites probably involved in unspecialized pollination.
	Licuala	Nectar secretion.
	Pritchardia	Flowers visited by bees, wasps, and other insects. Nectar secretion in *Pritchardia martioides* and *Pritchardia rockiana.*
	Serenoa	*Notolomus* sp. (Derelomini) and *Exoproso pafascipennis* (Bombyliidae) found on inflorescences.
	Corypha	*Chrysomyia* spp. (Diptera: Calliphoridae). Yellow flowers; odor offensive or of sour milk. Septal nectaries.
	Sabal	*Sabal palmetto* visited by a range of insects, initially bees and flies, later by Coleoptera and Lepidoptera. *Apis mellifera*, although exotic, thought to be the major pollinator. Continuous nectar and odor production throughout the day. Septal nectaries.
Phoeniceae	*Phoenix*	*Apis mellifera* attracted to *Phoenix reclinata* and many bees to *Phoenix caespitosa*. Male flowers scented, and nectaries possibly present.
Borasseae		Sepal nectaries.
Calamoideae		
Calameae	*Eugeissona*	Visited by a wide range of arthropods. Trigonid bees visit male flowers. Alcoholic smell resulting from fermentation of nectar.

(Continued)

TABLE 12.1 (*Continued*)
Insects Associated with Palm Pollination

Palm Subfamily and Tribe	Palm Genus	Insects associated with Palm Flowers and Floral Characteristics Suggestive of Entomophily
	Salacca	*Derelomus* species on male flowers of *Salaccazalacca* before visiting females, where they feed on nectar-like material. Females oviposit in female flowers, leaving the next day.
	Daemonorops	Musty odor, crowded flowers, many beetles present. *Calamus*. Musty or sour odor; crowded flowers, beetles, wasps, and flies.
	Calospatha	Musty odor, crowded flowers, many beetles present.
	Ceratolobus	Penetrating odor. Staphylinid beetles, ants, thrips, and Lepidoptera recorded on *Ceratolobus glaucescens*. Inflorescence enclosed, except for small slits to let pollinators through.
	Plectocomiopsis	Fragrant odors.
	Plectocomia	Fragrant odors. Visited by insects, especially weevils.
Nypoideae	*Nypa*	Visited by *Trigona* spp. and flies of the Drosophilidae. The bees were rarely found on female flowers, and the flies carried significant quantities of pollen. Dipteran larvae and occasional beetle larvae bred in the female flowers. Flower buds brightly colored; pollen orange and sticky, and flowers with a distinct odor.
Ceroxyloideae		
Cyclospatheae	*Pseudophoenix*	Many bees attracted by nectar.
Ceroxylyeae	*Ceroxylon*	Many melyrid beetles found on male flowers of *Ceroxylon mooreanum*.
	Ravenea	Curculionid and nitidulid beetles visit *Ravenea louvelii* and *Ravenea dransfieldii*. Inflorescence heating occurs, and musty aroma. Other species, for example, *Ravenea sambiranensis* and *Ravenea madagascariensis*, probably bee-pollinated.
Hyophorbeae	*Hyophorbe*	Intense fragrance; bright orange flowers and sepal nectaries.
	Synechanthus	Male flowers of *Synechanthus warscewiczianus* visited by many drosophilid flies, which later visited female flowers.
	Chamaedorea	The bees of *Trigona tataira* and *Cholus* spp. weevils visited the flowers of *Chamaedorea wendlandiana*, but only the former were recorded as having carried pollen. Visitors to other palms include trigonid and halictid bees, chrysomelid beetles, and drosophilid flies. Many palms of this genus have highly scented, brightly colored flowers with sticky pollen.
Arecoideae		
Caryoteae	*Arenga*	*Apis mellifera* visit scented male flowers of *Arenga tremula*. Sepal nectaries.
Iriarteae	*Iriartella*	Ants visit inflorescence.
	Iriartea	*Iriarteaventricosa* and *Iriartea gigantea* visited by *Trigona* sp. bees. *Melipona* and *Apis* species also recorded. Beetles of secondary importance.
	Socratea	*Socrateaexorrhiza* considered beetle-pollinated and visited by large numbers of *Phyllotrox* sp. (Derelomini) and *Mystrops* spp. (Nitidulidae), including *M. basalis*, *M. Dufaui*, and *M. corpulentus*. Sepal nectaries in *S. exorrhiza*.
	Catoblastus	*Phyllotrox* sp. and *Mystropsadustus* on *Catoblastuskalbreyeri*.
	Wettinia	*Mystrops basalis*, *Mystropsteapensis*, *Phyllotrox* sp., flies and bees on male flowers of *Wettiniahirsuta*.

TABLE 12.1 (*Continued*)
Insects Associated with Palm Pollination

Palm Subfamily and Tribe	Palm Genus	Insects associated with Palm Flowers and Floral Characteristics Suggestive of Entomophily
Podococceae	*Podococcus*	Sepal nectaries.
Areceae	*Manicaria*	Larvae probably breed in the flowers. Inflorescences with netlike bracts, which restrict entry of insects larger than pollinating nitulid beetles.
	Leopoldinia	Scented male flowers.
	Dypsis	Bee pollination in some *Dypsis* spp. Others have tiny specialized flowers for small pollinators not yet studied.
	Euterpe	Bees visited male and female flowers. Petals purple and nectar-like substance produced. Beetles of secondary importance.
	Prestoea	*Prestoeadecurrens* visited by *Trigona*, Neucorynura, and *Lasioglossum* species bees and one species of halictid bee. Small flies may also visit.
	Hyospathe	Few insect visitors, mainly ants, beetles, and bees.
	Archontophoenix	*Trigona Amalthea* considered major pollinator of *Archontophoenix cunninghamiana*
	Rhopalostylis	Flies (mainly) attracted to exudates of male and female lowers.
	Calyptrocalyx	Sweet scent, nectar produced.
	Ptychosperma	*Ptychosperma macarthurii*is visited by *Trigona, Nomia*, and *Homalictus* species bees and flies of the Syrphidae, Calliphoridae, and Drosophilidae. *Nomia* sp. considered most important pollinator. Nectar secreted; pollen heavy and sticky.
	Hydriastele	Bees of the genera *Nomia*, Homalictus, and *Trigona* visited male flowers and collected pollen but did not visit receptive female flowers. Drosophilid flies visited but considered unimportant. *Nodocnemus* spp. Weevils (Derelomini) present. Musky odor of flowers, and white sticky pollen.
	Nenga	*Nenga gajah* has a marked odor and is visited by nitulids and trigonids.
	Pinanga	Nitidulids and curculionids present on flowers of *Pinangac oronata*. Musty odor.
	Areca	Pollen of *Areca catechu* collected by bees and other insects, which did not visit female flowers. Sweet-scented flower.
	Iguanura	Male flowers attracted ants, flies, bees, wasps, and weevils. Fewer insects visited the female flowers.
	Marojejya	Inflorescence structures similar to those of palms known to be beetle-pollinated
	Cocoeae	*Butia leiospatha* pollinated by bees, wasps, flies, and curculionid and nitulid beetles.
	Cocos	*Cocos nucifera* inflorescences are visited by many insects, but only a few are common to both male and female flowers. *Derelomorphuseburneus*, apparently restricted to coconut flowers, may be an important pollinator. Nectaries present, and flowers sweetly scented. *Apis mellifera*.
	Attalea	Beetle-pollinated.
	Syagrus	Beetle or bee pollination predominates.
	Orbignya	*Mystrops* sp. found in numbers on *Orbignya martiana* flowers. Scented with nectaries.
	Maximiliana	*Maximiliana martiana* visited by *Melipona* sp. Bees.

(*Continued*)

TABLE 12.1 (*Continued*)
Insects Associated with Palm Pollination

Palm Subfamily and Tribe	Palm Genus	Insects associated with Palm Flowers and Floral Characteristics Suggestive of Entomophily
	Barcella	*Barcella odora* flowers are sweetly scented and visited by bees, flies, and wasps.
	Elaeis	Insect pollination important, especially *Elaedobius* spp. (Curculionidae). Female flowers strongly, but intermittently, scented.
	Acrocomia	Curculionids, nitulids, and scarabs implicated in pollination.
	Aiphanes	Bees, flies, and beetles implicated, depending on location.
	Bactris	Bee and beetle pollination important. Flowers with musky odor. Scarabs important.
	Astrocaryum	*Astrocaryum alatum* reported as beetle-pollinated. The scarab *Cyclocephala stictica* and *Mimeoma acuta* visited the flowers.
Geonomeae	*Welfia*	*Welfia georgii* visited by six *Trigona* species.
	Asterogyne	*Asterogyne martiana* is visited by many different insects. Weevils feed and breed in the flowers, but syrphids may be the major pollinators. Flowers cream- or white-colored, nectariferous, with a sweet smell. *Asterogyne spicata* has bright-orange male flowers with a strong scent which are visited by bees.
	Geonoma	Drosophilid flies may be important. Scented flowers, nectaries present.
Phytelephantoideae	*Phytelephas*	*Phytelephas microcarpa* visited by bees, flies, staphylinids, nitulids, and curculionids. One identified nitulid was *Mystropsteapensis*. The curculionids were thought to be the major pollinators. Strongly scented flowers. Scarabs also thought to be important pollinators.
	Palandra	Scarabs, *Cyclocephala aequatoria* and *Mystrops costaricensis* reported on male flowers of *Palandra aequatorialis*.
	Ammandra	Similar to *Phytelephas* and *Palandra* species. Curculionids, staphylinids, nitulids, and hydrophilid beetles found on male flowers.

According to Henderson (1986), the majority of palm tree pollination through entomophily can be categorized into three groups: beetle, bee, or fly pollination. Limited knowledge exists regarding pollination by Lepidoptera. Cantharophily, which involves beetles such as Nitidulidae and Curculionidae, is regarded as the most significant form of entomophily in palms, as stated by Henderson (1986).

Genera within these significant beetle families seem to show a specific affinity for palm flowers, and this level of specificity can be quite pronounced (Syed 1979, 1982). Other beetles, such as scarabs, may also be of importance, but without exhibiting the same degree of specificity (Faegri and van der Pijl 1979).

Cantharophily, just like bee pollination in temperate regions, might be a distinct characteristic of tropical areas (Faegri and van der Pijl 1979). Palms evolved early in evolutionary times and are at least as old as any other group of flowering plants (Corner 1966; Uhl and Dransfield 1987). *Sabal magothiensis* and *Palmoxylon cliffwoodensis* are the oldest palm species, whose fossil records have been definitively confirmed. These fossils date back approximately 84 million years to the Upper Cretaceous period, as documented by Uhl and Dransfield in 1987. Additionally, there are known Nypapollen fossils from the same period. In the Jurassic period, there exist fossilized beetles that bear resemblance to present-day *Floricolous* species, as noted by Crowson in 1981.

Evidence shows that palm and beetle association has long evolved in evolutionary history. Beetle pollination is widespread among the more primitive angiosperms. It may have promoted the development of the angiosperm-type flower. What are now pollination attractants may have developed first as anti-herbivore defenses (Vail 1992). It is suggested that some of the earliest beetles may have functioned as pollinators (Crowson 1981). However, fossil evidence also suggests that pollen-eating Lepidoptera and Diptera also appeared during the Cretaceous (Silberbauer-Gottsberger 1990).

Faegri and van der Pijl (1979) reported that beetles were often deceived by false attractants. In numerous palm species, the female flowers do not provide any nutritional reward, and pollination is accomplished through deceptive mechanisms (Olesen and Balslev 1990). Even when deceived into visiting flowers, beetles likely derive some benefits, such as finding shelter from predators or adverse weather conditions, and warmth from the actively metabolizing tissues within the flowers. Flowers pollinated by beetles typically have a limited number of visual cues, which is why they are often large in size or clustered together in large inflorescences. They are usually dull-colored or white. They have easy access, being open and actinomorphic.

These flowers possess strong odors and offer accessible food rewards, often having numerous stamens that produce abundant pollen (Faegri and van der Pijl 1979; Crowson 1981). According to Irvine and Armstrong (1990), beetle-mediated pollination can be categorized into two main types based on feeding behavior. The first type involves pollen and nectar feeders, which typically cause minimal to no damage to the flowers during their visitation.

The second type mainly consists of herbivores that feed on floral tissues, pollen, or nectar. These flowers typically have mature fruits or emit musky scents. While generalist foragers are responsible for a significant amount of pollination, there are also instances of highly specialized mutualism, such as the pollinators of African oil palm. Bees are the best-known pollinators, and melittophily dominates biotic pollination in many areas. Bees are considered to be important in the pollination of coconut palm (Sholdt 1966; Sholdt and Mitchell 1967). Stingless bees such as *Trigona* spp. have been observed on flowers of various palm species. In general, other than the ubiquitous occurrence in inflorescences of many palm, the value of bees as pollinators is presumed from such anecdotal information as the fact that tropical honey is largely dependent on palms (Corner 1966).

Myophily (pollination by Diptera) shows a range of forms, from the highly specialized to the opportunist. Several insects are attracted to simple, pale blossoms with easily accessible nectar. These flowers often have minimal scent. In oil palm, the process of fruit set formation relies directly on pollination. Pollen transfer from the male to the female inflorescences is crucial for fertilization and subsequent fruit development. Efficient pollination can lead to a remarkable increase in oil palm yield, with potential gains of up to 50% (Rajamohanan et al. 2017).

Pollination is affected by several factors, such as the abundance and diversity of pollinators, the quality and quantity of pollen, and the timing of pollination relative to flowering (Prada and Romero 2013). Efficient pollination results in greater proportions of normal or fertile fruits and increased oil production due to reduction in parthenocarpic fruits and aborted flowers (Corley and Tinker 2003). Timely and efficient pollination is very essential because of very short life span of less than 24 h in oil palm flowers (Klein et al. 2007). Pollination of oil palms in Cameroon was carried by a subfamily of weevils, the Derelominae (Henderson 1986; Syed 1979). These weevils feed on the soft parts of male flowers and on pollen and lay their eggs on male flowers; the larvae feed on the decaying flowers. An individual weevil remains for a very short time on the female inflorescence, which probably explains why their role was not recognized earlier. The genus *Elaeidobius*was absent from Southeast Asia, and in Malaysia, the main pollinator was *Thrips hawaiiensis* (Syed 1979); this species was usually absent from young plantings, explaining why assisted pollination was required. In Sabah, although *T. hawaiiensis* was present, it had apparently not adapted to the oil palm and did not contribute to pollination, but some pollen was dispersed by a moth, *Pyroderces* sp. (Syed 1979; Tan and Basri Wahid 1985). *T. hawaiiensis* was not a very efficient pollinator, and Syed (1982) considered that *E. kamerunicus* could usefully be introduced to Malaysia. Kang (1999) reported that the weevil was released on two plantations in Malaysia in February 1981, and population increase and spread were very rapid (Syed et al. 1982). Subsequently, releases were carried out in Indonesia, Papua New Guinea, and Colombia.

In Latin America, two pollinating beetles have been found on both *E. guineensis* and *E. oleifera*. These are the sap beetle, *Mystrops costaricensis*, which is common in areas of high, even rainfall, and the weevil, *Elaeidobius subvittatus*, a native of Africa, which is thought to have been fortuitously introduced, perhaps transported in pollen samples. However, the exact status of these two insects and the reasons for their uneven distribution are not fully understood.

12.2.3 Palm Pollinators

The number of insects visitors in the inflorescence varies from few to large numbers (Borchsenius 1997; Barfod et al. 1987). The number of visiting insect species and genera recorded for nine species of palms are as follows: The staminate inflorescence in Phytelephas species attracts thousands of insects. Barfod et al. (1987) thus recorded 6,000 stypyilinid beetles, 3,500 curculionid beetles (*Phyllotrox*), and 100 Nitidulid beetles (*Mystrops*) in *Phytelephas tenuicaulis.* Bernal and Ervik (1996), on *Phytelephass* eemannii, recorded 26 species of insects belonging to 13 genera. Borchsenius (1993) recorded 28 species of insects belonging to 24 genera on *Aiphaneserinacea*. Similarly, Borchsenius (1997) recorded 7 species belonging to 7 genera on *Geneomairena*. Búrquez et al. (1987), on the other hand, recorded 29 species belonging to 26 genera on *Astrocaryum mexicanum*. The number of insect species recorded on *Sabal etonia* was 30, belonging to 29 orders (Zona 1987). Khorsand Rosa and Koptur (2009) found 5 species belonging to 5 genera on *Cocothrinax argentata*. Listabarth (2001) found 60 species of insects belonging to 30 orders on *Hyospathe elegans*. Barfod et al. (2003) found *Licuala spinosa* attracting 29 species belonging to 29 orders. *Calamus rudentum* attracted 12 species belonging to 3 genera Bøgh (1996). New groups of visitors to palm inflorescences have also been recorded, such as Seba's short-tailed bat (*Carollia perspicillata*) and Commissaris's long-tongued bat (*Glossophaga commissarisi*) in *Calyptrogen eggiesbreghtiana* (Cunningham 1995; Tschapka 2003), pen-tailed trees hrew (*Ptilocercus lowii*) in *Eugeissona tristis* (Wiens et al. 2008), the Mexican mouse opossum (*Marmosa mexicana*) in *Calyptrogyne ghiesbreghtiana* (Sperr et al. 2009), and even the nectar-eating grapsid crab (*Armases* cf. *miersii*) in *Prestoea decurrens* (Ervik and Bernal 1996). Listabarth (1993*a*) reported that dioecious thrips and minute beetles of the family Ptiliidae play an important role in the pollination of *Chamaedorea pinnatifrons*. They enter the staminate flowers through small basal slits between the petals and disburse pollen by their activities.

As the female flowers are rarely visited by the same insects as the male flowers, Listabarth concluded that pollen grains are transferred by wind and suggested the name "insect-induced wind pollination" for the mechanism. Berry and Gorchov (2004) observed mixed insect-wind pollination in *Chamaedorea radicalis*. They discovered that fruit set increased from 3% in inflorescences covered with bags (to exclude insects) to 23% in inflorescences on palms that were not treated and allowed both insects and wind to contribute to pollination.

12.2.4 Beetle Pollinators

The most important beetle pollinators of palms are weevils (family Curculionidae), which comprise more than 48,000 species. The species that visit palm inflorescences are predominantly active during the nighttime. They seek shelter in the partially enveloping bracts of the peduncle and rachis during the daytime. These insects typically lay their eggs in the less-protected areas of the inflorescence. Among the visitors to palm inflorescences, members of the tribe Derelomi are commonly observed. A comprehensive study on *Elaeidobius kamerunicus*, the weevil species that visits African oil palm, has revealed its ability to carry a significant amount of pollen grains due to various types of setae (Dhileepan 1992). Other weevils belonging to subfamilies such as Dryophthorinae (formerly Rhynchophorinae) and Baridinae utilize palms as hosts for their offspring and are recognized as significant threats to palm health (Howard et al. 2001). Additionally, sap beetles, scarabid beetles, and rove beetles also play important roles in palm pollination.

One of the main pollinators of oil palm is *Elaeidobiusk amerunicus* (Syed 1982). This beetle is attracted by the scent of male flowers, which are stronger in aroma than female flowers. The pollen will stick on the body parts, and when moving on female flowers, the pollen is released to pollinate them (Syed 1982). *E. kamerunicus* beetles are very effective pollinators because of their shape, structure, and size, which match the size and structure of oil palm flowers, supported by high populations due to their proliferation in male oil palm flowers (Syed 1982), and have behavior that supports its function as a specialist pollinator on oil palm. Pollination of oil palm is strongly influenced by the conditions of *E. kamerunicus*, pollen conditions, and the environment that affects the pollination interaction process (Purba et al. 2010).

Several studies have investigated the floral biology and pollination efficiency of the subject (Sayed 1979; Pillai and Ponnamma 1992). Palms have been reported as wind-pollinated; however, Henderson (1986) reported that insects also play a significant role in palm pollination. According to Henderson (1986), three modes of pollination are most common in palms, such as canthrophily (beetle pollination), mellitophily (bee pollination), and myophily (fly pollination). In palms pollination phenomena, it is very complex, as palm inflorescences are borne laterally on the stem at various heights (Evans et al. 2001), and interactions between pollen vectors vary in time and space.

Interestingly, flowers within one inflorescence open in different sequences, and this variation is considered to be related to beetle pollination on one hand and fly and wasp on the other Henderson (2002). Barfod et al. (2003) reported that exceptions do occur, as *Licuala peltata* is pollinated by trigonid bees and not by beetles.

The number of insect visitors in the inflorescence varies from less than 10 Geonomairena (Borchsenius 1997) to as much as 10,000 (Barfod et al. 1987) in *Phytelephas tenuicaulis*. Beach (1984) recorded the highest number of weevils between 40,000 and 100,000 on a single inflorescence of *Bactris gasipaes* at the time of anthesis. Barfod et al. (1987) recorded 6,000 stypyilinid beetles, 3,500 curculionid beetles (*Phyllotrox*), and 100 Nitidulid beetles (Mystrops) in Phytelephas tenuicaulis. Bernal and Ervik (1996) recorded 26 species of insects belonging to 13 genera on Phytelephas seemannii. Borchsenius (1993) recorded 28 species of insects belonging to 24 genera on Aiphaneserinacea. Similarly, Borchsenius (1997) recorded 7 species belonging to 7 genera on *Geneomairena*. Búrquez et al. (1987), on the other hand, recorded 29 species belonging to 26 genera on *Astrocaryum mexicanum*. The number of insect species recorded on *Sabal etonia* was 30, belonging to 29 orders (Zona 1987).

Khorsand Rosa and Koptur (2009) found 5 species belonging to 5 genera on *Cocothrinax argentata*. Listabarth (2001) found 60 species of insects belonging to 30 orders on *Hyospathe elegans*. Barfod et al. (2003) found *Licuala spinosa* attracting 29 species belonging to 29 orders. Bøgh (1996) reported 12 species belonging to 3 genera visiting *Calamus rudentum* inflorescences.

Savolainen et al. (2006) demonstrated wind pollination in two sympatric species of Howea, endemic to the Lord Howe Islands. Listabarth (1993a) recorded a very noble mechanism of pollination in *Chamaedorea pinnatifrons*, where pollination is done by dioecious thrips (Thysanoptera) and minute beetles of the family Ptiliidae. Since female flowers are rarely visited by the same insects as male flowers, Listabarth (1993a) suggested that pollen grains are transferred by wind, referred to as "insect-induced wind pollination" mechanism.

Berry and Gorchov (2004) reported an increase of 3% fruit set in bagged inflorescences to 23% in palms in inflorescences that were treated to exclude insects, indicating the role of wind pollination in *Chamaedorea radicalis*. The most important pollinators are weevils, which constitute a very large group of living organism on earth.

Dhileepan (1992) conducted a detailed study on *Elaeidobius kamerunicus*, the main pollinator of the African oil palm, and found that this species has a very large pollen-carrying capacity due to a large number of setae.

Sap beetles constitute another group of visitors to palm inflorescences which has more than 3,000 species. These beetles mostly feed on decaying matter, but two species, *Mystrops* and *Epurea*, are frequent visitors to palm inflorescences Anderson et al. (1988).

Scarabid beetles, which comprise of more than 30,000 species, are yet another group of beetles which frequently visit palm inflorescences. They are renowned as borers of palms (Howard et al. 2001).

Another important group of beetles, the scarabid beetle *Cyclocephalaamazona*, was the most frequent visitor on the inflorescences of *Bactris gasipaes* in Costa Rica Beach (1984). A large number of scarabid beetles are attracted to inflorescences during pistillate anthesis and feed on staminate flowers. However, scarabid beetles have been considered to be of secondary importance in the pollination of *Bactris gasipaes* (Mora Urpí and Solis 1980).

The palm inflorescences are also found to attract rove beetles, which are a very large group, comprising of more than 46,000 species in 3,200 genera. They mostly prey on small invertebrates, and their role as pollinators is minor due to their moderate pollen-carrying capacity.

Bernal and Ervik (1996) found that staphylinid beetles were important pollinators of *Phytelephas seemannii* (Ashe 1984, 1987). Both staphylinid and curculionid beetles oviposit in the inflorescences to complete their life cycle in less than a week.

12.2.5 Bee and Fly Pollinators

Bee and flies represent a very predominant group of pollinators on palms in tropics and warm climates (Michener 2007). Bees constitute an important group as they rely on nectar and pollen to feed their offspring (Carrington et al. 2003). In case of bee-pollinated palms, conservation of bee nests is most important to ensure the reproductive success of plants.

Trigona bees are particularly the most important pollinators of palms attracted to inflorescences for nectar and pollen. They have pollen baskets (corbiculae) to collect pollen, which they use for mass provisioning of their brood.

Borchsenius (1997) reported that drosophilid flies are the primary pollinator of palms. Hoverflies and blow flies have been documented as important pollinators of palms according to studies conducted by Borchsenius (1997), Listabarth (2001), and Barfod et al. (2003). Siregar et al. (2016) found 497 individuals of insect pollinators belonging to 43 species, three orders, and seven families visiting palm flowers. Hymenopterans were the most abundant, followed by dipteral and Lepidoptera. Insect pollinators in oil palm plantations were dominated by giant honeybee (*Apisdorsata*) and stingless bee (*Trigona* sp. [= aff. *T. planifrons*]), whereas in rubber plantation, they were dominated by small carpenter bees (*Ceratinalieftincki* and *Ceratina simillima*) and, in jungle rubbers, were dominated by hoverfly (*Syrphid* sp.) and *Apis andreniformis*. Foraging activities of insect pollinators were highest during morning hours.

The studies reveal that several new groups of visitors are associated with palm inflorescences, such as bats, pen-tailed tree shrew, and Mexican mouse opossum and a decapod crustacean species (grapsid crab). Nevertheless, despite the significant number of weevils observed during the anthesis period in West Africa, beetles, bees, and flies continue to be the prevailing flower visitors.

12.2.5.1 Coleoptera

In the oil palm's native West African range, several weevil species of genus *Elaeidobius* (Coleoptera: Curculionidae) (Table 12.1) are found in oil palm inflorescences. Adults feed on flower parts and pollen in the male inflorescence, while females oviposit in male flower structures, on which the larvae develop and feed as it decomposes (Syed 1982; Mariau et al. 1991).

During male inflorescence anthesis, more than 100,000 flowers open gradually over two to four days, ceasing pollen production after five days (Corley and Tinker 2016). In West Africa, researchers have reported an observation of more than 8,000 weevils from various species per inflorescence during the anthesis period (Mariau et al. 1991). Similarly, in India, the introduction of weevils resulted in the presence of 2,000–3,000 individuals per inflorescence (Tandon et al. 2001). Weevils are attracted to receptive female flowers due to an aniseed-like scent that resembles the fragrance of male flowers (Adaigbe et al. 2011).

As female inflorescences have no resources, weevils only stay briefly; however, 5,000–12,000 visits can occur during the receptive period (Syed 1979; Chinchilla and Richardson 1991; Mariau et al. 1991).

Elaeidobius species differ in pollen-carrying ability and transported pollen viability (Table 12.1), as well as search capacity (Syed 1982). *Elaeidobius kamerunicus* was introduced to Southeast Asia in the 1980s due to its robustness in long rainy seasons, host specificity to oil palm, and high pollen-carrying capacity (Syed et al. 1982). This species was also introduced in tropical America, though *E. subvittatus* was already naturalized there accidentally (Chinchilla and Richardson 1991). A few other *Elaeidobius* species (Table 12.1) were introduced within limited areas of tropical America as well (Meléndez and Ponce 2016).

The Nitidulid beetle *Mistrop costaricensis* (Table 12.1) is a native pollinator in regions of tropical America that feeds on oil palm pollen (Syed 1984). While its occurrence on female inflorescences is less than 1% of the total population of 27,000–38,000 individuals found on male inflorescences, the beetle carries a highly viable pollen (70% viability) and continues to be active on female inflorescences beyond its diurnal foraging period. This behavior could potentially enhance pollination (Chinchilla and Richardson 1991).

Its rain tolerance may allow it to compensate for seasonal *E. kamerunicus* decreases in very wet regions (Syed 1984), though foraging time and rain resistance vary between regional sub-species, and it is still affected by heavy rains (Mariau and Genty 1988).

Other Coleopteran genera found on African oil palms include *Prosoestus* (Curculionidae), *Atheta* (Staphylinidae), and *Microporum* (Nitidulidae). However, the amount of pollen carried by these species is relatively limited and lower in quality (Mariau et al. 1991; Kouakou et al. 2014).

12.2.5.2 Thysanoptera

In Southeast Asia, the Thysanopteran *Thrips hawaiinensis* (Table 12.1) was seen as the main pollinator before *E. kamerunicus* was introduced (Syed 1979). Syed (1979) found that a male flower spikelet could contain up to 1,000 individuals (with up to 200 spikelets per inflorescence). Many thrips may swarm female inflorescences (up to seven per flower), making continuous brief visits. On average, thrips may carry four to five pollen grains, with high (76%) viability. However, this species is absent from young plantations, perhaps due to difficulty flying in open and windy areas (Syed 1979).

12.2.5.3 Lepidoptera

The moth *Pyroderces* sp. (Lepidoptera: Cosmopterygidae) is another pollinator endemic to Southeast Asia (Table 12.1). This species oviposits on male inflorescences but also visits female inflorescences, often depositing pollen on them (Syed 1979). It is only active for 2–3 hr around sunset, which may reduce contribution to fruit set (Tan and Basri 1984). Syed (1979) estimated visitors to a female inflorescence in one night did not exceed 500 individuals.

12.2.5.4 Other Visiting Groups

True flies (Diptera) can also be found in oil palm inflorescences, but their role in pollination has not yet been assessed (Mayfield 2005; Hala et al. 2012).

12.2.6 Role of Insects

In the oil palm, insects have also been observed to be involved in pollination (Syed 1979; Abraham 1994), but their efficiency varies with the height of the canopy (Jyoti and Saradamani 1994). Generally in aged tall plants, pollination is successfully done by airborne pollen, whereas in the young palms, insects could enhance the pollination to a larger extent, especially in deeply borne flowers within the female inflorescences.

Several species of insects are associated with the male and female inflorescences of the oil palm (Syed 1979), and these insects are believed to assist in transferring pollen to the lower parts of the female inflorescence, which is protected by persistent bracts, and are, as such, not easily accessible to airborne pollen.

Syed (1979) reported that a large number of insects visited female inflorescences throughout the period of their receptivity, mainly during the daytime. These insects were observed to arrive in intermittent storms. The main insects observed by Syed (1979) on both male and female inflorescences were *Elaeidobius* spp. and *Atheta* spp.

The introduction of the insect *Elaeidobius kamerunicus* into Malaysia led to the cessation of the practice of assisted pollination, saving the plantations millions of dollars (Abraham 1994). Higher bunch weight and greater yield of fruits were also observed with the introduction of the pollinator insects (Syed 1994). Similarly, Donough and Law (1987) reported an increase in yield and profitability with the introduction of pollinator insects in Johore and Sabah. Basri (1984) had earlier reported that the introduction of *E. kamerunicus* into Malaysia led to a 36% improvement in fruit set, a 12% increase in fruit-to-bunch ratio, and a 28% increase in the mean bunch weight. He further reported a 9% increase in the oil-to-bunch ratio and an improvement of 43% in the kernel-to-bunch ratio.

12.3 INSECT POLLINATION AND EFFECTS ON PRODUCTION

Once *E. kamerunicus* was introduced to Southeast Asia in the early 1980s, insect-pollinated oil palms achieved higher fruit sets than previous hand-pollinated rates (Syed et al. 1982). The weevil's ability to penetrate deeper into the inflorescence compared to manually applied pollen, combined with its continuous visitation throughout the asynchronous blooming period, may explain the observed phenomenon (Harun and Noor 2002). In a Malaysian plantation, Syed (1984) conducted a comparison before (1981) and shortly after (1982) the introduction of *E. kamerunicus* and found an increase in fruit set from 48% to 71%. This was accompanied by improvements in fruit-to-bunch ratio, bunch weight, and a 48% increase in total oil production per bunch, equivalent to approximately 1 kg. Continued observations in Malaysia by Donough and Law (1987) confirmed these positive effects even five years after the introduction. Similar increases in yield were also observed in Indonesia (Hutauruk et al. 1984), India (Dhileepan and Nampoothiri 1989), and tropical America (Mariau and Genty 1988; Chinchilla and Richardson 1991; Moura et al. 2012). To achieve acceptable fruit set levels, certain weevil population parameters have been proposed, such as a minimum population density of 20,000 individuals per hectare, resulting in a 55% fruit set (Donough et al. 1996).

However, the relationships between pollinator population size, pollination services, and yield are not altogether linear or monotonic. While bunch weight and fruit set generally increased with weevil introduction, bunches produced per year decreased in some cases, though this was still offset by bunch weight (Taniputra and Muluk 1989; Harun and Noor 2002). Though oil content is closely related to fruit set, the ratio of oil to bunch weight only increases with fruit set up to about 75% and decreases above that (Harun and Noor 2002).

Appiah and Agyei-Dwarko (2013) reported that Iinsects belonging to the *Elaeidobius* spp. (*Coleoptera: Curculiondae; Derelominae*) are the main insects involved in the pollination of the oil palm. Other insects listed are the *Prosoestus* spp., *Microporium* spp. (*Coleoptera: Nitidilidae*), and *Atheta spp.* (*Coleoptera: Staphylinidae*). Insects generally carry pollen on their hairy bodies and wings. Larger and more hairy insects are generally more effective pollinators, leading to sufficient fruit set and increased crop yield. Sustainable production and conservation of pollinator insect populations can be achieved through the judicious application of insecticides, strategic timing of their use, and adopting suitable modes of application.

TABLE 12.2
Insect Visitors on Oil Palm Inflorescence

S. No.	Insect	Order: Family	Insect Visit on Inflorescences	
			Male	Female
1.	African oil palm weevil (*Elaeidobius kamerunicus*)	Coleoptera: Curculionidae	✓	✓
2.	Earwig Dermaptera	Forficulidae	✓	✓
3.	Moth Lepidoptera	Cosmopterigidae	✓	✓
4.	Leaf caterpillar (*Elymnias* sp.)	Lepidoptera: Nymphalidae	✓	x
5.	Citrus butterfly (*Papilio* sp.)	Lepidoptera: Papilionidae	✓	x
6.	Butterfly—yellows (*Eurema hecabe*)	Lepidoptera: Pieridae	✓	x
7.	Indian honeybee (*Apiscerana indica*)	Hymenoptera: Apidae	✓	✓
8.	Dammer bee (*Trigona iridipennis*)	Hymenoptera: Meliponidae	✓	x
9.	Black ant (*Camponotus* sp.)	Hymenoptera: Formicidae	x	✓
10.	Leaf-cutting bee (*Megachile* sp.)	Hymenoptera: Megachilidae	✓	x
11.	Cuckoo wasp	Hymenoptera: Chrysididae	✓	x
12.	Carpenter bee (*Xylocopa* sp.)	Hymenoptera: Xylocopidae	✓	x
13.	Giant hornet (*Vespa* sp.)	Hymenoptera: Vespidae	✓	x
14.	Mealy bug (*Paracoccus marginatus*)	Hemiptera: Pseudococcidae	x	✓
15.	Hoverfly	Diptera: Syrphidae	✓	x

Source: Sambathkumar and Ranjith (2011).

Sambathkumar and Ranjith (2011) studied the insect pollinators of oil palm in Kerala, India, and found that more than 15 insect species visit the inflorescence of oil palm (Table 12.2). Among them, population of the African oil palm weevil, *Elaeidobius kamerunicus* (Coleoptera: Curculionidae), was found to be most predominant. They further found that weevils of *E. kamerunicus* were congregating on the male inflorescence during the entire period of anthesis. Earlier, Syed (1984) reported that *E. kamerunicus* is highly specific to the oil palm and breed on spent male inflorescences. This specificity is due to the presence of some anise seed, like odor. Hussein et al. (1991) found that the host specificity of *E. kamerunicus* was due to presence of estragole (5, 4—allylanisole), a plant chemical in male flowers of oil palm. The presence of specialized hair-like structures of weevils (on the periphery of abdomen) and their frequent visit to female inflorescence could make them an ideal pollinating agent in oil palm.

The interaction between palms and their visiting insects represents a trade-off between the two. Some weevils completely depend on one palm species to complete their life cycle. Taking all this evidence into consideration, 29% of all palm species studied have been referred to as beetle-pollinated, 26% as bee-pollinated, 8% as fly-pollinated, 7% wind-pollinated, and about 3% as pollinated by mammals.

REFERENCES

Abraham, V. K. 1994. Insect pollination: Revolution by weevil. *Indian Oil Palm Journal* 3(9), 5–10.

Adaigbe, V., Odebiyi, J., Omoloye, A., Aisagbonhi, C., and Iyare, O. 2011. Host location and ovipositional preference of Elaeidobius kamerunicus on four host palm species. *Journal of Horticulture and Forestry* 3, 163–166.

Adam, H., Jouannic, S., Escoute, J., Duval, Y., Verdeil, J. L., and Tregear, J. W. 2005. Reproductive developmental complexity in the African oil palm (Elaeis guineensis, arecaceae). *American Journal of Botany* 92, 1836–1852.

Anderson, A. B., Overal, B., and Henderson, A. 1988. Pollination ecology of a forest dominant palm (*Orbignya phalerata* Mart.) in Northern Brazil. *Biotropica* 20, 192–205.

Appiah, S. O. and Agyei-Dwarko, D. 2013. Studies on Entomophil pollination towards sustainable production and increased profitability in the oil Palm: A review. *Elixir Agriculture* 55, 12878–12883.

Ashe, J. S. 1984. Major features in the evolution of relationships between gyrophaenine staphylinid beetles (Coleoptera: Staphylinidae: Aleocharinae) and fresh mushrooms. In: Wheeler, Q., and Blackwell, M. (eds.) *Fungus Insect Relationships; Perspectives in Ecology and Evolution*. Columbia University Press, New York, pp. 227–255.

Ashe, J. S. 1987. Egg chamber production, egg protection and clutch size among fungivorous beetles of the genus *Eumicrota* (Coleoptera: Staphylinidae) and their evolutionary implications. *Zoological Journal of the Linnean Society* 90, 255–273.

Barfod, A. S., Burholt, T., and Borchsenius, F. 2003. Contrasting pollination modes in three species of *Licuala* (Arecaceae: Coryphoideae). *Telopea* 10, 207–223.

Barfod, A. S., Henderson, A., and Balslev, H. 1987. A note on the pollination of *Phytelephasmicrocarpa* (Palmae). *Biotropica* 19, 191–192.

Basri, M. W. 1984. Developments of the oil palm pollinator, *Elaeidobius kamerunicus* in Malaysia. *Palm Oil Developments* 2, 1–3.

Beach, J. H. 1984. The reproductive biology of the peach or 'Pejibayé' palm (*Bactris gasipaes*) and a wild congener (*B. porschiana*) in the atlantic lowlands of Costa Rica. *Principes* 28, 107–119.

Bernal, R. and Ervik F. 1996. Floral biology and pollination of the dioecious palm *Phytelephasseemannii* in Colombia: An adaptation to staphylinid beetles. *Biotropica* 28, 682–696.

Berry, E. J. and Gorchov, D. I. 2004. Reproductive biology of the dioecious understorey palm Chamaedorearadicalis in a Mexican cloud forest: Pollination vector, flowering phenology and female fecundity. *Journal of Tropical Ecology* 20, 369–376.

Bertin, R. I. 1989. Pollination biology. In: Abrahamson, W. G. (ed.) *Plant—Animal Interactions*. McGraw-Hill, New York, pp. 23–86.

Bøgh, A. 1996. The reproductive phenology and pollination ecology of four Calamus (Arecaceae) species in Thailand. *Principes* 40, 5–15.

Borchsenius, F. 1993. Flowering biology and insect visitation of three Ecuadorean *Aiphanes* species. *Principes* 37, 139–150.

Borchsenius, F. 1997. Flowering biology of Geonomairena and G. cuneata var. sodiroi (Arecaceae). *Plant Systematics and Evolution*, 208187–208196.

Búrquez, A., Jose Sarukhán, K., and Pedroza, A. L. 1987. Floral biology of a primary rain forest palm, Astrocaryum mexicanum Liebm. *Botanical Journal of the Linnean Society* 94, 407–419.

Carrington, M. E., Gottfried, T. D., and Mullahey, J. J. 2003. Pollination biology of saw palmetto (Serenoa repens) in Southwestern Florida. *Palms* 47, 95–103.

Chinchilla, C. M. and Richardson, D. 1991. *Pollinating Insects and the Pollination of Oil Palms in Central America*. ASD Oil Palm Papers (Costa Rica), No. 2: 1–18.

Corley, R. and Tinker, P. 2003. *The Oil Palm*. 4th edn. Blackwell Science, Oxford, U.K., p. 608.

Corley, R. H. V. and Tinker, P. B. 2016. *The Oil Palm*. Wiley-Blackwell, Hoboken, NJ.

Corner, E. J. H. 1966. *The Natural History of Palms*. Weidenfeld and Nicolson, London.

Crowson, R. A. 1981. *The Biology of the Coleoptera*. Academic Press, New York, p. 802.

Cunningham, S. A. 1995. Ecological constraints on fruit initiation by Calyptrogyne ghiesbreghtiana (Arecaceae): Floral herbivory, pollen availability and visitation by pollinating bats. *American Journal of Botany* 82, 1527–1536.

Dhileepan, K. 1992. Pollen carrying capacity, pollen load and pollen transferring ability of the oil palm pollinating weevil Elaeidobiuskamerunicus Faust in India. *Oléagineux* 47, 55–61.

Dhileepan, K. and Nampoothiri, K. U. K. 1989. Pollination potential of introduced weevil, Elaeidobius kamerunicus, in oil palm (Elaeis guineensis) plantations. *Indian Journal of Agricultural Sciences* 59, 517–521.

Donough, C. R., Chew, K. W., and Law, I. H. 1996. Effect of fruit set on OER and KER: Results from studies at Pamol Estates (Sabah) Sdn Bhd. *Planter* 72, 203–219.

Donough, C. R. and Law, I. H. 1987. The effect of weevil population on yield and profitability at Pamol Plantations. In: Proceedings of the International Oil Palm Conference, Kuala Lumpur, Malaysia.

Ervik, F. and Bernal, R. 1996. Floral biology and insect visitation of the monoecious palm Prestoea decurrens on the Pacific coast of Colombia. *Principes* 40, 86–92.

Essig, F. B. 1971. Observations on pollination in Bactris. *Principes* 15, 20–24, 35.

Essig, F. B. 1973. Pollination of some New Guinea palms. *Principes* 17, 75–83.
Evans, T. D., Sengdala, K., Viengkham, O. V., and Thammavong, B. 2001. *A Field Guide to the Rattans of Lao PDR*. Royal Botanic Gardens, Kew, UK.
Faegri, K. and van der Pijl, L. 1979. *The Principles of Pollination Ecology*. 3rd edn. Pergamon Press, Oxford, UK.
F.A.O. of the U.N. (FAO). Why Bees Matter. Available at www.fao.org/3/I9527EN/i9527en.PDF (accessed on 20 February 2023).
Hala, N., Tuo, Y., Akpesse, A. A. M., Koua, H. K., and Tano, Y. 2012. Entomofauna of oil palm tree inflorescences at La Mé experimental station (Côte d'Ivoire). *American Journal of Experimental Agriculture* 2, 306–319.
Harborne, J. B. 1998. Textbook of phytochemical methods. In: *A Guide to Modern Techniques of Plant Analysis*, 5th edn. Chapman and Hall Ltd, London, pp. 21–72.
Hardy, O. J., Fernandez, E., and Nordine, O. 2020. Elaeisguineensis (oil palm). In: *Fruit Trees and Oil Producing Trees*. Springer, Berlin/Heidelberg, Germany, pp. 17–27.
Hartley, C. 1988. *The Oil Palm*, 3rd edn. Editorial Longman, London, p. 958.
Hartley, C. 2003. Oil palm breeding: Genetics and genomics. *Ann. Appl. Biol.* 142, 107–120.
Harun, M. H. and Noor, M. R. M. 2002. Fruit set and oil palm bunch components. *Journal of Oil Palm Research* 14, 24–33.
Henderson, A. 1986. A review of pollination studies in the Palmae. *The Botanical Review* 52, 221–259.
Henderson, A. 2002. *Evolution and Ecology of Palms*. New York Botanical Garden Press, New York.
Hoppe, L. E. 2005. The pollination biology and biogeography of the mangrove palm NypafruticansWurmb (Arecaceae). MSc Thesis. wAarhus University Denmark Howard.
Howard, F. W., Moore, D., Giblin-Davis, R. M., and Abad, R. G. 2001. *Insects on Palms*. CABI Publications, Wallingford, UK.
Hussein, M. Y., Lajis, N. H., and Ali, J. H. 1991. Biological and chemical factors associated with the successful introduction of Elaeidobius kamerunicus Faust, the oil palm pollinator in Malaysia. *Acta Horticulture* 288, 81–86.
Hutauruk, C., Sipayung, A., Ps, S., and Simangunsong, G. 1984. Population build up and dispersal of Elaeidobius kamerunicus Faust in Indonesia. *Buletin, Pusat Penelitian Marihat* 4, 8–29.
Irvine, A. K. and Armstrong, J. E. 1990. Beetle pollination in tropical forests of Australia. In: Bawa, K. S. and Hadley, M. (eds.) *Reproductive Ecology of Tropical Forest Plants*. UNESCO, Paris, pp. 135–149.
Jyoti, P. V. and Saradamani, N. 1994. Pollinating agents of oil palm: A study on Plantations on Andra Pradesh. *Indian Oil Palm Journal* 4(22), 24–26.
Kang, S. M. 1999. The Elaeidobius kamerunicusstory. *Planter, Kuala Lumpur* 75, 143–150.
Khorsand Rosa, R. and Koptur, S. 2009. Preliminary observations and analyses of pollination in Coccothrinaxargentata: Do insects play a role? *Palms* 53, 75–83.
Klein, A. M., Vaissie, B., Cane, J., Steffan-Dewenter, I., Cunningham, S., Kremen, C., and Tscharntke, T. 2007. Importance of pollinators in changing landscapes for world crops. *Proc. R. Soc. B Biol. Sci.* 274, 303–313.
Knudsen, J. T., Tollsten, L., and Ervik, F. 2001. Flower scent and pollination in selected Neotropical palms. *Plant Biology* 3, 642–653.
Kouakou, M., Hala, N., Akpesse, A. A. M., Tuo, Y., Dagnogo, M., Konan, K. E., and Koua, H. K. 2014. Comparative efficacy of Elaeidobius kamerunicus, E. plagiatus, E. subvittatus (Coleoptera: Curculionidae) and Microporum spp. (Coleoptera: Nitidulidae) in the pollination of oil palm (Elaeis guineensis). *Journal of Experimental Biology and Agricultural Sciences* 2, 538–545.
Listabarth, C. 1993a. Pollination in Geonomamacrostachys and 3 congeners, G. acaulis, G. gracilisand G. interrupta. *Botanica Acta* 106, 496–506.
Listabarth, C. 2001. Palm pollination by bees, beetles and flies: Why pollinator taxonomy does not matter. *Plant Species Biology* 16, 165–181.
Mariau, D. and Genty, P. 1988. IRHO contribution to the study of oil palm insect pollinators in Africa, South America and Indonesia. *Oléagineux* 43, 233–240.
Mariau, D., Houssou, M., Lecoustre, R., and Ndigui, B. 1991. Oil palm pollinating insects and fruitset rates in West Africa. *Oleagineux* 46, 43–51.
Mayfield, M. M. 2005. The importance of nearby forest to known and potential pollinators of oil palm (Elaeis guineensis Jacq.; Areceaceae) in southern Costa Rica. *Economic Botany* 59, 190–196.
Meléndez, M. R. and Ponce, W. P. 2016. Pollination in the oil palms Elaeis guineensis, E. oleifera and their hybrids (OxG), in tropical America. *Pesquisa Agropecuária Tropical* 46, 102–110.

Michener, C. D. 2007. *The Bees of the World*, 2nd edn. Johns Hopkins University Press Mogea, Baltimore, MD.

Mora Urpí, J. and Solis, E. M. 1980. Polinazaciónen Bactris gacipaes H. B. K. (Palmae). *Revista de Biologia Tropical* 30, 153–174.

Moura, J. I. L., Cividanes, F. J., Pires, J. L., dos Santos, L. P., and Valle, R. R. M. 2012. Visitation and transportation of pollen grains by pollinator beetles of oil palm. *Agrotrópica* 24, 109–114.

Olesen, J. M. and Balslev, H. 1990. Flower biology and pollinators of the Amazonian monoecious palm, Geonoma macrostachys: A case of bakerian mimicry. *Principes* 34, 181–190.

Pellmyr, O., Tang, W., Groth, I., Bergstro¨m, G., and Thien, L. B. 1991. Cycad cone and angiosperm floral volatiles: Inferences for the evolution of insect pollination. *Biochemical Systematics and Ecology* 19, 623–627.

Pillai, R. S. N. and Ponnamma, K. N. 1992. Botany. In: Nair, M. K. and Nampoothiri, K. U. K. (eds). *Oil Palm Production Technology*. Central Plantation Research Institute, Kasargod, pp. 13–20.

Prada, F. and Romero, H. M. 2013. Muestreo y Ana´ lisis de RacimosenelCultivo de la Palma de Aceite. In: *Tecnologı´as para la Agroindustria de la Palma de Aceite: Guı´a para Facilitadores*. Cenipalma, Bogota´, p. 149.

Purba, R. Y., Harahap, I. Y., Pangaribuan, Y., and Susanto, A. 2010. Menjelang 30 tahunkeberadaanseranggapenyerbukkelapasawit Elaeidobiuskamerunicus Faust di Indonesia. *J. l. Pen. KelapaSawit* 18(2), 73–85.

Purseglove, J. W. 1975. *Tropical Crops: Monocotyledons*. Vols 1 & 2. Longman Group Ltd, London.

Rajamohanan, P. R., Kushairi, A., Chee, S., Ong-Abdullah, M., Leslie, C., and Singh, R. 2017. Effect of pollination on oil palm yield. *J. Oil Palm Res.* 29, 220–227.

Rethinam, P. 1992. History of oil palm cultivation and potential in India. In: Nair, M. K. and Nampoothiri, K. U. K. (eds.) *Oil Palm Production Technology*. Central Plantation Research Institute, Kasargod, Kerala, India, pp. 7–11.

Sambathkumar, S. and Ranjith, A. M. 2011. Insect pollinators of oil palm in Kerala with special reference to African weevil, *Elaeidobius kamerunicus* Faust. *Pest Management in Horticultural Ecosystems* 17(1), 14–18.

Savolainen, V., Anstett, M. C., Lexer, C. et al. 2006. Sympatric speciation in palms on an oceanic island. *Nature* 444, 210–213.

Sayed, R. A. 1979. Studies on oil palm pollination by insects. *Bulletin of Entomological Research* 69, 213–224.

Schmid, R. 1970a. Notes on the reproductive biology of Asterogyne martiana (Palmae). I. Inflorescence and floral morphology; phenology. *Principes* 14, 3–9.

Schmid, R. 1970b. Notes on the reproductive biology of Asterogyne martiana (Palmae). II. Pollination by syrphid flies. *Principes* 14, 39–49.

Sholdt, L. L. 1966. Insects associated with the flowers of the coconut palm, Cocos nucifera L., in Hawaii. *Proceedings of the Hawaiian Entomological Society* 19, 293–296.

Sholdt, L. L. and Mitchell, W. A. 1967. The pollination of Cocos nucifera L. in Hawaii. *Tropical Agriculture of Trinidad* 44, 133–142.

Silberbauer-Gottsberger, I. 1990. Pollination and evolution in palms. *Phyton* 30, 213–233.

Silberbauer-Gottsberger, I., Webber, A. C., Küchmeister, H., and Gottsberger, G. 2001. Convergence in beetle-pollinated Central Amazonian Annonaceae, Araceae, Arecaceae, and Cyclanthaceae. *Dissertationes Botanica* 346B, 165–183.

Siregar, E. H., Atmowidi, T., and Kahono, S. 2016. Diversity and abundance of insect pollinators in different agricultural lands in Jambi, Sumatera. *Hayati Journal Of Biosciences* 23(1), 13–17.

Soh, A. C. 2012. Breeding and genetics of the oil palm. *Palm Oil*. Elsevier. Boca Raton, FL, pp. 31–58.

Sperr, E. B., Fronhofer, E., and Tschapka, M. 2009. The Mexican mouse opossum (*Marmosa mexicana*) as a flower visitor at a neotropical palm. *Mammalian Biology* 74, 76–80.

Syed, R. A. 1979. Studies on oil palm pollination by insects. *Bull. Entomol. Res.* 69(2), 213–224.

Syed, R. A. 1982. Insect pollination of oil palm: Feasibility of introducing elaeidobius spp. [Species] into Malaysia [From Africa]. Proc. of the International Conference on Oil Palm in Agriculture in the Eighties, Push-parajah, E., Chew, P. S. (eds.)-Kuala Lumpur (Malaysia): PPP .ISP, 1982, pp. 263–289.

Syed, R. A. 1984. Insect pollinators of African palm. *Palmas* 5, 19–64.

Syed, R. A. 1994. Insect pollination of the Oil palm: A proven success. *Indian Oil Palm Journal* 3(19), 12–21.

Syed, R. A., Law, I. H., and Corley, R. H. 1982. Insect pollination of oil palm: Introduction, establishment and pollinating efficiency of Elaeidobiuskamerunicus in Malaysia. *Planter Kuala Lumpur* 58, 547–561.

Tan, Y. P. and Basri, M. W. 1984. Another pollinating insect of oil palm [Elaeis guineensis] Pyroderces sp. PORIM.

Tan, Y. P. and Basri, M. W. 1985. Another pollinating insect off oil palm—Pyroderces spp. Proceeding of Seminar on Impact of Weevil on the Malaysian Industry 21–24 February 1984. Palm Oil Research Institute of Malaysia, Kuala Lumper, pp 50–57.

Tandon, R., Manohara, T., Nijalingappa, B., and Shivanna, K. 2001. Pollination and pollenpistil interaction in oil palm. *Elaeisguineensis. Ann. Bot.* 87, 831–838.

Taniputra, B. and Muluk, C. 1989. The influence of Elaeidobius kamerunicus on the yield pattern of Elaeis guineensis at Bukit Sentang Estate, North Sumatra, Indonesia. *Planter* 653, 493–499.

Tschapka, M. 2003/2009. The Mexican mouse opossum (*Marmosa mexicana*) as a flower visitor at a neotropical palm. *Mammalian Biology* 74, 76–80.

Turner, P. D. and Gilbanks, R. A. 1974. Oil palm cultivation and management. The Incorporated Soc. of Planter, Kuala Lumpur, Malaysia. pp. 247–276.

Uhl, N. W. and Dransfield, J. 1987. Genera Palmarum: A classification of palms based on the work of Harold E. Moore Jr. The L. H. Bailey Hortorium and the International Palm Society, Lawrence, Kansas, U.S.A.

Uhl, N. W. and Moore, H. E. 1977. Correlation of inflorescence, flower structure, and floral anatomy with pollination in some palms. *Biotropica* 9, 170–190.

Vail, S. G. 1992. Selection for overcompensatory plant-responses to herbivory—a mechanism for the evolution of plant-herbivore mutualism. *American Naturalist* 139, 1–8.

Wiens, F., Zitzmann, A., Lachance, M. A. et al. 2008. Chronic intake of fermented floral nectar by wild treeshrews. *Proceedings of the National Academy of Science* 105, 10426–10431.

Wood, B. J. 1874. Research in relation to natural resources-oil palm. *Planter Kuala Lumper* 54, 414–442.

Zona, S. 1987. Phenology and pollination biology of Sabal etonia (Palmae) in Southeastern Florida. *Principes* 31, 177–182.

13 Coconut

13.1 INTRODUCTION

The coconut palm, scientifically known as *Cocos nucifera* L., is a perennial crop that undergoes cross-pollination, facilitated by wind and insects. Bees greatly benefit from the abundant nectar and pollen provided by coconuts, making it an excellent source throughout the year. Found along tropical seashores worldwide, coconuts are also cultivated in inland regions. They offer a wide range of resources to humans, including food, beverages, fuel oil, and various other products. The global coconut acreage is approximately 8.5 million acres, with significant portions located in the Philippines (2.45 million acres), India (1.59 million acres), Indonesia (1.5 million acres), Ceylon (1.07 million acres), Malaya (0.6 million acres), other South Sea Islands (0.6 million acres), and 0.7 million acres elsewhere (Menon and Pandalai 1958; Woodruff 1970). Apacible (1968) noted an increase from 2.4 million acres in 1958 to 4.5 million acres in 1967. Additionally, coconut production has experienced a consistent annual growth rate of 5% over the past 50 years. In the United States, coconuts can be found in Florida, Hawaii, and Puerto Rico.

The largest coconut plantation in the United States, located in Key Biscayne, Florida, comprises approximately 30,000 trees (Woodruff 1970). The trunk of the coconut tree is typically inclined and devoid of branches, reaching heights of up to 100 ft. However, there are now cultivated variations of dwarf coconut plants that can be as short as 6 ft. Nonetheless, Wrigley (1969) noted that dwarf coconuts have a shorter life span and yield inferior copra production.

The inflorescence of the coconut tree produces anywhere from 1 to 20 nuts, although typically around 6 nuts, with each nut weighing several pounds. The nut is enveloped by a thick fibrous husk which, upon removal, reveals the familiar brownish fiber-coated coconut. Inside the hard shell lies the edible meat and milk. A single tree can produce around 100 fruits per year, and an average of 90 trees per acre is utilized (Woodruff 1970).

The coconut tree is a monoecious plant, featuring both staminate and pistillate florets on the same extensively branched inflorescence, known as the spadix or fleshy panicle, which measures 2 to 4 ft in length and is found in the leaf axil (Figure 13.1 and 13.2). The inflorescence is primarily composed of up to 8,000 staminate flowers, while the base typically hosts 1 to 30 pistillate flowers (Aldaba 1921a; Ochse et al. 1961).

13.2 FLOWER ECOLOGY

Flowering initiation in larger plants typically occurs between five and eight years of age, as stated by Chandler in 1958. However, on dwarf plants, it commences in the third or fourth year, as noted by Woodruff in 1970. Throughout the year, the plants experience continuous flowering.

A specific type of male flower, as described by Juliano and Quisumbing in 1931, opens for only one day and is primarily active from 0600 h to noon. These flowers are small, measuring just a few millimeters, and consist of three cream-colored petals and six stamens. The stamens release a substantial amount of pollen, some of which is shed even before the flower opens. In total, each inflorescence can produce as much as 6.1 g of pollen, according to Whitehead in 1963. Additionally, there is a non-functional pistil in these flowers, which possesses three parts, each containing an active nectar gland.

In contrast, the pistillate flowers, which are significantly larger, ranging from 1/2 to 1 in in size, have an oval shape. They possess three stigmas attached to a short style and three ovules, out of which two consistently fail to develop. Sholdt and Mitchell (1967) noted that honeybees collect nectar from the nectary openings and the region surrounding the stigmas. Menon and Pandalai (1958) reported that the secretion of nectar is most abundant between the stigma and the base of the ovary.

DOI: 10.1201/9781032656724-14

Whitehead (1965) reported that a significant amount of nectar is produced by three nectaries in the pistillate flower. Patel (1938) observed that a clear sweet fluid is abundantly secreted in four locations when the stigma is receptive, including the base of the stigma and three pores on the pericarp near the top of the ovary. It should be noted that not all pistillate flowers develop into fruit. Lever (1961) reported two types of shedding in the plant: a natural shedding similar to the "June drop" in fruit trees, and a shedding caused by harmful insects. Normally, only one spadix opens at a time on a plant. Moreover, the staminate flowers typically complete their flowering three to six days before the pistillate flowers open. As a result, it is unlikely for cross-pollination to occur between flowers on a spadix or even within the same plant, although there is some overlap in flowering periods in newer dwarf varieties (Woodruff 1970; Ochse et al. 1961).

The duration of staminate flowering on a spadix can range from 18 to 38 days, while the pistillate phase lasts from 2 to 12 days. The interval between spadices varies from 10 to 57 days, with an average of 18 days (Kidavu and Nambiyar 1925). The extent of overlapping phases on a plant ranges from "rarely" to 20% of the time (Sholdt and Mitchell 1967; Ochse et al. 1961).

A pistillate flower has the potential to receive pollen either from staminate flowers on the same spadix or from a subsequent spadix on the same plant. However, if there is no overlap between spadices, the pollen must originate from a different plant (Chapman 1964). According to Free (1970), staminate flowers of tall plants typically start blooming approximately one month earlier than pistillate flowers,

FIGURE 13.1 Female flowers.

FIGURE 13.2 Male flowers of coconut.

whereas pistillate flowers of dwarfs tend to open about a week after the staminate flowers. These flowers attract honeybees and various other insects due to the presence of nectar and pollen (Sholdt 1966).

Nectar production, in terms of honey stored by a colony of honeybees, is not great (Pellett 1947; Sholdt and Mitchell 1967), and the amount stored by a colony varies with the time of the year (Wolfenbarger 1970). Whitehead (1965) stated that nectar is produced in considerable quantity from the three nectaries in the female flower.

During a 30 min interval, he observed 103 instances of bees gathering nectar from a single flower, and the nectar was quickly replenished after each visit. Ochse et al. (1961) also mentioned the abundant amount of nectar that oozes from the flower.

13.3 FLORAL BIOLOGY

The coconut tree is a monoecious plant, with the sequence of flower clusters occurring approximately on a monthly basis, although it may be more rapid during the dry season compared to the wet season. The male phase lasts approximately 20 days, but this duration can vary depending on the variety and season.

The female phase spans from four to seven days.

13.4 POLLINATION REQUIREMENTS

For coconuts to develop, pollen must be transferred from male flowers to female flowers. Sholdt and Mitchell (1967) demonstrated that the source of the pollen is not significant in terms of fruit set, as good results were obtained whether the pollen originated from the same plant or a different one. Pollen can come from the same flower cluster, another cluster on the same tree, or a different tree altogether.

The first day the stigma is exposed is when the pollen is most effective, and theoretically, only one pollen grain per female flower is enough to fertilize a single ovule. Aldaba (1921b) calculated that one flower cluster produces 272 million pollen grains. Whitehead (1965) conducted a study on coconut flowering in Jamaica and reported various factors affecting pollination requirements. He believed that it was unreliable to conclude that the plants are exclusively self-pollinated or cross-pollinated, as the extent of cross-pollination depended on the importance of wind, insects, proximity of other trees, efficiency of self-pollination, presence of nectaries on male and female flowers, and frequency of insect visits, particularly by bees.

Copeland (1931) stated that the succession of flower clusters is usually timed in a way that necessitates pollen from another tree, ensuring cross-pollination. However, Tammes and Whitehead (1969) stated that this applies only to tall coconut palms. In dwarf palms, except for the "Niu Leka" variety, female flowers become receptive before male flowers cease, allowing pollen to come from the same flower cluster. Wrigley (1969) mentioned that self-fertilization between flower heads on a dwarf coconut tree is typical.

13.5 POLLINATORS

There has been significant inquiry regarding the agents responsible for the transfer of pollen from the male to the female flowers, a crucial process, irrespective of the plant's flowering patterns. Self-pollination is frequently mentioned, but this only refers to the source of the pollen, whether from the same inflorescence on which the stigma is located or another inflorescence. The flower cannot fertilize itself.

The coconut tree relies on various agents for cross-pollination, such as wind, birds, mites, and insects, including ants, bees, earwigs, flies, and wasps. Previous studies by Davis (1954) and Kidavu and Nambiyar (1925) have mentioned these agents. The efficiency of each agent is likely influenced by specific local conditions. Furtado (1924) expressed skepticism regarding the contribution of birds

FIGURE 13.3 *Apis dorsata* and wasps foraging on coconut male and female flowers.

FIGURE 13.4 *Apis mellifera* foraging on male flowers.

FIGURE 13.5 *Apis florea* visiting female flowers.

FIGURE 13.6 Coconut flowers foraging by *Apis cerana indica* and ant.

to pollination, while Sampson (1923), Tammes (1937), and Whitehead (1965) asserted that insects play a key role in the pollination process.

Huggins (1928) regarded honeybees and various other hymenoptera as significant, while considering ants as insignificant. Hunger (1920), Patel (1938), and Ochse et al. (1961) acknowledged the importance of both insects and wind. In Hawaii, Sholdt (1966) gathered 51 insect species from coconut inflorescences, with ants, bees, earwigs, flies, and wasps being the most frequently found. Sholdt and Mitchell (1967) recognized the importance of both wind and insects, with honeybees being the most crucial among the insects. The recognition of honeybees' value in coconut pollination is not a recent revelation, as an anonymous author in 1916 mentioned their significant role in Fiji.

Inflorescences that attracted bees while in bloom resulted in a high nut yield, leading to the suggestion of introducing bee colonies into coconut plantations (Figure 13.3–13.6). According to Sampson's findings in 1923, estates with a large population of bees for various purposes experienced remarkably high nut yields. Huggins (1928) believed that insufficient cross-pollination often caused a decrease in nut production.

In 1958, Haldane proposed the use of honeybees to enhance yields. However, Tammes and Whitehead (1969) disagreed with this viewpoint, asserting that there was already adequate natural pollination by wild bees, as observed in trials. They did not specify the required population of wild bees for optimal pollination.

Apart from bees, other insects such as houseflies, wasps, and beetles (Thampan 1982) also contribute to pollination. The hybrids attracted 36% more honeybees compared to the varieties.

Adoro (1982) reported that there were 13 flower-visiting insects in coconut, and from these, 3 species were considered to be effective pollinators, namely, *Apis (Sigmatapis) indica*, the most significant; *Halictus* sp.; and *Trigona* sp. Sholdt (1966) found that 51 different species of insects were collected from the flowers of the coconut palm.

13.6 POLLINATION RECOMMENDATIONS AND PRACTICES

Apart from the previously mentioned sources, bees have not been recommended for coconut pollination. In 1967, Sholdt and Mitchell suggested that introducing bee colonies could be beneficial in boosting yields, but they did not specify the required number of colonies per acre or bees per inflorescence. The available evidence suggests that having a sufficient number of honeybees could enhance coconut production. However, it is unclear what constitutes an adequate population on the flowers. It is worth considering the significantly lower coconut production per acre in the Philippines, where the bee population is comparatively low (Morse and Laigo 1969), in contrast to other coconut-producing regions worldwide.

The concentration of honeybees, even if it meant the development of a strong apicultural industry in the Philippines, might considerably improve the coconut industry.

REFERENCES

Adoro, L. C. 1982. Insect pollinators of coconut (Cocos nucifera Linn.) [study conducted in the Philippines] University of the Philippines at Los Banos, College, Laguna [Corporate Author].

Aldaba, V. C. 1921a. The pollination of coconut. *Philippine agr.* 10(5), 195–208.

Aldaba, V. C. 1921b. The pollination of coconut. *Philipp Agriculturist* 10, 195–210.

Anonymous. 1916. Bees and pollination. *Planters' chron.*, Bangalore 9(46), 572.

Apacible, A. R. 1968. The philippine coconut. *Sugar News [Manila]* 44(10), 599–606.

Chandler, W. H. 1958. *Evergreen Orchards.* 2nd ed. Henry Kimpton, London.

Chapman, 1964. G.P. Chapman. Pollination and the yields of tropical crops: an appraisal. *Euphytica*, 13, 187–197.

Copeland, E. B. 1931. *The Coconut*, 3rd edn. Macmillan, London, p. 225.

Davis. J. A. 1954. Mysteries of cross-pollination. *Indian Cent. Coconut Com. Ernakulam, Bul.* 7, 226–227.

Free, J. B. 1970. *Insect Pollination of Crops.* Academic Press, London.

Furtado, C. X. 1924. A study of the coconut flower and its relation to fruit Production. *Gard. Bul. [Singapore]* 3(7–8), 261–273.

Haldane, J. B. S. 1958. Some suggestions for coconut research. *Indian Coconut Jour.* 12, 1–9.

Huggins, H. D. 1928. Pollination and crop production (concluded). *Agr. J. Br. Guiana* 1, 90–94, 164–169.

Hunger, F. W. T. 1920. Cocos nucifera 518 pp. Scheltema and holkema's boekhandel, amsterdam.

Juliano, J. B. and Quisumbing, E. 1931. Morphology of the male flower of cocos nucifera linn. *Philippine Jour. Sci.* 45, 449–458.

Kidavu, M. G. and Nambiyar, E. K. 1925. Pollination in coconut. *Madras Dept. Agr. Yearbook* 1925, 43–49.

Lever, R. J. A. W. 1961. Immature nutfall of coconuts; the war of the ants. *World Crops* 13(2), 60–62.

Menon, K. P. V. and Pandalai, K. M. 1958. *The Coconut Palm—A Monograph.* ICAR, Ernakulam, Kerala, India, pp. 65–68.

Morse, R. A. and Laigo, E. M. 1969. The potential and problems of beekeeping in the philippines. *Bee World* 50(1), 9–14.

Ochse, J. J., Soule Jr., M. J, Dijkman, M. J., and Wehlburg, N. C. 1961. *Tropical and Subtropical Agriculture*, Vol. 1. The Macmillan Company, New York, London, 760 p.

Patel, J. S. 1938. *The Coconut—A Monograph.* Government Press, Madras, p. 262.

Pellett, F. C. 1947. *American Honey Plants.* Hamilton, IL: Dadant & Sons.

Sampson, H. C. 1923. *The Coconut Palm.* J. Bale, sons, and danielson, ltd., London, p. 262.

Sholdt, L. L. 1966. Insects associated with the flowers of the coconut palm, *Cocos nucifera* L. in Hawaii. *Proc. Hawaii. Ent. Soc. Proc.* 19(2), 293–296.

Sholdt, L. L. and Mttchell, W. A. 1967. The pollination of *Cocos nucifera* l. In Hawaii. *Trop. Agr. [Trinidad]* 44(2), 133–142.

Tammes, P. M. L. 1937. On the inflorescence and pollination of the coconut. *Landbouw, buitenz.* 13, 74–89.

Tammes, P. M. L and Whitehead, R. A. 1969. Coconut. In: Ferwerda, E. P. and Wit, F. (eds.) *Outlines of Perennial Crop Breeding in the Tropics.* H. Veenman and Zonen, n. V. Wageningen, the Netherlands, pp. 175–188.

Thampan, P. K. 1982. *Handbook on Coconut Palm.* Oxford and IBH Publ. Co., New Delhi, India.

Whitehead, R. A. 1963. The processing of coconut pollen. *Euphytica* 12, 167–177.

Whitehead, R. A. 1965. The flowering of cocos nucifera l. In jamaica. *Trop. Agr. [Trinidad]* 42(1), 19–29.

Wolfenbarger, D. O. 1970. Notes on pollen dispersers and pollination of tropical plants and on attractancy of aluminum mulched plants for Honey bees: In the indispensable pollinators, ark. *Agr. Ext. Serv. Misc. Pub.* 127, 150–156.

Woodruff, J. G. 1970. *Coconuts: Production, Processing, Products.* A.V.I. Publishing co., inc., Westport, conn, p. 241.

Wrigley, G. 1969. *Tropical Agriculture.* Frederick a. Praeger, n. Y. And Washington, p. 376.

14 Olive

14.1 INTRODUCTION

Olive pollination is important to improve the fruit setting rates and genetic improvement. Olive is very widely grown in different parts of the world, covering an area of 10 million hectares. During recent years, consumption of olive oil has increased due to improvements in living standards of the people.

The olive tree is usually 15–20 ft tall but sometimes reaches 30–35 ft when fully developed and properly nurtured, with oval 1 to 3 in gray-green leaves and gray branches. Its beauty, sturdiness, and symmetrical growth make it a prized ornamental as well as commercial fruit tree. It will live hundreds of years in mild, arid climates. It blossoms profusely in the spring, producing the well-known oval one-seeded green to blue-black fruit about an inch long. It is cultivated in similar manner as other warm-weather fruit or nut trees. In the grove, the trees are spaced well apart (35 to 40 ft) so the sunlight can reach the tree on all sides. Olive (*Olea europaea*) trees, though, produce an enormous number of flowers, but it is the proportion of complete or perfect flowers which determines the final crop volume. However, a negative relationship exists between the number of flowers to the proportion of complete flowers and with fruit set.

The process of pollination in olive is further complicated by the small size of the flowers and scanty pollen production in many cultivars, which often lead to an inadequate pollination, and consequently, an exceptionally higher intensity of flower abscission is noticed soon after full bloom. It is therefore imperative to understand various processes that are related to pollination and fruit set.

Alternate arid erratic bearing behavior of olives has further restricted the proliferation of this crop in many areas. This tendency becomes more conspicuous when olive trees are grown under rainfed areas which are prone to prolonged droughts. Thus, it becomes most essential to understand various processes related with the reproductive cycle of the tree, so as to ensure a regular succession of optimum crops.

14.2 FLORAL BIOLOGY

The cluster of one to two dozen cream-colored to white fragrant flowers (4 mm) that develop in the axil of the leaf is usually shorter than the leaf itself. The individual flower has four valvate corolla lobes, a short four-toothed calyx, and two stamens that produce pollen copiously and little, if any, nectar (Figure 14.1). The flower opens before the discharge of pollen from the anthers, enabling cross-pollination before self-fertilization occurs within the same flower. The flower can exist in two forms: either as a complete and potentially productive structure, exhibiting a robust green pistil, a short style, and a green ovary, or solely as a male flower with a yellow non-functional pistil (Condit 1947). It does not produce purely female flowers.

Most cultivars possess self-fertility; there are some that exhibit self-fertility, and others fall in between (Crider 1922; Morettini 1958; Mort 1952; Pierce 1896). On occasion, a limited fruit yield occurs due to an abundance of staminate flowers during flowering (Hartmann and Opitz 1966). Despite the plentiful presence of pollen during the flowering period, honeybees tend to collect it rather sparingly from olive trees.

Sometimes, an olive honey flow is reported by beekeepers. In olives, flowers are borne on paniculate inflorescences arising in the leaf axils of oppositely arranged leaves, borne on the previous season's shoots. However, occasionally, they may arise from dormant buds of 1 to 2-year-old shoots. In areas where olive trees fail to put forth their growth flush in autumn, inflorescence may arise from axillary buds borne by 1-year-old shoots (Sharma 1971; Singh et al. 1979; Cimato and

DOI: 10.1201/9781032656724-15

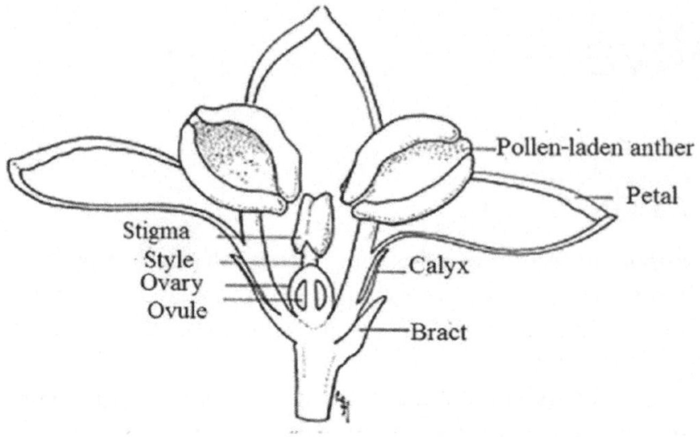

FIGURE 14.1 Floral structure of the olive flower.

Bartolini 1987). A typical olive flower consists of four-lobed gamosepalous calyx, four corolla, yellowish white in color, alternating with sepals, stamens (usually two), single carpel, ovary superior (two-celled), with one syncarpous style and syncarpous ovary with one or two ovules in each locul (Kar and Singh 1982). In olives, the size of panicle and number of flowers borne on a panicle are influenced by the genetic constitution of the tree and environmental conditions prevailing during the period of bud development/ontogeny of different floral parts.

The prevalence of higher temperatures, coupled with acute moisture stress during this period, tends to inhibit the size of flowering panicle and number of flowers per panicle. The proportion of ovary degeneration has been depicted in some olive cultivars. Principally, olives bear two types of flowers, viz, (1) complete or bisexual flowers, containing functional stamen and pistil, and (2) incomplete or unisexual flowers, containing functional stamens but aborted pistil.

However, in arid areas with frequent periods of moisture stress during late winters or spring, intermediary flowers containing rudimentary but non-functional pistils may develop (Singh et al. 1979; Kar and Singh 1982; Thakur 1987). Such flowers are incapable of bearing fruits. The ability of perfect flowers to bear fruits is further affected by adequate pollination and post-pollination abscission (Morettini 1941).

The extent of pistil abortion is cultivar-dependent and on prevailing weather conditions. Histological studies have attributed this condition as the manifestation of incomplete or restricted development of embryo sac and its allied tissues, leading to lack of fertilization. A higher proportion of such type of flowers results in a marked reduction of fruit set.

Each olive cultivar has been found to show a specific extent of pistil abortion even under normal or ideal weather conditions. Thus, environmental factors, genetic constitution, and cultural practices seem to have a pronounced bearing on the expression of this phenomenon.

14.2.1 Flowering Season

The time of onset of flowering in olives is influenced by environmental conditions prevailing between time of panicle emergence until beginning of flowering. Flowering occurs during the month of March to April, whereas in cooler climates it is in the month of May.

Usually, olive plants bloom profusely, wherein flower buds appear on 1-year-old twig in terminal position of the leaves. However, only 10% to 15% fruit set has taken place out of about 5 million flowers produced by a mature olive tree (Martin 1990; Cuevas and Polito 2004).

Between five and seven weeks after full bloom, there is a significant decline in the number of flowers and fruits (Rallo and Fernandez-Escobar 1985). Shortly after full bloom, the shedding of male flowers initiates (Cuevas et al. 1995) and partially overlaps the abscission of unfertilized perfect flowers triggered by pollination taking place in the days after petals drop (Rosati et al. 2011). After petals fall, about 25% of the ovaries are retained. However, only a small percentage of fruits out of 25% reaches maturity.

A shift in the time of flowering has also been noticed at the same location from year to year. The duration of flowering varied between 6 and 13 days in the mid-hills of Himachal Pradesh. In Israel, flowering in olives occurs in the month of April (Shulman and Lavee, 1979), whereas in California, it takes place in May (Hartmann 1949).

It was also found that the prevalence of cool weather due to a light shower prior to flowering tends to delay the time of flowering, whereas the occurrence of such conditions during blooming resulted in extended duration of flowering. Thus, time and duration of flowering depend upon the cultivar and season. Thus, the occurrence of lower temperature coupled with a moderate relative humidity is conducive for anthesis of olive flowers.

The peak period of dehiscence of anthers was observed between 0200 and 0060 h. The pollen grains released during early morning hours can be conveniently used for pollination immediately after their dehiscence from pollen sacs. The increased time lag between these two processes will tend to lower down the extent of pollen viability.

14.2.2 Pollen Morphology

Fresh pollen grains appear as a mass of golden-yellow colour with the naked eye, but they appear oblong and pale yellow in color under a low-power microscope. In aqueous medium, pollens attain a circular shape. The extent of pollen germination varied with the cultivar and season. However, a pollen germination of 28.0 to 48.0% has been reported in olive cultivars in various investigations conducted so far (Sharma 1971; Khurana 1982; Thakur 1987).

14.2.3 Stigma Receptivity

Various investigations have revealed that peak period of stigma receptivity varied between one day before and one day after anthesis. Thus, pollination should be carried out within a specified period in order to obtain satisfactory fruit set in controlled cross-pollination experiments.

14.2.4 Floral Sterility

Three types of floral sterility observed by Cimato and Bartolini (1987) are: (a) morphological sterility (lack of or insufficient development of sexual organs), which is a frequent characteristic of some cultivars (Morchiaio, 70% of aborted flowers; Moraiolo, 20–30%; and Frontoio, 10%) and can also be attributed to nutritional factors; (b) cytologic sterility (anomalies occurring during micro- and macro-sporogenesis), which involves the formation of pollen with poor germinating capacity and the degeneration of the female gametophyte; and (c) factorial sterility or incompatibility (pollen unable to germinate or to develop a pollen tube in a sufficiently short time to allow the fertilization before the egg cell degenerates). This type of sterility can affect the flowers of different cultivars (incompatibility between plants) or flowers belonging to the same cultivar (self-incompatibility within the same plant). The self-incompatible cultivars can neither be used for self-pollination nor for pollination of other trees of the same cultivar.

In such a case, it would be worthwhile to find out a suitable compatible cultivar.

14.3 POLLINATION

14.3.1 Pollination Requirements

Different cultivars of olive are reported to have different pollination requirements. Crider (1922) listed two self-sterile, one partly self-sterile, and five self-fertile cultivars. In greenhouse experiments conducted by Bradley et al. (1961), it was demonstrated that even in cultivars capable of self-pollination, the pollen tubes from different cultivars exhibited faster growth down the style compared to self-pollen tubes when subjected to the same temperature conditions. The researchers observed that when the growth of pollen tubes was excessively slow, the embryo sac began to deteriorate before the tube could reach it, leading to a lack of fertilization. Consequently, they concluded that cross-pollination had a higher likelihood of fertilization compared to self-pollination, as evidenced by the greater percentages of pistils in which a pollen tube successfully reached the embryo sac.

The concept of cross-pollination was first introduced in olive by Miller (1895), who reported that in Chino Valley (Southern California), most of the olive cultivars yielded better crops when two or more of them were interplanted. But it seems that pollination requirement is also significantly influenced by prevailing weather conditions. Condit (1947), Bradley et al. (1961), and Griggs et al. (1975) were of the opinion that for obtaining commercial yields from olive orchards, an adequate provision for cross-pollination should be kept.

This is also essential to compensate for an annual variation found in the self-pollination capability of different cultivars. In cross-pollination, pollen tubes grow faster in stylar tissue, and effective pollination period (EPP) is much shorter as compared to self-pollination; hence, a higher fruit set is obtained (Bradley et al. 1961; Griggs et al. 1975).

14.3.2 Self-Pollination

Pendulino and Cornicobra showed a successful fruit set through self-pollination, while Frontoio and Coratina exhibited a moderate fruit set when selfed, as reported by Thakur (1987) and Singh and Awasthi (1993). This revealed that Frontoio and Coratina appeared to be partially self-fruitful. The lower fruit set recorded in Cornicobra Attica, Leccino, and Aglandau under self-pollination emphasized the need for their cross-pollination with compatible cultivars.

In Ascolano, no fruit set could be achieved under self-pollination (Thakur 1987; Singh and Awasthi, 1993, which emphasized the need of inter-planting this cultivar with other cultivars having synchronizing flowering periods. In self-pollination, fruits never reached maturity but remained small, were misshapen, and appeared as "shot berries." Olive cultivar Coratina is more prone to this malady. But production of shot berries is not entirely attributed to lack of cross-pollination.

14.3.3 Natural or Open Pollination

A remarkable improvement in fruit set of Ascolano, Leccino, Cornicobra Attica, and Aglandeau was observed in natural/open pollination over self-pollination by Sharma (1971), Singh and Kar (1980), Thakur (1987), and Singh and Awasthi (1993) Wind pollination has also been considered to be the main mode of pollination in olives by Lavee (1986). The distribution of pollen grains through wind has also been ascertained by Lavee and Datt (1978) so as to determine the appropriate placement of pollinizers for effective pollination.

14.3.4 Placement of Pollinizers

In earlier studies, Molinari and Nicolea (1947) emphasized that for adequate pollination of olives, every other tree in every other row should be of pollinizer cultivar. However, keeping in view the time lag between fruit ripening of olive cultivars, it would be appropriate to keep every third row of the pollinizing cultivar, and it would be advisable to use two pollinizing cultivars instead of one so as to ensure the availability of pollen mass during the entire period of flowering of the main cultivar.

14.3.5 Cross-Pollination

In general, cross-pollination results in a higher fruit set than either self- or open or natural pollination (Sharma 1971; Singh and Kar 1980; Khurana 1982; Thakur 1987; Singh and Awasthi 1993). A remarkable improvement in fruit set of Coratina was recorded by Kar and Singh (1984) when cross-pollinated with Cornicobra. An exceptionally higher fruit abscission in olive cultivar Biancolilla was recorded following self-pollination, but cross-pollination of this cultivar resulted in a higher fruit set. Gentile (1951) and Debertoldi and Fiorinto (1968) also emphasized the need of cross-pollination, improving the extent of fruit set of Dritta, Castiglionese, Olivoce, and Indorso cultivars under Italian agro-climatic conditions.

In India, too, the need for cross-pollination of olive cultivars has been emphasized by several workers (Sharma 1971; Singh and Kar 1980; Khurana 1982). Further investigations have indicated that Cornicobra, Aglandeau, and Frontoio are good pollinizers for Coratina, Aglandeau, Cornicobra Attica, and Leccino.

14.3.6 Role of Pollinators

Canale and Loni (2010) conducted study in a Tuscan olive grove, where they gathered and classified a grand total of 380 insect specimens. These insects represented two orders, three families, eight genera, and ten species, all of which were observed visiting flowers of *Olea europaea*. Among Hymenoptera, workers of *Apis mellifera* L. were frequently seen to collect pollen from flowers at different times of the day. Three Halictidae species, *viz, Halictus maculatus* Smith, *Lasioglossum nitidulum* (F.), and *Lasioglossum villosulum* (Kirby), and six Syrphidae species, both Syrphinae and Milesiinae, were identified as foragers on olive flowers. Further, they suggested that *O. europea* is a potential source of pollen for a number of insect species in late spring when blossom of other flora is scanty.

The olive flowers are generally pollinated by wind. Honeybees, though, visit olive flowers to collect the pollen but are not essential for pollination (Morettini 1941; Griggs et al. 1975). However, wind pollination is supplemented by the activity of honeybees. In fact, honeybees do not collect olive pollen as activity as in other plants. It requires relatively heavy concentration of honeybee colonies in or near the grove in order to increase the visits of bees to olive flowers.

In the absence of adequate wind pollination, honeybee pollination leads to better yield, but extensive research is still required in this direction. There is no information on the concentration that might be desired. Studies in this area would be productive.

REFERENCES

Bradley, M. V., Griggs, W. H., and Hartmann, H. T. 1961. Studies on self- and cross-pollination of olives under varying temperature conditions. *Calif. Agr.* 15(3), 4–5.

Canale, A. and Loni, A. 2010. Insects visiting olive flowers (*Olea europaea* L.) in a Tuscan olive grove. *Redia* 93, 95–98.

Cimato, A. and Bartolini, G. 1987. *A Handbook of Olive Cultivation*. Agrotec, Rome.

Condit, I. J. 1947. Olive culture in california. *Calif. Agr. Ext. Serv. Cir.* 135, 36 pp.

Crider, F. J. 1922. The olive in arizona. *Ariz. Agr. Expt. Sta. Bul.* 94, 491–528.

Cuevas, J. and Polito, V. S. 2004. The role of staminate flowers in the breeding system of *Olea europaea* (Oleaceae): An andromonoecious, wind-pollinated taxon. *Annals of Botany* 93(5), 547–553.

Cuevas, J., Rapoport, H. F., and Rallo, L. 1995. Relationships among reproductive processes and fruitlet abscission in Arbequina olive. *Advances in Horticultural Science* 9(2), 92–96.

Debertoldi, M. and Fiorinto, O. 1968. Test on Foliar feeding, artifical pollination, barkringing in olive variety Ascolano. *Teneva. Tech. Agrer.* 8, 12.

Gentile, R. 1951. Investigations on floral biology and experiments in cross pollination with some olive varieties grown at pescora (Italy). *Olivicultura* 9. In *Bot. Olercult.* 5, 67–68.

Griggs, W. H., Hartmann, H. T., Bradley, M. V., Iwakiri, B. T., and Whistler, J. E. 1975. Olive pollination in California. *California Agricultural Experimental Station, Bulletin* 869, 1–49.

Hartmann, H. T. 1949. Growth of the olive fruit. *Proceedings of the American Society of Horticultural Science* 54, 86–94.

Hartmann, H. T. and Opitz, K. W. 1966. Olive production in California. California Agricultural Experimental Station and Extension Service. *Circular*, 540, 63 pp.

Kar, P. L. and Singh, R. P. 1982. Pollination and compatibility studies in some olive cultivars. *Himachal Journal of Agricultural Research* 10(1), 55–57.

Kar, P. L. and Singh, R. P. 1984. Pollination and compatibility studies in some olive cultivars. *Himachal Journal of Agricultural Research* 10, 55–57.

Khurana, M. L. 1982. Studies on flowering, pollination, fruit set and fruit drop in olive. M.Sc. Dissertation, HPKVV, Solan, India.

Lavee, S. 1986. Olive. In: Monselise, S. P. (ed.). *Handbook of Fruit Set and Development.* CRC Press, Boca Raton, FL, pp. 261–276.

Lavee, S. and Datt, Z. 1978. The necessity of cross pollination for fruit set of Manzanillo olives. *Journal of Horticultural Science* 53, 261–266.

Martin, G. C. 1990. Olive flowers and fruit population dynamics. *Acta Horticulturae*, 286, 141–153.

Miller, J. W. 1895. *Experiments on Pollination of Olives.* California Agricultural Experiment Station. Report 1894–1895.

Molinari, C. and Nicolea, O. Y. 1947. *Tratado general de olivicultura.* El Ateneo, Buenos Aires Argentina.

Morettini, A. 1941. L'incremento produttivo negli olivi Moraiolo e Frantoio con I'impiego di adatte variété impollinatrici. *L'Italic Agricola* 9, 631–639.

Morettini, A. 1958. The biology of fertilization in olive cultivars and its practical importance. *Italian. Agr.* 94, 1103–1116. [In Italian.] Plant Breed. *Abstracts* 28(3), 586587.

Mort, C. H. 1952. Fruitfulness in olives. *Agr. Gaz. N.S. Wales* 63, 371–372.

Pierce, N. B. 1896. Olive culture in the united states. *U.S. Dept. Agr. Yearbook*, 371–390.

Rallo, L. and Fernandez-Escobar, R. 1985. Influence of cultivar and flower thinning within the inflorescence on competition among olive fruit. *Journal of the American Society for Horticultural Science*, 110, 303–308.

Rosati, A., Caporali, S., Paoletti, A., and Famiani, F. 2011. Pistil abortion is related to ovary mass in olive (*Olea europaea* L.). *Scientia Horticulturae* 127(4), 515–519.

Shulman, Y. and Lavee. S. 1979. Fruit development and maturation of olives affected by treatments with auxins. *Rivista Ortoforofrullicoltura Italiana* 63, 31–40.

Singh, R. P. and Awasthi, R. P. 1993. Effect of tree vigour and blooming intensity on olive production. *OLEA* 22, 96.

Singh, R. P. and Kar, P. L. 1979. Compatibility studies in some olive cultivars. *Himachal Pradesh Journal of Agriculture*, 8–15.

Singh, R. P. and Kar, P. L. 1980. Compatibility studies in some olive cultivars. *Progressive Hort.* 12, 9–15.

15 Planned Pollination

15.1 INTRODUCTION

Edible vegetable oils are a high-value agricultural commodity, and demand for high-quality seed oils continues to grow as the world population increases (Wittkop et al. 2009). The decline in insect pollinators enhancing yields in almost 70% of crops worldwide poses a genuine threat to global food security (Klein et al. 2007). It has been found that insect pollination increases yield in oilseed rape by 25%. The varieties with higher nectar secretion are highly preferred by bees and also good for enhanced agricultural production. Rapeseed crop is important for beekeepers as it is a good source of nectar (Bommarco et al. 2012).

Within intensively managed agricultural landscapes, the presence of natural or semi-natural elements plays a crucial role in offering valuable nesting and foraging areas for wild pollinators. When agricultural areas are located close to these habitats, there is an increase in the diversity of pollinator species, the frequency of crop visitation, and ultimately, the success of pollination (Garibaldi et al. 2011).

Maintaining pollinator diversity can ensure resilience of pollination as an ecosystem service due to species' showing differential response to environmental change (i.e., response diversity). Although pollination positively impacts the productivity of numerous crop species and accounts for a significant portion of global crop production, it raises concerns in the context of intensified agriculture due to the depletion of floral resources and declining pollination services (Klatt et al. 2014; Vanbergen et al. 2013; Wratten et al. 2012).

When bees are employed in a planned manner to facilitate the natural process of pollination, it can have a substantial impact on the output and value of crop yields. This deliberate intervention has become a vital element of contemporary farming practices as it enhances crop productivity and preserves the quality and quantity of food resources.

Oilseed crops hold particular importance in India as they significantly contribute to the country's economy. India has a prominent position as regards to acreage and production of oilseeds. The cultivation of oilseeds constitutes the second-largest agricultural commodity after cereals in India, sharing 14% of the country's gross cropped area. These crops are cultivated on about 16.5 million hectares, with total production of 10 million tons (Abrol and Shankar 2012).

Rapeseed and mustard, sesame, linseed, safflower, sunflower, niger, and linseed are among the significant oilseed crops cultivated in India (Guidry 1964). The oilseed crops are either fully dependent upon cross-pollinating agents or benefit greatly from seed production. Inadequate pollination results not only in reduced yields but also in delayed yield and a high percentage of culls or inferior fruits. Though self-pollination can set fruits, if insect pollination is allowed, there can be a number of advantages as far as quality or quantity of seed/fruit production is concerned. Insect pollination increases either (a) the proportion of fruits set or (b) the quality of fruits set, because fruit size depends on the number of seeds set or the size of seeds (Charlesworth and Charlesworth 1987; Richards 1997). This group of pollination systems includes various oilseed crops, like oilseed rape (Free and Ferguson 1983; Williams et al. 1987), flax- and linseed, as well as sunflowers (Free and Williams 1977).

The level of productivity of oilseeds in India is far behind the average productivity in the world. One of the factors responsible for low productivity of oilseed crop is the failure of proper seed setting due to lack of pollination (Rao et al. 1981; Free 1993; Abrol 2007, 2008, 2009). The cross-pollination of entomophilous crops by honeybees is regarded as a highly efficient and cost-effective approach to increase crop yields. This is because insect pollinators not only improve crop productivity but also promote consistent and early fruit or seed development.

Recently, there has been a widespread adoption of various practices, such as applied pollination, pollinator management, and managed pollination interventions. These approaches aim

DOI: 10.1201/9781032656724-16

to optimize and maximize production in cross-pollinated crops by attracting pollinators to the desired crops.

Insect pollination plays an important role in increasing the yield of oilseed crops (Mishra et al. 1988). A brief description of the pollination requirements, pollinator types, their behavior, and effectiveness in enhancing the production of various important oil crops, such as rapeseed-mustard, sunflower, safflower, sesame, niger, taramira, and Linseed, is given next.

15.2 RAPESEED, MUSTARD, AND CANOLA (*BRASSICA* SPP.)

The rapeseed-mustard group is mostly cross-pollinated, although some varieties are self-fruitful. Selfing has been observed to reduce seed yield, seed size, and yield in subsequent generation (Delaplane and Mayer 2000; Wallace et al. 2002). Pollination interventions in the form of planned pollination have been found to greatly benefit quality and quantity of seed production in crucifers (Figure 15.1).

For example, honeybee pollination (Westcott and Nelson 2001) resulted in higher yields for male sterile oilseed rape (*B. napus*) plants. Similarly, in sarson (*B. campestris*), pollination by insects increased the seed yield, caused formation of well-shaped, larger grain, and produced more viable seed (Abrol 2009). The impact of insect pollination on the production of different oil crop is presented in Table 15.1.

FIGURE 15.1 Honeybee colonies in *Brassica* field.

TABLE 15.1
Impact of Animal Pollination on the Production of Oil Crops

Crop species	Commodity	World production (Mt)	Animal pollination	References
Brassica napus	Rapeseed, Oilseed rape	46 770 903	increase	Free 1993; Adegas and Nogueira Couto 1992; Abel and Wilson 1999; Bürger 2004; Manning and Boland 2000; Abel et al. 2003; Morandin and Winston 2005
Helianthus annuus	Sunflower	26 460 824	increase	Bichee and Sharma 1988; Crane 1991; Free 1993; DeGrandi-Hoffman and Martin 1993; Moreti et al. 1996; Heard 1999; DeGrandi-Hoffman and Watkins 2000; Dag et al. 2002; Greenleaf and Kremen 2006

Source: Data compiled from different sources.

Many Brassica species, including rape (*Brassica napus*), sarson (*Brassica campestris* var. sarson), toria (*Brassica campestris* var. toria), Indian mustard or rai (Brassica juncea), white mustard (Brassica alba), and black mustard (Brassica nigra), are extensively cultivated worldwide as oilseed crops. These plants produce vibrant yellow flowers with a pleasant fragrance in long terminal racemes, which attract honeybees and other natural insect pollinators. In a study conducted by Mohammed (1935) in West Punjab at Lyallpur, a total of 117 insect species from seven orders was observed visiting the crops. Among them, *Andrena ilerda*, *Apis florea*, and *Halictus* spp. constituted 82% of the insect population and were particularly significant. Additionally, Mohammed noted that the stigma of the crops remained receptive for two days after opening, and within five days, their fertility declined. However, the pollen remained viable for up to seven days.

It was observed in Australia that among *Apis mellifera* foragers, 72% collected nectar, only 25% both nectar and pollen, and 3% collected pollen only (Langridge and Goodman 1975). The pollen was collected between 1000 and 1400 h.

Mohammed (1935) reported that by enclosing individual plants in muslin bags, seed set was 12.3% in toria, 20.3% in brown sarson, and 91.0% in yellow-seeded sarson. He further reported that in case of hand-pollinated toria flowers, there was 100% pod formation.

In a similar study, Sihag (1986) examined the seed production in toria plots under different conditions, caged and uncaged, allowing for natural cross-pollination. The results showed that in the caged plots, there were 25 pods per plant, with an average of 6.3 seeds per pod and a total of 4 seeds per plant (Figure 15.2). In the uncaged plots, the numbers increased significantly to 1,060 pods per plant, with an average of 18.2 seeds per pod and a total of 385 seeds per plant. Similar findings were observed in sarson crops, where Mishra et al. (1988) reported that 59% and 95% of the pods contained an average of 3.5 and 12.7 seeds per pod, respectively. Further observations were conducted in field plots for both sarson and toria crops. When the crops were caged to exclude insects, the yield was 68 g. However, when caged with *Apis cerana* (honeybees), the yield increased to 219 g. In open-field conditions near honeybee colonies, where other insect pollinators were also present, the yield further increased to 244 g (Latif et al. 1960).

In a similar study, Sihag (1986) examined the seed production in toria plots under different conditions, caged and uncaged, allowing for natural cross-pollination. The results showed that in the caged plots, there were 25 pods per plant, with an average of 6.3 seeds per pod and a total of 4 seeds per plant. In the uncaged plots, the numbers increased significantly to 1,060 pods per plant, with an average of 18.2 seeds per pod and a total of 385 seeds per plant. Similar findings were observed in sarson crops, where Mishra et al. (1988) reported that 59% and 95% of the pods contained an average of 3.5 and 12.7 seeds per pod, respectively. Further observations were conducted in field plots for both sarson and toria crops. When the crops were caged to exclude insects, the yield was 68 g.

FIGURE 15.2 Caged experiment in (a) Gobhi sarson, (b) mustard.

However, when caged with *Apis cerana* (honeybees), the yield increased to 219 g. In open-field conditions near honeybee colonies, where other insect pollinators were also present, the yield further increased to 244 g (Latif et al. 1960).

In general, there is quite a high degree of self-incompatibility in these crops, and good yields are obtained by cross-pollination with the aid of natural fauna in most places, where good weather prevails and probably yield can be further increased with the help of honeybee colonies to the extent of 10–30% (Rahman 1940; Singh 1954), covering for inclement weather. Singh and Singh (1992) made an observation that plants pollinated by bees within field cages yielded pods that were three times heavier compared to plants that were self-pollinated within individual bags. Furthermore, these bee-pollinated plants produced four times more seeds per pod, 50 times more seeds overall, and 84 times higher seed yield per plant.

Singh and Singh (1992) studied the impact of bee pollination on seed yield, carbohydrate composition, and lipid composition of mustard seed *Brassica campestris* L. and found that bee-pollinated plants produced 3 times heavier pods, 4 times more seeds per pod, 50 times more seeds per plant, 11 times more pods per plant, and 84 times more seed yield per plant than self-pollinated plants. Carbohydrate content was inversely proportional to lipid content. The total amount of carbohydrates in self-pollinated plant seeds was approximately twice as high as the carbohydrate content found in seeds of plants pollinated by bees. Triglycerides constituted the majority of neutral lipids. Sterol was the least abundant of all the lipids, and phosphatidylcholine was absent from all seeds. High lipid content was directly related to high seed yield, and concentration of total lipid increased according to the type of pollination in the order SP < OP < HP < BP.

Deodikar and Suryanarayana (1972) found enhanced yield in oilseed crops when pollinated by bees. Plots receiving unrestrained visits had significantly higher plant density, seed yield, seed/plant, weight of 1,000 seeds, weight of seeds per plant, germination, and percentage of oil content as compared to those isolated from insect visits.

Mishra et al. (1988) reported that percent pod setting was significantly higher in plants which had access to insect pollination (95.21%) than those which were net-caged (70.27%) or muslin-bagged (58.81%). Similarly, the number of seeds per pod was also significantly higher in open-pollinated flowers (912.62 seeds/pod) in comparison to nylon-caged (5.55 seeds/pod) and muslin-bagged (3.05 seeds/pod).

Pods obtained from open-pollinated flowers exhibited a significantly higher percentage of healthy seeds (84.78%) compared to those from flowers that were net-caged (41.45%) or bagged with muslin (48.37%). However, the latter two did not vary significantly. Furthermore, the average weight of 100 healthy dry seeds from muslin-bagged flowers (542.20 mg) was significantly greater than that of seeds from open-pollinated (478.50 mg) and net-caged flowers (459.40 mg). In terms of oil content, the percentage of oil in healthy seeds from muslin-bagged flowers (41.64%) was significantly higher than that from open-pollinated flowers (39.02%). However, there was no significant difference in oil content between seeds from muslin-bagged and net-caged flowers. Taking all yield parameters into account for calculating oil content per pod, the total oil yield increase from open-pollinated and net-caged plants was 9.76 and 1.55 times higher, respectively, compared to muslin-bagged plants.

Singh et al. (2000) found that there was significant improvement in qualitative and quantitative parameters of rape when pollinated by *Apis cerana Himalaya*. Atmowidi et al. (2007) analyzed the diversity of pollinator insects and its effect to seed set of mustard (*Brassica rapa*) planted in agricultural ecosystem in West Java. They found 19 species of insects visiting *Brassica* flowers. Chhuneja et al. (2007) observed a significant increase in various factors, including the average number of siliquae per plant, the number of seeds per 20 siliquae, the weight of a thousand seeds, the seed yield per plant, and the oil content, when the distance from bee colonies was within 50 and 75 m. It was further reported that one colony of *Apis mellifera* with ten-bee strength was sufficient for effective pollination of 1 ha of crop area. Araneda et al. (2010) evaluated the yield of *B. napus* cv. Artus pollinated by *A. mellifera* in Chile and found that the parameter least affected by bee intervention was the grains-per-silique variable.

TABLE 15.2
Seed Yield and Yield Parameters in Toria as Influenced by Varied Mode of Pollination

Yield Attribute	Bee Pollination	Open Pollination	Closed Pollination
Siliqua per plant	258.33 ± 0.25	276.33 ± 0.16	122.67 ± 0.32
Seed per siliqua	22.10 ± 0.11	29.50 ± 0.22	15.67 ± 0.17
Pod length (cm)	6.23 ± 0.006	7.37 ± 0.18	5.26 ± 0.01
1,000 seed weight (g)	5.65 ± 0.01	5.97 ± 0.04	2.15 ± 0.01
Seed yield (q/ha)	12.77 ± 0.02	13.73 ± 0.01	7.50 ± 0.02

Source: Sharma and Abrol (2014).

Sharma and Abrol (2014) found that the number of siliqua per plant in toria under open pollination were twice (279.33 siliqua/plant) than those formed under caged condition (128.99 siliqua/plant). Seed weight was lower under caged conditions than under open pollination, and more and better-quality seeds were obtained from open pollination. Munawar et al. (2009) reported significant increases in a range of plant parameters when caged with bees as compared to plants without bees.

Abrol (2009) studied that pollination requirement of different oilseed crops and expected increase in yield due to bee pollination was found to be between 13 and 222% in mustard and from 21 to 3,400% in sunflower.

In rapeseed, while wind plays a significant role in transporting pollen over long distances (Hoyle et al. 2007), bees are the primary pollinators within the cultivars and varieties of rapeseed used for seed production, as they are the most abundant insects (Westcott and Nelson 2001; Pordel et al. 2007). Therefore, the introduction of beehives of *A. mellifera* for controlled pollination can effectively enhance the crop yield of rapeseed (*Brassica napus*) (Sabbahi et al. 2006). Rapeseed plants produce ample pollen and high-quality nectar with relatively high sugar concentrations, and their flowers exhibit attractive colors and structures specifically appealing to insects, particularly bees (Smith 2002; Sabbahi et al. 2006).

The studies clearly establish that most of the oilseed crops are cross-pollinated, and adequate pollination is vital for any significant seed production. Mustard (Mohammed 1935; Latif et al. 1960) and oilseed rape (Langridge and Goodman 1975; Kisselhagen 1977) have been reported to result in an increase in seed yield due to insect pollination. Latif et al. (1960) observed that the presence of *Apis cerana* colonies near sarson and toria fields resulted in a 60% increase in seed yield. Similar findings were observed in *Brassica campestris* var Jambuck, where Kubisova et al. (1980) reported a significant increase in seed yield. Bisht et al. (1980) conducted a study and observed that rapeseed flowers visited by *Apis species* exhibited a higher pod set, greater number of seeds per pod, and heavier seed weight compared to those deprived of pollinator visits.

Mishra et al. (1988) found that in *Brassica campestris* percent pod setting, number of pods per plant and proportion of healthy seeds were significantly higher in open-pollinated flowers than in net-caged and muslin-bagged ones. Similarly, open-pollinated flowers exhibited higher average seed weight and oil content. *Apis cerana* was the most common pollinating species. The other pollinators observed included *A. mellifera*, syrphid flies, etc. In a similar study, Prasad et al. (1989) found that pollination *of Brsassica juncea* by *A. cerana* resulted in more siliqua setting, increased length of siliqua, seed weight increased yield, and had a pronounced effect on oil contents and germination.

Mishra and Kaushik (1992) found that honeybee-pollinated crops exhibited higher yield and oil content in comparison to self-pollinated. The results revealed that open-pollinated crops of *Brassica nigra*, *B. campestris*, *B. carinata*, *B. napus*, and *Eruca sativa* exhibited a higher percentage of successful pod formation, with rates of 80.00%, 84.42%, 71.90%, and 70.30%, respectively. In terms of seed characteristics, the average weight of 1,000 healthy seeds ranged from 3 to 5.1 g when open-pollinated, whereas it varied from 1.8 to 3.7 g when self-pollinated.

Langridge and Goodman (2003), in a trial on oilseed rape (*Brassica napus*) cv. Midas in Northern Victoria, found that no increase in yield of seed, oil content, or percentage germination was obtained from plots where bees and larger insects had access compared with plots enclosed in cages of 2.5 mm mesh to exclude these insects. The weight of 1,000 seeds was slightly greater in the enclosed plots.

Insect pollination is one of the most important factors enhancing crop production. Among the various pollinating agents, insect pollinators play a predominant role in increasing the yield of oilseed crop. Almost one-third of total cropped area under oilseeds has been reported to be entomophilous (Mishra et al. 1988). Pollination by insects can effectively double the yield of rapeseed. In addition to boosting crop yield, pollinators play a crucial role in promoting uniform and early pod setting. Being a cross-pollinated crop, rapeseed attracts a large number of insect pollinators.

15.3 SUNFLOWER (*HELIANTHUS ANNUS* L., FAMILY COMPOSITAE)

In case of sunflower (*Helianthus annuus* L.), there is strong evidence that insufficient pollination can significantly minimize the yields (Free 1999). The pollination of sunflowers by honeybees (*Apis mellifera* L.) results in a 30% increase in seed yield and a more than 6% rise in oil content in hybrid varieties (Furgala et al. 1979; Jyoti and Brewer 1999). Honeybees are widely recognized as the primary pollinators of sunflowers across the globe. Recent research conducted by Greenleaf and Kremen (2006) clearly demonstrates the beneficial role of non-Apis bees in sunflower pollination, particularly through their behavioral interactions with honeybees. When multiple non-Apis bees forage simultaneously on the same floral heads, honeybees are compelled to move among numerous flowers. This enhances the efficiency of honeybees in pollinating sunflowers.

In an attempt to determine the effect of honeybee visits on sunflower seed production in Manitoba, Furgala (1954) placed 2.5 colonies per hectare along the east border of sunflower fields. The seed production per hectare in these two fields was as much as recorded in four fields located 4.8 km away from the honeybee colonies. In addition, he observed a gradual decrease in seed yield as the distance extended from the east to the west in the farmers' field. Manzoor-ul-Haq and Fiaz (1980) studied insect pollination of sunflower in Pakistan. Wakhle et al. (1978), on the basis of chemical analysis of sunflower seeds obtained from different pollination treatment, showed that there was a significant increase in oil (6.5%) as also oil and protein contents together (7%) in seeds resulting from bee pollination when compared with self-pollinated seeds.

It is evident that bee pollination brings about not only a quantitative but also a qualitative enhancement in sunflower seeds. In a similar study conducted by Singh et al. (2000), it was observed that seeds in bee pollination and controlled pollination treatments had higher oil percentages and higher oil and protein contents compared to those in self-pollination or even open-pollination treatments. Seeds under OP, BP, and CP treatments showed nearly 6, 7, and 8.5% increase in oil and protein contents over SP treatment, respectively.

Rajasri et al. (2012), in a two-year study on sunflower hybrid NDSH1, found that the crop covered with insect-proof net without honeybees recorded significantly lower seed setting percent in both the years of study with a mean of 6.33% compared to other methods.

Among the various pollination methods that involved the introduction of honeybees, all were observed to be considerably more effective than open pollination, with a difference of 16.52%. The combination of bee pollination and hand pollination yielded the highest average seed setting percentage of 74.13%, followed by bee pollination with eight frames at 50.89% and four frames at 33.26%, respectively.

Honeybees + hand pollination gave significantly higher seed yield (14.4 q/ha) than open pollination (6.4 q/ha), and with crop caged with net (2.3 q/ha), thereby the yield was increased about 526% with honeybee + hand-pollination and 178% with open pollination compared to control plots. The introduction of honeybee colonies, consisting of eight frames and four frames, respectively, resulted in a significant increase in sunflower yield of approximately 330% and 191%, respectively.

The supplemental honeybee pollination + hand-pollination significantly increased the percentage of seed setting and seed yield compared with open pollination and crop in cages without honeybees.

Hybrid sunflower seeds produced through a combination of honeybee pollination and hand-pollination showed remarkable superiority in terms of germination rates. The percentage of germination reached 99% for this method, surpassing both open pollination (96%) and exclusive bee pollination (95%). These results highlight the significant impact of bees on seed germination compared to the control plot, where bees were absent, resulting in a germination rate of 93%.

15.4 SAFFLOWER (*CARTHAMUS TINCTORIUS* L., FAMILY ASTERACEAE)

Safflower has been reported to benefit greatly from insect pollination (Knowles 1969).

Classen (1950) observed a range of cross-pollination rates from 0 to 100%. The majority of plants exhibited cross-pollination rates varying between 5 and 40% (Kadam and Patankar 1942; Levin and Butler 1966; Butler et al. 1966; Levin et al. 1967). In majority of the plants, cross-pollination ranged from 5 to 40%. Safflower is usually considered to be a self-pollinated crop. Kumari and Pandey (2005) found that there was 85.99% seed setting in the open-pollinated heads, followed by 38.15% (in muslin cloth) and 35.54% (in butter paper) bagged condition.

15.5 SESAME (*SESAME INDICUM* L., FAMILY PEDALIACEAE)

Sesame plants exposed to insects had a 25% higher yield compared to those covered with cages. Cages containing *A. cerana* bees yielded 2.6 kg of okra per plot, while cages with regular bees produced 1.6 kg (Phadke et al. 1967; Panda et al. 1988). In open areas, *A. cerana*, *A. dorsata*, and *A. florea* were the most common insects observed on the crops, and they started foraging at 0600, 0700, and 0800 h, respectively. They visited 6.3, 5, and 1.9 flowers per minute, respectively, and the flowers pollinated by them produced 51, 53, and 4 seeds per pod. Additionally, the pollinated flowers had a heavier seed weight compared to the control group (Rao et al. 1981). During foraging, 72% of *A. cerana* bees collected nectar, 23% collected pollen only, and 4% collected both nectar and pollen. At some point, they touched anthers, facilitating cross-pollination and increasing the overall yield (Phadke et al. 1967). *A. dorsata* was also identified as an important pollinator of sesame. Abrol (1991) observed that bee activity in the field was more intense at higher temperatures, light intensity, and radiation. Extensive studies on the genetics and pollination requirements of sesame were conducted in Punjab by Sikka and Gupta in 1949.

15.6 LINSEED/FLAX (*LINUM USITATISSIMUM* L., FAMILY LINACEAE)

Flax is generally considered to be a self-pollinated crop, although small degree of natural cross-pollination (ranging from 1 to 6%) also occurs, particularly among the larger-flowered varieties (Dillman 1938; Dillman and Goar 1937; Masuo 1958). Rubis (1970) conducted research using a male sterile line with disk-form flowers and observed minimal cross-pollination between the male sterile lines with tubular flowers.

Several studies have demonstrated that bee pollination significantly enhances seed yields in fiber flax. Bezdenezhnykh (1956) reported that honeybees confined in cages increased seed production in fiber flax by 22.5% compared to plants in cages without bees. Similarly, Gubin (1945) reported an increase in seed production ranging from 22.5 to 38.5% due to bees' presence. Luttso (1957) reported that bee pollination resulted in a 29% increase in seed production, 18% increase in the number of seeds per capsule, and 11% increase in the weight per seed when compared to fields without bee pollination. Similarly, Smirnov (1954) observed a 19% increase in the number of seeds per capsule, 22% increase in the total weight of seeds, and a 2.2% increase in the weight per seed due to bee visitation. He also noted that bee-visited plants initiated crop development earlier and ceased blooming sooner than plants from which bees were excluded.

Kozin (1954) reported a significant increase in the number of seeds per boll and seed weight when 40 colonies were placed near a fiber flax field, although the exact size of the field was not specified. Overall, the studies on crop pollination provide clear evidence that cross-pollination not only enhances the quantity but also improves the quality of oil crops. Even self-compatible varieties demonstrate increased yields and improved quality when subjected to cross-pollination. Therefore, pollination plays a vital role in boosting oil crop production, particularly in a world facing a shortage of oil and oilseeds.

Clearly, most of the oilseed crops are cross-pollinated and exclusively depend upon bees and other pollinators for pollination services.

15.7 POLLINATION MANAGEMENT

15.7.1 Problems Associated With Bee Pollination

Pollinator–plant interaction has many overlapping effects. The protection of pollinators, including honeybees, is as essential as the protection of crops from the insect pest damage. The use of pesticides for pest control and pollination for enhanced production are important components of modern agriculture. Unfortunately, these two practices are not always compatible, as honeybees are susceptible to many of commonly used pesticides (Johansen 1977; Russell et al. 1998; Cunningham et al. 2002; Sundararaju 2003), used for the control of insect pests (Poehling 1989; Stark et al. 1995). The indiscriminate and excessive use of pesticides for controlling insect pests is a major constraint (Bisht et al. 1980; Rana and Goyal 1991; Zhong et al. 2004). The loss of honeybees directly affects beekeeping through the loss of honey production, and indirectly the crop production, due to inadequate pollination. Reduction of population of these beneficial insects due to insecticides, therefore, incurs significant environmental, ecological, and economic costs (Bai and Reddy 1977; Pimentel et al. 1980a, 1980b; Crane and Walker 1983; Parkash and Kumaraswami 1984; Khan and Dethe 2004).

Furthermore, modern farming methods have led to the decline of indigenous insect pollinators and disrupted the intricate connection between insects and flowers due to the disappearance of unused lands and uncultivated areas, the elimination of specific food sources through weed management, and general alterations in the surrounding ecosystem.

Wild bees are also damaged by pesticides. Poisoning may result from contaminated food as well as from florets, leaves, soil, or other material used by the bees in nesting. The toxicity of a specific insecticide to honeybees and wild bees is not always the same, and even among wild bees, some materials are more toxic to one species than to another. Poisoning can result from florets, leaves, soil, or other material used by the bees in nesting. The following practices need to be adopted to reduce bee poisoning:

1. Apply pesticides only when needed.
2. The pesticide selected should be less hazardous to bees and with the lowest residual effect.
3. Use liquid or granule formulations and avoid dusts. Avoid microencapsulated forms of pesticides.
4. Ground application is less hazardous than aerial application, particularly when applied in close distances to apiaries. Fine sprays are less toxic than coarse sprays.
5. Apply pesticides when bees are not foraging on the crop. Early morning, late afternoon, or early evening.
6. The time of day when a pesticide is applied should be chosen to minimize the risk of spray drift occurring either over apiaries or over plants being foraged by bees. Treatments can and should be applied only when bees are not foraging for nectar or pollen. If bee-attractive plants need to be treated while in bloom, they should be treated at night or in the early morning or late evening when the bees are not flying.

7. Where practical, beekeepers should be given prior notice, preferably a minimum of 48 h, of a pesticide application to allow apiaries to be moved from the area or their entrance closed.

15.7.2 Number of Colonies Required for Pollination

Several investigators have attempted to determine the number of colonies of honeybees required for increased yields in rape, and their recommendations vary from place to place and crop. Hammer (1963, 1966) suggested a density of three colonies per hectare, while Radchenko (1964) recommended two colonies, Downey and Bolton (1961) proposed one colony, White (1970) suggested two colonies, and Vesely (1962) suggested three to four colonies per hectare. While hoverflies may have some involvement in rape pollination, we regard honeybees as the more effective pollinators. It is necessary to establish the optimal population of pollinators and the appropriate distribution of colonies to achieve the most efficient pollination for different crops.

15.7.3 Pollination Recommendations

Honeybees are the most effective agents involved in cross-pollination of rapeseed-mustard and other oil crops. The excessive use of pesticides and fungicides often results in the indiscriminate killing of numerous pollinators. In some instances, cultivating a single crop over a vast area further diminishes the population of wild honeybee colonies in those regions. Consequently, recognizing the significance of managed and migratory beekeeping in the field has become crucial to enhance the production of oilseed crops. To achieve effective pollination and maximize oilseed crop yields, the efficacy of bee colonies as pollinators relies on various factors, including colony strength, the number and timing of colony placements, their distribution in the field or orchards, and prevailing weather conditions.

15.8 CONCLUSIONS AND FUTURE STRATEGIES

This chapter clearly demonstrates that supplementing oilseed fields with honeybee colonies can greatly enhance oilseed production through pollination interventions. However, a more comprehensive approach to managing crop pollination is required to ensure a dependable supply of pollinators, both from managed colonies and native pollinators.

The honeybee pollination is not only responsible to enhance the yield in self-incompatible varieties, but even the self-fertile varieties give manifold increased yield due to intervention of managed pollinators. This requires understanding of the biology and ecology of pollinating insects, as well as providing appropriate nesting habitat and ensuring the availability of alternative sources of "forage" to sustain them when target crops are not available. Additional role can be played by native or wild pollinators, provided that attention is given to curtailing of population losses caused by both inadvertent insecticide poisoning and habitat destruction (Kevan et al. 1990). Pollinator management and managed pollination are the common efforts recently practiced for maximization of production in cross-pollinated crops and to bring the pollinator to the target crop. Kluser and Peduzzi (2007) have presented strong evidence indicating a decline in pollinators due to the degradation of both local and global environments.

Recent declines in both wild and domesticated pollinators and parallel declines in the plants which rely upon them possess a serious threat to global food security and sustainable agriculture (Potts et al. 2010). As the preference for cultivating crops that rely on pollinators persists, the demand for the essential role fulfilled by dwindling pollinators will substantially rise in the upcoming times.

Kevan et al. (1990) highlighted the pivotal role played by managed and native insect pollinators in the sustainability of agricultural practices. However, they have declined in some regions (Biesmeijer et al. 2006; Potts et al. 2010). The suitability of farm environments for insect pollinators can be restricted due to various factors linked to intensified farming practices aimed at accommodating larger human populations.

These factors include the reduction of natural areas, habitat fragmentation, and a scarcity of flowering food and nesting resources, as noted by Carvell et al. in 2006. The planting of monocultures, which lack floral diversity, can also restrict the availability of resources for pollinators throughout the entire season.

The presence of a greater variety of landscapes fosters higher insect pollinator diversity, whereas agricultural landscapes with limited resources may experience a decline in this diversity (O'Toole 1993). Moreover, the diminished presence of wild bees in crop pollination can be a potential consequence (Potts et al. 2010). However, by introducing habitats that include flowering plants and grasses, which enhance structurally deficient environments, it is possible to provide support for beneficial insects in agricultural landscapes (Long et al. 1998; Kells et al. 2001; Sheffield et al. 2008).

These agricultural restoration programs are expected to provide the greatest support for bees in simple landscapes with the greatest floral contrast to the background landscape (Scheper et al. 2013). Wildflower plantings can offer pollen and nectar sources during periods when the crop is not flowering. Furthermore, depending on the biological characteristics of the bee species, these plantings may serve as nesting habitats (Carreck and Williams 2002; Kremen et al. 2004; Heard et al. 2007).

Hoverflies play a valuable role in pollinating various crops, including mango (Dag and Gazit 2001) and oilseed rape (Jauker and Wolters 2008). The larvae of hoverflies that feed on aphids act as effective biological agents for controlling numerous soft-bodied arthropods (Bugg et al. 2008; Smith et al. 2008). This dual contribution of hoverflies provides an extra ecosystem service to agricultural crops.

REFERENCES

Abel, C. A. and Wilson, R. L. 1999. The use of diverse plant species for increasing *Osmia cornifrons* (Hymenoptera: Megachilidae) in field cages. *J. Kansas Entomol. Soc.* 71, 23–28.

Abel, C. A., Wilson, R. L., and Luhmann, R. L. 2003. Pollinating efficacy of *Osmia cornifrons* and *Osmia lignaria* subsp. *lignaria* (Hymenoptera: Megachilidae) on three Brassicaceae species grown under field cages. *J. Kans. Entomol. Soc.* 38, 545–552.

Abrol, D. P. 1991. Path analysis of environmental factors influencing daily flight activity of *Apis dorsata* F. *Acta Ecologia, Oecologica Generalis* 12(6), 619–826.

Abrol, D. P. 2007. Foraging behaviour of Apis mellifera and Apis cerana as determined by the energetics of nectar production in different cultivars of Brassica campestris var. toria. *Journal of Apicultural Science* 51(2), 5–10.

Abrol, D. P. 2008. Plant insect interaction in crucifers. In: Gupta, S. K. (ed.) *Biology and Breeding of Crucifers*. CRC Press, Taylor and Francis, Boca Raton, FL, pp. 129–149.

Abrol, D. P. 2009. *Bees and Beekeeping in India*, 2nd edn. Kalyani Publishers, New Delhi, p. 720.

Abrol, D. P. and Shankar, U. 2012. Pollination in oil crops: Recent advances and future strategies. In: Gupta, S. K. (ed.) *Technological Innovations in Major World Oil Crops*, vol. 2. Springer, New York, pp. 221–267.

Adegas, J. E. B. and Nogueira Couto, R. H. 1992. Entomophilous pollination in rape (*Brassica napus* L. var. *oleifera*) in Brazil. *Apidologie* 23, 203–209.

Araneda, X. D., Ulloa, R. B., Carrillo, J. A., Contreras, J. L., and Bastidas, M. T. 2010. Evaluation of yield component traits of honeybee pollinated (*Apis mellifera* L.) rapeseed canola (*Brassica napus* L.). *Chilean Journal of Agricultural Research* 70(2), 309–314.

Atmowidi, T., Buchori, D., Manuwoto, S., Suryobroto, B., and Hidayat, P. 2007. Diversity of pollinator insects in relation to seed set of mustard (*Brassica rapa* L.: Cruciferae). *HAYATI Journal of Biosciences* 14(4), 155–161.

Bai, A. R. K. and Reddy, C. C. 1977. Laboratory toxicity of some insecticides to *Apis cerana indica*. *Journal of Apicultural Research* 16(3), 142, 161–162.

Bezdenezhnykh, S. M. 1956. Pollination of fiber flax and training of bees. In: Krishchunas, I. V. and Gubin, A. F. (eds.) *Pollination of Agricultural Plants*. Moskva, Gos. Izd-vo. Selkhoz Lit-ry, pp. 21–24. [In Russian.]

Bichee, S. L. and Sharma, M. 1988. Effect of different modes of pollination in sunflower *Helianthus annuus* L. (Compositae) sown on different dates by *Trigona iridipennis* Smith. *Apiacta* 23, 65–68.

Biesmeijer, J. C. et al. 2006. Parallel declines in pollinators and insect-pollinated plants in Britain and the Netherlands. *Science* 313, 351–354. doi:10.1126/science.1127863

Bisht, D. D., Naim, M., and Mehrotra, K. N. 1980. Studies on the role of honeybees in rapeseed production. In: *Proc. 2nd Int. Conf. Apiculture in Tropical Climate*. Indian Agricultural Research Institute, New Delhi, pp. 491–496.

Bommarco, R., Marini, L., and Vaissiere, B. E. 2012. Insect pollination enhances seed yield, quality, and market value in oilseed rape. *Oecologia* 169(4), 1025–1032.

Bugg, R. L., Colfer, R. G., Chaney, W. E., Smith, H. A., and Cannon, J. 2008. *Flower Flies (Syrphidae) and Other Biological Control Agents for Aphids in Vegetable Crops* (book online). University of California, Division of Agriculture and Natural Resources, California. Available at https://escholarship.org/uc/item/37k3d5g8

Bürger, C. 2004. Effects of landscape structure on bee diversity and pollination at different spatial scales. PhD thesis. Institute of Agroecology, University of Göttingen, Germany.

Butler, G. D., Jr., Werner, E. G., and Leven, M. D. 1966. Native bees associated with safflower in South Central Arizona. *Kans. Ent. Soc. Jour.* 39(3), 434–436.

Carreck, N. and Williams, I. H. 2002. Food for insect pollinators on farmland: Insect visits to flowers of annual seed mixtures. *Journal of Insect Conservation* 6(1), 13–23.

Carvell, C., Roy, D. B., Smart, S. M., and Pywell, R. F. 2006. Declines in forage availability for bumblebees at a national scale. *Biological Conservation* 132(4), 481–489.

Charlesworth, D. and Charlesworth, B. 1987. Inbreeding depression and its evolutionary consequences. *Annual Review of Ecology and Systematics* 18, 237–268.

Chhuneja, P. K., Singh, J., Gatoria, G. S., and Blossom. 2007. Assessing the role of honey bee (*Apis mellifera* L.) in the seed production of *Brassica campestris* var. toria in the Punjab. *Indian Journal of Crop Science* 2(2), 34–40.

Classen, C. E. 1950. Safflower production in the western part of the northern great plains, University of Nebraska Agriculture Experiment Station Circular No. 87 rev.

Crane, E. 1991. *Apis* species of tropical Asia as pollinators and some rearing methods for them. *Acta Hort.* 288, 29–48.

Crane, E. and Walker, P. 1983. Pollination of tropical and subtropical crops by bees. In: Crane, E. and Walker, P. (eds.) *The Impact of Pest Management on Bees and Pollination*. IBRA, Cardiff, pp. 5–21.

Cunningham, S. A., FitzGibbon, F., and Heard, T. A. 2002. The future of pollinators for Australian Agriculture. *Agriculture* 53(8), 893–900.

Dag, A., and Gazit, S. 2001. Mango pollinators in Israel. *J. Appl. Hortic.* 2, 39–43.

Dag, A., Lior, E., and Afik, O. 2002. Pollination of confection sunflowers (*Helianthus annuus* L.) by honey bees (*Apis mellifera*). *Am. Bee J.* 142, 443–445.

DeGrandi-Hoffman, G. and Martin, J. H. 1993. The size and distribution of the honey bee (*Apis mellifera* L.) cross-pollinating population on male sterile sunflowers (*Helianthus annuus* L.). *J. Apic. Res.* 32, 135–142.

DeGrandi-Hoffman, G. and Watkins, J. C. 2000. The foraging activity of honey bees *Apis mellifera* and non-*Apis* bees on hybrid sunflowers (*Helianthus annuus*) and its influence on cross-pollination and seed set. *J. Apic. Res.* 39, 37–45.

Delaplane, K. S. and Mayer, D. F. 2000. *Crop Pollination by Bees*. CABI Publishing, New York.

Deodikar, G. B. and Suryanarayana, M. C. 1972. Crop yields and bee pollination. *Indian Bee J.* 34, 53–64.

Dillman, A. C. 1938. Natural crossing in flax. *J. Am. Soc. Agron.* 30, 279–286.

Dillman, A. C. and Goar, L. G. 1937. Flaxseed production in the far western states. *U.S. Dept. Agr. Farmers' Bul.* 1792, 22 pp.

Downey, R. K. and Bolton, J. L. 1961. Production of [Polish and Argentine] rape in Western Canada. *Canada Dept. Agr. Res. Br. Pub.* 1021, 19 pp.

Free, J. B. 1993. *Insect Pollination of Crops*. 2nd edn. Academic Press, Harcourt Brace Jovanovich Publication, London, 684 pp.

Free, J. B. 1999. Pollination in the tropics. *Beekeeping and Development* 51, 6–7.

Free, J. B. and Ferguson, A. W. 1983. Foraging behaviour of honeybees on oilseed rape. *Bee World* 64, 22–24.

Free, J. B. and Williams, I. H. 1977. The Pollination of Crops by Bees. Apimondia and International Bee Research Organisation, Bucharest, Romania.

Furgala, B. 1954. Honey bees increase seed yields of cultivated sunflowers. *Glean. Bee Cult.* 82, 532–534.

Furgala, B., Noetzel, D. M., and Robinson, R. G. 1979. Observations on the pollination of hybrid sunflowers. *Proc. IVth Int. Symp. on Pollination, Md. Agric. Exp. Sta. Spec. Misc. Publ.* 1, 45–48.

Garibaldi, L. A., Aizen, M. A., Klein, A. M., et al. 2011. Global growth and stability of agricultural yield decrease with pollinator dependence. *Proc. Natl. Acad. Sci. USA* 108, 5909–5914.

Greenleaf, S. A. and Kremen, C. 2006. Wild bees enhance honey bees' pollination of hybrid sunflower. *Proc. Nat. Acad. Sci. USA* 103, 13890–13895. doi:10.1073/pnas.0600929103

Gubin, A. F. 1945. Cross pollination of fibre flax. *Bee Wld.* 26, 30–31.

Guidry, N. P. 1964. Graphic summary of world agriculture. *U. S. Dept. Agr. Misc. Pub.* 705, 64 pp.

Hammer, O. 1963. Summer rape as a competitor affecting the pollination of clovers. *Dansk Froavl.* 14, 7 pp.

Hammer, O. 1966. Some problems of competition between summer rape and clover, in relation to pollination. *Bee World* 47(Suppl. 1), 99–106.

Heard, M. S., Carvell, C., Carreck, N. L., Rothery, P., Osborne, J. L., and Bourke, A. F. G. 2007. Landscape context not patch size determines bumble-bee density on fower mixtures sown for agri-environment schemes. *Biol. Lett.* 3, 638–641.

Heard, T. A. 1999. The role of stingless bees in crop pollination. *Ann. Rev. Entomol.* 44, 183–206.

Hoyle, M., Hayter, K., and Cresswell, J. E. 2007. Effect of pollinator abundance on self-fertilization and gene flow: Application to GM Canola. *Ecol. Appl.* 17, 2123–2135.

Jauker, F. and Wolters, V. 2008. Hover flies are efficient pollinators of oilseed rape. *Oecologia* 156, 819–823.

Johansen, C. A. 1977. Pesticides and pollinators. *Ann. Rev. Ent.* 22, 177–192.

Jyoti, J. and Brewer, G. J. 1999. Effect of honeybee (Hymenoptera: Apidae) pollination on sunflower hybrids. Proc. 21st Sunflower Research Workshop. January 14–15, Nat. Sunflower Assoc, pp. 103–107.

Kadam, B. S. and Patankar, V. K. 1942. Natural cross-pollination in safflower. *Indian Jour. Genet. Plant Breed* 2, 69–70.

Kells, A. R., Holland, J. M., and Goulson, D. 2001. The value of uncropped field margins for foraging bumblebees. *J. Insect Conservation* 5, 283–291.

Kevan, P. G., Clark, E. A., and Thomas, V. G. 1990. Insect pollinators and sustainable agriculture. *Am. J. Altern. Agric.* 5, 12–22.

Khan, R. B. and Dethe, M. D. 2004. Median lethal time of new pesticides to foragers of honeybee. *Pestology* 28(1), 28–29.

Kisselhagen, S. 1977. Seed yield is increased if there are bees on the rape. *Tidsskrift for Biavl* 111, 66–67.

Klatt, B. K., Holzschuh, A., Westphal, C., Clough, Y., Smit, I., Pawelzik, E., and Tscharntke, T. 2014. Bee pollination improves crop quality, shelf life and commercial value. *Proc. R. Soc. B* 281, 20132440.

Klein, A. M., Vaissiere, B. E., Cane, J. H., Steffan-Dewenter, I., Cunningham, S. A., Kremen, C., and Tscharntke, T. 2007. Importance of pollinators in changing landscapes for world crops. *Proc. R. Soc. B Biol. Sci.* 274, 303–313.

Kluser, S. and Peduzzi, P. 2007. *Global Pollinator Decline: A Literature Review.* UNEP/GRID, Geneva.

Knowles, P. F. 1969. Modification of quantity and quality of safflower oil through plant breeding. *Journal of the American Oil Chemists Society* 46, 130–132.

Kozin, R. B. 1954. Effect of bees on the seed crop of flax. *Pchelowdstvo, Mosk.* 6, 41–43.

Kremen, C., Williams, N. M., Bugg, R. L., Fay, J. P., and Thorp, R. W. 2004. The area requirements of an ecosystem service: Crop pollination by native bee communities in California. *Ecology Letters* 7(11), 1109–1119.

Kubisova, S., Nedbalova, V., and Plesnik, R. 1980. Cinnost vcely medonosne *(Apis mellifera* L.) na repce *(Brassica napus* L. var. *napus). Pol'nohospoddrstvo* 26, 744–754.

Kumari, A. and Pandey, A. K. 2005. Pollination mechanism and behaviour of pollinators in safflower (*Carthamus tinctorius* L.). *J. Ind. Bot. Soc.* 82, 56–61.

Langridge, D. F. and Goodman, R. D. 1975. A study on pollination of oilseed rape (*Brassica campestris*). *Austral. J. Exptl. Agric. Anim. Husband.* 15, 285–288.

Langridge, D. F. and Goodman, R. D. 2003. Honeybee pollination of oilseed rape, cultivar Midas. *Australian Journal of Experimental Agriculture and Animal Husbandry* 22(115), 124–126.

Latif, A., Qayyum, A., and Abbas, M. 1960. The role of *Apis indica* in the pollination of toria and sarson (*Brassica campestris* var. Toria and dichotoma). *Bee World* 41, 283–286.

Levin, M. D., Butler, G. D., and Rubis, D. D. 1967. Pollination of safflower by insects other than honeybees. *Jour. Econ. Ent.* 60, 1481–1482.

Long, R. F., Corbett, A., and Lamb, C. 1998. Beneficial insects move from flowering plants to nearby crops. *Cal. Ag.* 52(5), 23–26.

Luttso, V. P. 1957. The pollination of flax by honeybees. *Dokl. TSKhA* 30, 327–331.

Manning, R. and Boland, J. 2000. A preliminary investigation into honey bee (*Apis mellifera*) pollination of canola (*Brassica napus* cv. Karoo) in Western Australia. *Aust. J. Exp. Agr.* 40, 439–442.

Manzoor-ul-Haq and Fiaz, M. 1980. Insect pollination of sunflower *(Helianthus annuus)* in Pakistan. *J. Apic. Res.* 19, 83–88.

Masuo, Y. 1958. On natural crossing in flax. *Crop Sci. Soc. Japan Proc.* 27, 321–323. [In Japanese.] *Abstract in Plant Breed* 29(4), 817–818. Oct. 1959.

Mishra, R. C., Gupta, J. K., and Kumar, J. 1988. Effect of modes of pollination on fruit characteristics of okra, *Abelmoschus esculentum* (L.) Moench. *Proc. Indian Natl. Sci. Acad.* 53B(2), 157–160.

Mishra, R. C. and Kaushik, H. D. 1992. Effect of cross-pollination on yield and oil content of *Brassica campestris* L. var. sarson with pollination efficiency of honeybees, *Apis* spp. *Ann. Entomol.* 10, 33–37.

Mohammed, A. 1935. Pollination studies in tara *(Brassica napus* L. var. dichotsma Prain) and sarson *(Brassica campestris* L. var. sarson, Prain). *Indian J. Agric. Sci.* 5, 125–154.

Morandin, L. A. and Winston, M. L. 2005. Wild bee abundance and seed production in conventional organic, and genetically modified canola. *Ecol. Appl.* 15, 871–881.

Moreti, A. C. de C. C., da Silva, R. M. B., da Silva, E. C. A. et al. 1996. Aumento na produção de sementes de girassol (Helianthus annuus) Pela ação de insetos polinizadores. *Sci. Agric.* 53, 280–284.

Munawar, M. S., Sarwar, G., Raja, S., Waghchoure, E. S., Iftikhar, F., and Mahmood, R. 2009. Pollination by honeybee (Apismellifera) increases seed setting and yield in black seeds (Nigella sativa). *Int. J. Agric. Biol.* 11(2009), 611–615.

O'Toole, C. 1993. Diversity of native bees and agroecosystems. In: La Salle, J. and Gauld I. D. (eds) *Hymenoptera and Biodiversity.* CAB International, Wallingford, UK, pp. 169–196.

Panda, P., Sontakke, B. K., and Sarangi, P. K. 1988. Preliminary studies on the effect of bee (*Apis cerana indica* F.) pollination on yield of sesamum and niger. *Indian Bee J.* 50, 63–64.

Parkash, R. and Kumaraswami, T. 1984. Toxicity of some insecticide spraying on hymenopertrans pollinators in new Brunswick. *Canadian Entomologist* 112, 259–269.

Phadke, K. G., Pant, N. C., and Sinha, R. B. P. 1967. The foraging behaviour of *Apis indica* F ((Hymenoptera: Apidae) visiting flowers of sesame *(Sesamum orientate* L.). *Indian J. Ent.* 29, 157–163.

Pimentel, D. (ed.). 1980a. *CRC Handbook of Energy Utilization in Agriculture.* CRC Press, Boca Raton Florida.

Pimentel, D. et al. 1980b. Environment and pesticides: A preliminary assessment. *Oikos* 34(2), 126–140.

Poehling, H. M. 1989. Selective application strategies for insecticides in agricultural crops. In: *Pesticides and Non Target Invertebrates.* Intercept, Wimborne, Dorset, UK, pp. 151–175.

Pordel, M. R., Hatami, B., Mobli, M., and Ebadi, R. 2007. Identification of insect pollinators of three different cultivars of winter canola and their effect on seed yield in Isfahan. *J. Sci. Technol. Agric. Natur. Resour.* 10(4B), 413–426.

Potts, S. G. et al. 2010. Declines of managed honeybees and beekeepers in Europe? *J. Apic. Res.* 49, 15–22.

Prasad, D., Hameed, S. F., Singh, R., Yazdani, S. S., and Singh, B. 1989. Effect of bee pollination on the quantity and quality of rai crop (Brassica juncea Coss). *Indian Bee Journal* 51(2), 45–47.

Radchenko, T. H. 1964. The influence of pollination on the crop and the quality of seed of winter rape. *Bdzhil'nitstvo* 1, 68–74.

Rahman, K. A. 1940. Insect pollinators of toria (*Brassica napus* L. *var dichotoma) and Sarson* (brassica campestris L. var Sarson) at Lyallpur. *Indian J. Agric. Sci.* 10, 422–447.

Rajasri, M., Kankasurga, K., Anuradha, Ch., and Durga, V. 2012. Honey bees-potential pollinators in hybrid seed production of sunflower. *International Journal of Applied Biology and Pharmaceutical Technology* 3(2), 216–221.

Rana, B. S. and Goyal, N. P. 1991. Field toxicity of methyl demeton and dimethoate to the foragers of honeybee (*Apis cerana indica* F.) visitors to *Brassica chinensis* L. *Indian Bee Journal* 53(1–4), 73–77.

Rao, G. M., Lazar, G. M., and Suryanarayana, M. C. 1981. Foraging behavior of honeybees in sesame (Sesamum indicum L.). *Indian Bee J.* 43, 97–100.

Richards, A. J. 1997. *Plant Breeding Systems*, 2nd edn. London: Chapman and Hall.

Rubis, D. D. 1970. Breeding insect pollinated crops. *Arkansas Agricultural Extension Service* 127, 19–24.

Russell, D., Meyer, D., and Bukowski, J. 1998. Potential impact of micro encapsulated pesticides on New Jersey Apiaries. *American Bee Journal* 38(3), 207–210.

Sabbahi, R., De Oliveira, D., Marceau, J., and Houle, E. 2006. Floraison du canola (Crucifera: Brassicaceae) et le rendement des colonies d'abeilles domestiques (Hymenoptera: Apidae). Available at http://www.agrireseau.qc.ca/apiculture/documents/Rendem_apicole-Rachid_revise.pdf

Scheper, J., Holzschuh, A., Kuussaari, M., Potts, S. G., Rundlöf, M., Smith, H. G., and Kleijn, D. 2013. Environmental factors driving the effectiveness of European agri-environmental measures in mitigating pollinator loss—a meta-analysis. *Ecology Letters* 16, 912–920.

Sharma, D. and Abrol, D. P. 2014. Effect of insecticides on foraging behaviour and pollination role of Apis mellifera L. (Hymenoptera: Apidae) on toria (Brassica campestris var. toria) crop. *Egyptian Journal of Biology* 16, 79–86.

Sheffield, C. S., Westby, S. M., Smith, R. F., and Kevan, P. G. 2008. Potential of bigleaf lupine for building and sustaining *Osmia lignaria* populations for pollination of apple. *The Canadian Entomologist* 140, 589–599.

Sihag, R. C. 1986. Insect pollination increases seed production in cruciferous and umbelliferous crops. *J. Agric. Res.* 25, 121–126.

Sikka, S. M. and Gupta, N. P. 1949. Pollination studies in *Sesamum orientale* L. *Indian J. Genetics* 9, 33–191.

Singh, M. P., Singh, K. I., and Devi, C. S. 2000. Role of Apis cerena Himalaya pollination on yield and quality of rape seed and sunflower crops. Proceedings of 4th Asian Apicultural Association, International Conference, March 23–28, 1998, Kathmandu, Nepal.

Singh, R. P. and Singh, P. N. 1992. Impact of bee pollination on seed yield, carbohydrate composition and lipid composition of mustard seed. *Journal of Apicultural Research* 31, 128–133.

Singh, S. 1954. Insect pollination and breeding of fruit varieties. *Indian J. Hort. Sci.* 11, 123–125.

Smirnov, V. M. 1954. Cross-pollination of flax by bees. *Pchelovodstvo, Mosk.* 9, 53–55.

Smith, H. A., Chaney, W. E., and Bensen, T. A. 2008. Role of syrphid larvae and other predators in suppressing aphid infestations in organic lettuce on California's Central Coast. *Journal of Economic Entomology* 101, 1526–1532.

Smith, W. 2002. *Honey Bees on Canola*. New South Wales Agriculture, Department of Primary Industries, Orange, New South Wales, Australia.

Stark, D. J. D., Jepson, P. C., and Mayer, D. F. 1995. Limitation to use of topical toxicity data for production of pesticide side effects in the field. *Journal of Economic Entomology* 88(5), 1081–1088.

Sundararaju, D. 2003. Occurrence of bee fauna and extent of pollination in insecticide sprayed ecosystem of cashew. *J. Palynol.* 39, 121–125.

Vanbergen, A. J., Biesmeijer, J. C., Britton, N. F., and Brown, M. 2013. Threats to an ecosystem service: Pressures on pollinators. *Frontiers in Ecology and the Environment* 11, 251–259.

Vesely, V. 1962. The economic effectiveness of bee pollination on winter rape (*Brassica napus* l., var. Oleifera metz.). Min. Zemedel. Lesn. A vodniho hospodar. *Ust. Vedtech. Inform. Zemedel. Ekon.* 8, 659–673.

Wakhle, D. M., Shakuntala Nair, K., and Phadke, R. P. 1978. Effect of bee-pollination on the oil and protein content in the seeds of sunflower (*Helianthus annuus* L.). *Indian Bee J.* 40, 1–2.

Wallace, H. M., Maynard, G. V., and Trueman, S. J. 2002. Insect flower visitors, foraging behaviour and their effectiveness as pollinators of *Persoonia virgata* R. Br. (Proteaceae). *Austral J. Entomol.* 41, 55–59.

Westcott, L. and Nelson, D. 2001. Canola pollination: An update. *Bee World* 82, 115–129.

White, B. 1970. Pollination of commercial rape seed crops. *Australasian Beekeeper* 72, 99–100.

Williams, I. H., Martin, A. P., and White, R. P. 1987. The effect of insect pollination on plant development and seed production in winter oil-seed rape (*Brassica napus* L.). *J. Agric. Sci. Camb.* 109, 135–139.

Wittkop, B., Snowdon, R. J., and Friedt, W. 2009. Status and prospective of breeding for enhanced yield and quality of oilseed crops for Europe. *Euphytica* 170, 131–140.

Wratten. S. D., Gillespie, M., Decourtye, A., and Mader, E. 2012. Pollination habitat enhancement: Benefits to other ecosystem services. *Agriculture Ecosystem & Environment* 159, 112–122.

Zhong, H. E., Mark, L., Sheve, P., and Cate, B. 2004. Minimizing the import if the mosquito adulterine noted or honeybees, *Apis mellifera* (Hymenoptera: Apidae): Aerial Ultra volume application using a high-pressure nozzle system. *Journal of Economic Entomology* 97(1), 1–7.

16 Protecting Honeybees From Pesticides

The importance of insect pollination, especially by bees, for pollinator-dependent crops and their potential to increase yields is well established. However, further research is needed to understand how the interaction between insect pollination and agricultural inputs affects crop yields and farm economics, in order to strike a balance between food production and biodiversity conservation.

Honeybees are required for the pollination of many vegetable and fruit crops. The production of crops and increase thereof is not imaginable without adequate population of bees. Almost each crop is attacked by pests, which require treatment, which adversely affects the beneficial insects, like honeybees and natural enemies. Hence, modern agriculture is faced with two different but complementary approaches which require controlling pests on the one hand and providing safety to pollinators on the other. Without either of the two, global food production shall be adversely impacted. These two approaches are not always compatible, as most of the pesticides used are toxic to honeybees (Johansen 1977; MacKenzie and Winston 1989; Russell et al.1998; Cunningham et al. 2002; Sundararaju 2003), employed for management of insect pests (Poehling 1989; Stark et al. 1995). The excessive and indiscriminate use of insecticides is a major constraint in pollinator–plant interaction (Bisht et al. 1980; Rana and Goyal 1991; Zhong et al. 2004). The loss of honeybees shall result in poor pollination and loss in honey production.

The decline in the population of these beneficial insects caused by insecticides results in substantial environmental, ecological, and economic consequences (Bai and Reddy 1977; Pimental et al. 1980; Crane and Walker 1983; Parkash and Kumaraswami 1984; Khan and Dethe 2004). In India, 50 million hectares of area is under entomophilous crops, cross-pollinated by different abiotic and biotic agents. About 90% pollination is carried out by insects, 85% of which comprise the bees (Singh et al. 1989).

Honeybees are major pollinators of a wide variety of crops, resulting in their increased yield and improvement in quality (Melnichenko 1976). Scientists and farmers are taking into account an economic factor when it comes to insecticidal applications while managing pests. Our primary aim should be to assess how crop pests can be kept under control without killing insect pollinators and to ensure optimum pollination by these insects.

While controlling pests, the scientists and farmers are looking on to one aspect of the economic considerations in insecticidal applications. Our primary aim should be to assess how crop pests can be kept under control without killing insect pollinators and to ensure optimum pollination by these insects. Some of the pesticides have been evaluated for their toxicity against honeybees.

The first study in this field was carried out by Cherian and Mahadevan (1946) with DDT and Gammexane against *Apis cerana indica*. Hameed allowed the worker bees of *Apis mellifera* to collect nectar from mustard flowers that had been treated with both systemic and contact insecticides. Bai and Reddy (1977) also examined the toxicities of various insecticides through contact and oral exposure. Bai conducted a study on cholinesterase inhibition caused by insecticides in Indian honeybees and observed that signs of poisoning in *Apis cerana indica* became apparent when acetylcholinesterase inhibition exceeded 35%. Death occurred when inhibition reached 96% or higher. Bee poisoning poses a significant challenge for beekeepers, with most incidents of bee mortality happening when pesticides are applied or unintentionally drift onto blossoming crops or weeds (Mayer 2003).

Hameed allowed the worker bees of *Apis mellifera* to forage on cut flowers of mustard, to which systemic and contact insecticides had been sprayed. Bai and Reddy (1977) also assessed the contact

DOI: 10.1201/9781032656724-17

and oral toxicities of some insecticides. Cholinesterase inhibition by insecticides in Indian honeybee was studied by Dale Bai, who reported that signs of poisoning in *Apis cerana indica* were first observable when acetylcholinesterase inhibition exceeded 35%, and death occurred at 96% or more inhibition. Bee poisoning is a serious problem for beekeepers.

Most bee kill occurs when pesticides are applied or allowed to drift on to blooming crops or weeds (Mayer 2003). Most (99%) bee kills result from bees picking up the pesticides when foraging. A prolong repellent effect will deprive the flowers of pollination benefits of insect visits, while a short repellency will deter the insect pollinators from visiting the treated bloom for a brief period and thereafter allow them to resume their foraging activity (with minimum residual hazards) without compromising the yield potential of the crop.

In regions with insufficient wild pollens and nectar plants to support the necessary number of bee colonies for crop pollination, bee poisoning tends to become a chronic issue. Bees are primarily kept in agricultural areas, where they face the risk of pesticide exposure, leading to mortality (Mayer and Johansen 1988; Johansen and Daniel 1990; Caron 2000; Mayer 2003).

Conservation of honeybees for crop pollination is vital to agricultural production (Bisht et al. 1980; Sorthia and Chari 1985; Smirle et al. 1990; Gupta 1994; Dhuley et al. 1996; Russell et al. 1998; Kremen et al. 2002).

Though the monetary loss to the beekeeping is large, the value of the seed and fruit lost through the lack of pollination is estimated to be 50–100 times greater (Atkins 1975).

Pesticides applications also change the physiology of nectar pollen–producing plants and their attraction to bees (Eaton 1963; Church and Williams 1978; Fell et al. 1983; Kuriachen et al. 1992; Sharma 1993). The use of all these pesticides has serious impact on honey production and pollination potential of honeybees.

The immature stages of honeybees are vulnerable to insecticidal poisoning that result in hidden damage to honeybee colony (Davis and Shuel 1985; Naumann and Isman 1996; Davis and Shuel 1988). The detrimental impact of insecticide exposure on brood and newly emerged bees may be more significant compared to the loss of foragers, as the latter can be replaced more quickly and contribute less overall value to the colony in comparison to the emerging workers (Abrol and Kumar 2000).

Bee poisoning has been a longstanding issue dating back to the 1870s, when Thompson (1881) accidentally discovered the killing of bees due to the application of Paris green on blooming pear trees. During that time, various bee ailments were observed around beehives. The period between 1870 and 1888 contributed to the growing understanding of the toxicity of substances like Paris green and London purple. Troop (1918) and Hoskins and Harrison (1934) further documented the high toxicity of inorganic compounds such as arsenics to honeybees. Similarly, Kingsmill (1917) reported significant accidental bee mortality resulting from a mixture of Paris green and molasses bait.

The detrimental impact of insecticide exposure on brood and newly emerged bees may be more significant compared to the loss of foragers, as the latter can be replaced more quickly and contribute less overall value to the colony in comparison to the emerging workers (Abrol and Kumar 2001; Henry et al. 2012).

16.1 FACTORS INFLUENCING BEE POISONING

The extent of damage to the colony by a pesticide application is influenced by various factors. These factors include the toxicity level of pesticide, the quantity and techniques used during application, the time of day, and the prevailing weather conditions. Besides the number of bees from the colony that visit flowers in the treated area, the specific type of food (nectar or pollen) they gather, the kind of flowers they collect from, the season when the damage transpires, and even the availability of forage for the bees in the weeks prior to and following the application, all play a role in influencing the extent of damage.

Wild bees are also damaged by pesticides. Bees may be poisoned from contaminated food as well as nesting materials, such as leaves and/or soil. Different factors influencing bee poisoning are as given:

Pesticide formulation. Dusts are highly hazardous to bees because of their tendency to drift to considerable distance, and particles remain adhered to plant surface for long. The emulsifiable concentrates are relatively safer (Kapil 1970). The granular insecticides are safest. However, granular insecticides with systemic action may contaminate nectar and may result in losses to bees, foraging upon them. The insecticides with fumigant effect and microencapsulated granules are also toxic and stored in the hives, where they may be beaten by adult bees or fed to brood, causing high mortality.

Selectivity of pesticides. Some insecticides are less toxic to honeybees when used as spray, viz, endosulfan, phosalone, pirmicarb, fluvalinate, or trichlorfon. High toxicity of carbamates to honeybees has been considered due to their very low level of phenolase enzymes.

Period of application. Insecticides when applied to crops in bollom are more harmful to bees. In mango orchards, bees attracted to honey dew secreted by mango hoppers are killed in large numbers when insecticides have been applied to the trees.

Time of application. Application of pesticides early in the morning or late evening is safe to bees as there is little or no foraging during this period. Species variation has also been noticed. Whereas *Apis cerana* became active at 7–10°C ambient temperature, *Apis mellifera* never became active unless the ambient temperature reached 13–16°C or above.

Attractiveness of crop. Some crops, notably rapeseed and mustard, are extremely attractive to bees, which will forage from colonies up to 3 km away or more. These crops remain attractive until the very end of flowering. It remains attractive even in cool and dull weather, when they do not visit other crops.

Weather. Application of pesticides during dull weather is safer to bees as foraging is considerably low.

Temperature. Immediate effects of pesticides may be much higher at higher temperature. However, residual effects are likely to be less due to quick breakdown of materials.

Method of application. Aerial application of pesticides has been regarded as more hazardous than ground application. Bees get less time to escape the drift of insecticides. Systemic insecticides applied on the blooming crop may cause hazards to the bees. Fine sprays are safer than coarse sprays.

Colony strength. Populous colonies always suffer greater losses than small colonies, because a greater number of foragers are exposed to insecticides.

Age and body size of honeybees. Newly emerged bees are more susceptible to insecticides than older bees. Smaller bees likewise are more susceptible to insecticides because their body surface area is larger in relation to their body weight.

Distance of colonies. Honeybee colonies are less affected due to a larger distance from the treated field.

16.2 HOW BEES ARE EXPOSED TO PESTICIDAL HAZARDS

Many of the crop plants need cross-pollination. Despite being in the flowering stage, these plants are often plagued by pests, leading to significant losses that necessitate the implementation of pest control methods. Pesticidal applications present a significant risk and also result in the elimination of a substantial number of insect pollinators. Some of the crops benefitting from honeybees are heavily attacked by pests, which include cotton, **brassica and vegetable seed crops**, sunflower, sesame, seed crops, pulses, cucurbits, tobacco, pome, stone fruit and other fruit crops. In case of cotton, flowering continues for two months, and at least 15 to 20 pesticidal applications are required to control the pests.

Hence, it is advisable to minimize the application of insecticides during this timeframe in order to facilitate the relocation of bees to the crop. Further, the nectar present in flowers and extrafloral nectaries is depleted by midday, resulting in fewer bees actively foraging in the afternoon. This presents an opportune window for applying insecticides with reduced risks to bees. Aerial spraying has picked up for cotton crops.

In such a situation, the colonies should be located away from the flight path of the plane.

Brassica and vegetable seed crops. These are attacked by aphids, caterpillars, and bugs during flowering and pod-formation stage. These crops include oilseeds Brassica, seed crops of cauliflower, cabbage, radish, turnip, carrot, fennel, and coriander. In these crops, too, the flowering is greatly extended, lasting for about l–l 1/2 months. These crops need insecticidal applications during flowering periods.

Bees, which serve as valuable pollinators for these crops, also show great preference in foraging for them. To safeguard the bees, it is advisable to conduct spraying in the early morning or late evening. In India, it was observed that Endosulfan posed lower toxicity to honeybees compared to Fenthion, Carbaryl, or Parathion. Moreover, the application of Endosulfan did not negatively impact seed set or yield, since bee activity remained unaffected in the plots where Endosulfan was sprayed (Ramakrishna et al. 1974; Bhattacharya et al. 1982).

Sesame. It is self-pollinated crop, but honeybees also visit the flowers. The crop is attacked by several sucking insects, which require chemical application at flowering time, causing great hazard to bees.

Seed crops. Seed crops are a good source of forage for honeybees. Under semi-arid tropics, the legume flowers usually close in the afternoon, and it allows time for safe application of pesticides afterward against caterpillar pests.

Pulses. Like soybean, cajanus and others are self-pollinated crops, but yield increases by bee pollination have been observed. Application of pesticides causes great harm to bees.

Cucurbits. Honeybees visit flowers of cucurbit for nectar and pollen. Pesticidal applications are required to control fruit flies and aphids at flowering period.

Tobacco. In this case, self-pollination is normal, but honeybees and other insects visit the flowers for nectar, affecting some cross-pollination. Insecticidal applications need to be avoided during flowering.

Pome and stone fruits. Apple, peach, plum, apricot, and almond are attacked by caterpillars at blooming time. Insecticidal use has been suggested by economic entomologists against blossom thrips, though economic losses by thrips have not been ascertained. The caution is required during application of pesticides so that honeybees are not killed.

Other fruit crops. Citrus, litchi, olive, grapes, coconut, and cocoa are also visited by bees for floral rewards.

16.3 PHYTOTOXICITY TO PLANTS

The impact of applying an insecticide may extend beyond harming pollinators for a different crop or eliminating pollinators for the intended crop. Another aspect that has been underestimated in relation to the pesticide is its potential to reduce the productivity of plants. Sedivy (1970) discovered that when pollen grains were treated with Melipax (a toxaphene like chlorinated camphene), only 10.5% of them germinated, whereas the control pollen had a germination rate of 62.1%.

When the pollen grains were treated with 0.3% Fribal emulsion, another apparently toxaphene-like compound, only 28.2% germinated as compared to 81.5% of the control pollen. Carbaryl and methomyl had little or no deleterious effect on pollen, and xylene was non-injurious. The separation of the toxic or repelling effect of the presence of the insecticide on the plant from the possible less attractiveness of affected pollen is difficult.

16.4 INTENSITY OF DAMAGE TO BEES BY PESTICIDES

Pesticides have been reported to cause severe losses to honeybees; 500,000 colonies were killed in USA in 1967 (Levin 1970). Few industries can tolerate such losses and survive. However, the effect of these losses on the adequacy of crop pollination is unknown.

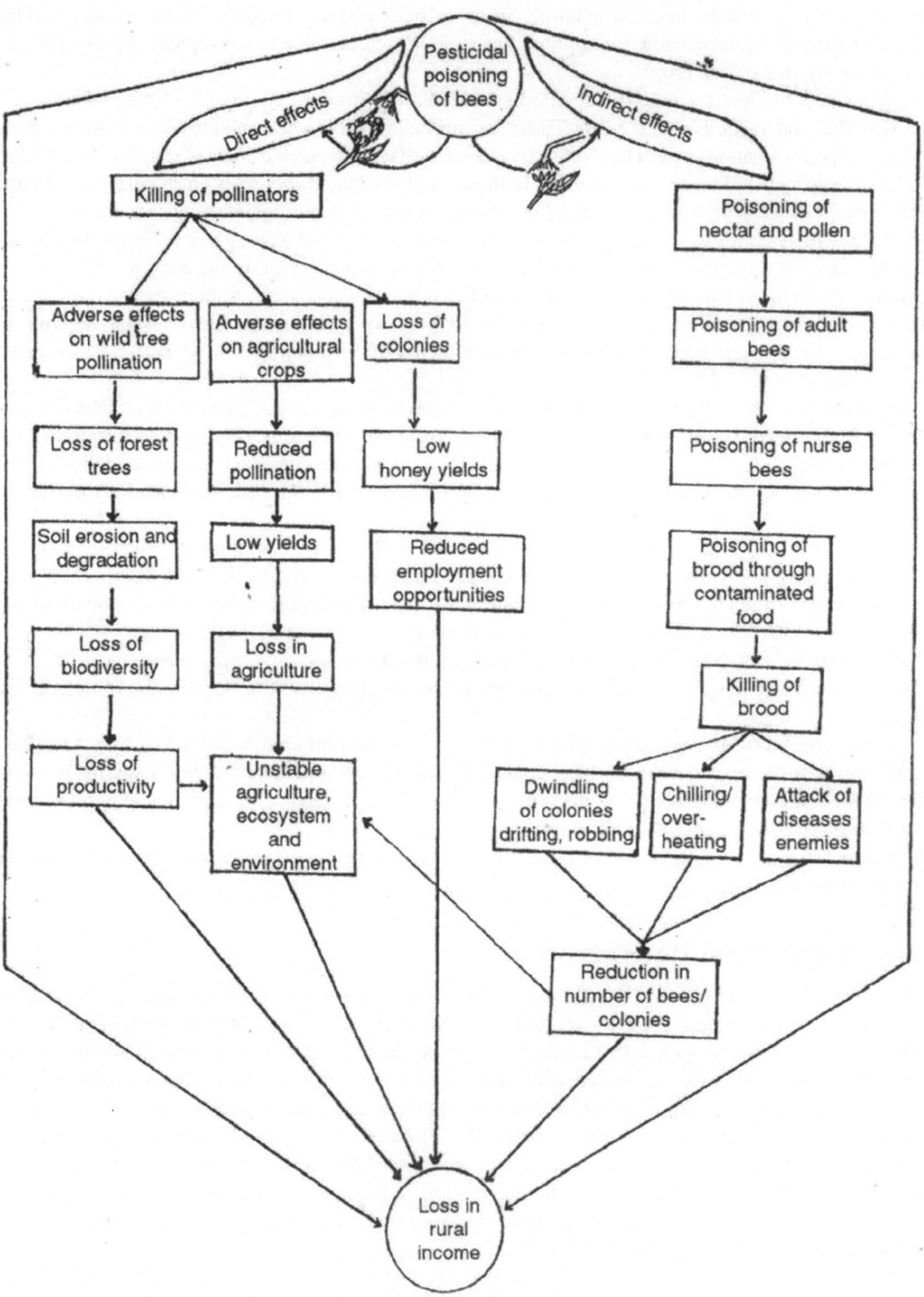

FIGURE 16.1 Schematic model exhibiting the direct and indirect effects of pesticides on honeybee adults and brood.

16.5 INDIRECT EFFECTS OF PESTICIDES ON BEES

Pesticides not only kill bees directly but also indirectly cause several adverse effects:

- Reduced foraging due to repellency or reduced foraging due to killing of foragers due to poisoning.
- Sub-lethal doses can also influence other behaviors, such as orientation, dance rhythm, dance velocity, walking speed, wing beat frequency.
- Physiological injury resulting into reduced longevity.
- Pesticide generally accumulates in combs as a result of absorption from stored pollen, honey/nectar, and may cause chronic paralysis under stress conditions.
- Reduced egg laying and brood rearing due to small doses of pesticides.
- Amorphogenic effects on developing brood and delayed and abnormal development.
- Pesticide application changes physiology of plants, affecting nectar, pollen production, pollen viability; consequently, bee behavior, nectar pollen collection, honey storage, and pollination.

A schematic model exhibiting the direct and indirect effects of pesticides on bee behavior, ecosystem, and crop productivity is given in Figure 16.1.

16.6 PESTICIDES INVOLVED: BASIC TYPES AND CLASSES

The following information is presented to help the beekeeper better understand pesticides and to successfully meet the challenge of pesticides killing honeybees.

16.6.1 Classes of Pesticides

The need of human beings to effectively control their environment is most evident in their agricultural pursuits. The materials can be grouped according to general chemical properties and modes of action.

16.6.2 Insecticides

Insecticides can be categorized as stomach poisons, contact materials, and fumigants. Arsenicals are stomach poisons, pyrethrum as contact insecticide, and hydrogen cyanide, paradichlorobenzene, and carbon disulfide as fumigants.

Botanicals. These are the insecticides derived from plants, such as cube, derris, nicotine, pyrethrins, ryania, sabadilla, and tephrosia. They present no or minor hazard to bees.

Occasionally, sometimes, bees are poisoned by feeding on nectar or pollen of certain plants, such as California buckeye (*Aesculus californica* (Spach) Nutt.), locoweed (*Astragalus* spp.), or mountain laurel (*Kalmia latifolia* L.).

Inorganics. They are arsenicals, fluorides, mercury compounds, and sulfur. They pose a hazard to bees when used in the field; present slight repelling action.

Organics. The chlorinated hydrocarbons, organophosphates, and carbamates vary in their toxicity to bees from relatively less hazardous to highly hazardous, depending upon the individual material or combination of materials.

16.6.3 Chlorinated Hydrocarbons

They include BHC, toxaphene, and chlordane. These chemicals are slow-reactive and persist in the environment for longer periods. These chemicals degrade very slowly, enter fatty and muscle tissues, get concentrated, and enter the food chain. They are classified as neuroactive agents, which block the transmission of nerve impulses, for example, DDT.

16.6.4 Organophosphorus Insecticides

Organophosphates also are neuroactive agents. They include parathion, Systox, DDVP, and malathion. They are quite reactive chemically and are not regarded as persistent in our environment.

16.6.5 Carbamate Insecticides

These also are inhibitors of cholinesterase. A sublethal dose can result in symptoms of nerve poisoning, such as tremors, loss of muscular control, and incontinence.

Pathogens: bacteria, protozoans, and viruses. None of these that are currently recommended or that have been tested for biological control pose a hazard to bees (Cantwell et al. 1972).

16.6.6 Defoliants, Desiccants, and Herbicides

They are nonhazardous to bees and were also highly toxic when sprayed onto older bees in small cages (Moffett et al. 1972).

16.6.7 Diluents, Synergists, and Activators

There is little information on the influence of these agents on the toxicity on honeybees.

16.6.8 Fungicides

Fungicides such as copper compounds, mercury compounds, pentachlorophenol, sulfur, and zineb are not toxic when applied to crops on which bees may be foraging.

16.6.9 Sex Lures, Attractants, and Other Hormones

These usually cause no problems to bees (Hamilton et al. 1970).

16.6.10 Biological Control Agents (Parasitic and Predatory Insects)

Biological control is most useful for controlling pests and providing safety to bees.

16.6.11 Nonchemical Control

Use of IPM that utilizes all techniques and methods in controlling pests, including pesticides, provides safety to bees. The various tactics include insect attractants, repellents, traps, insect-resistant plants, insect pathogens (disease), insect predators and parasites, time of planting, cultivation, time of harvest, sterilized insects, quarantines, and other practices. Not all these techniques are utilized at the same time in controlling a pest; however, all the control methods are considered non-injurious to honeybees.

16.7 TOXICITY OF A PESTICIDE

The toxicity of a particular pesticide results from a combination of its physical and chemical characteristics, the way it is formulated, and the inherent capacity of honeybees to internally process the substance. In the case of a highly volatile pesticide like TEPP, for instance, the chemical can be absorbed by honeybees through their spiracles or respiratory system. Ingestion of contaminated pollen and nectar may result in paralyzing the alimentary canal.

16.7.1 How Bee Poisoning Occurs

Bee poisoning occurs when pesticides are applied on crops on which bees are foraging. Bees collect contaminated nectar, pollen, or collect water from the areas near treated areas. Pesticides dirt may also result in poisoning of bees.

16.7.2 Symptoms of Bee Poisoning

Some of the most important and common symptoms of pesticides poisoning in bees include the presence of large numbers of dead bees in front of the beehive. Individual bees may be seen crawling on the ground near the hive entrance. Foraging bees die in the field. Bees become paralytic, lose orientation; legs, wings, and digestive tract stop functioning, and abdomen becomes distended. Dead bees have unhooked wings, proboscis extended.

Regurgitation of contents of gut is seen in case of organophosphate insecticidal poisoning.

Chlorinated hydrocarbons and organophosphorus insecticides are frequently responsible for inducing stupefaction, paralysis, and abnormal behaviors in bees. By examining the observed symptoms, one can deduce the specific group of insecticides that have been applied to the crop in close proximity to the beekeeping area.

In addition to this, a scientist or a beekeeper as well as farmers should also know the degree of toxicity of different insecticides to the bees so that a person can spray such an insecticide, which is quite effective against pests but least toxic to the bees.

16.7.3 Groups of Insecticides Based on Their Toxicity to Bees

Toxicity is life property of a substance (insecticides) that causes any adverse effect in an organism. The toxicity may be due to single short-term exposure period (acute toxicity) or due to repeated/continuous exposure within less than half a lifetime of that animal (sub-acute toxicity) or repeated exposure over a period of at least half the lifetime of that organism (chronic toxicity). On the basis of toxicity to bees, insecticides can be grouped into the following three categories:

1. **More-Toxic Insecticides**
   The insecticides which have adverse effect on bees even after 90 hr (about four days) come under this category. Such insecticides are carbofuran, dichlorovous fenthion, monocrotophos, fenitrothion, lindane, malathion, carbaryl, methyl parathion, chlorpyriphos, dimethoate, phosphamidon, diaginon, etc.

2. **Less-Toxic Insecticides**
   Those insecticides whose residue is less persistent in nature and remain toxic to bees up to 90 hr (less than four days) are in this category. Such insecticides are endosulfan, phosalone, etc. Besides this, all systemic insecticides also fall in this category if used properly.

   In order to minimize harm to bees, it is advisable to apply these insecticides after 1500 h and allow bees to access the crop plants a little bit the next morning. This time delay ensures that the majority of the systemic insecticide have entered the plant system, reducing the potential harm to bees. Besides, there are non-toxic insecticides available that effectively control insect pests without posing a threat to bees.

3. **Non-Toxic Insecticides**
   There are some chemicals which are very effective for control of insect pests but DOD-toxic to bees. They are some bacteria (BT formulations), viruses (NPV), insect growth regulators, etc. Besides this, fungicides, weedicides, and some plant growth regulators have also been found safe for ten bees.

16.7.4 Relative Toxicity of Pesticides

The kill of bees can be classified as:

<100 per day—normal die-off rate
200–400 per day—low kill
500–900 per day—moderate kill
>1,000 per day—high kill

The relative toxicity of different insecticides to honeybees has been given in Tables 16.1–16.6.

LD50 (μg/bee)	Classification
>100	Virtually non-toxic
11–100	Slightly toxic
2–10.99	Moderately toxic
<2.0 (i.e., 0.001 to 1.99 μg/bee)	highly toxic

TABLE 16.1
Classification of Insecticides on the Basis of Their Mode of Action

Group	Subgroup	Example
Physical poisons	-	Heavy mineral oils, inert dusts
Protoplasmic poisons	-	Heavy metals, for example, mercury, acids
Metabolic inhibitors	Respiratory poisons	HCN, CO, H2S, rotenone, dinitrophenols
	Inhibitors of mixed-function oxidase	Pyrethrin synergists
	Inhibitors of carbohydrate metabolism	Sodium fluoroacetate
	Inhibitors of amine metabolism	Chlordimeform
	Insect hormones	Juvenile hormone analogues
	Inhibitors of chitin synthesis	Diflubenzuron
Neuroactive agents (nonmetabolic)	Anticholinesterases	Organophosphorus compounds, carbamates
	Effects on ion permeability	DDT analogues, pyrethroid
	Agents for nerve receptors	Neonicotinoids, Nereistoxin analogues, GABA cyclodienes, HCH, Avermectins (Spinosoids) Octopamine Formamidines
Hormone mimics	-	Methoprene
Stomach poisons	-	*Bacillus thuringiensis* toxin, Thiodicarb

TABLE 16.2
Pesticide Risk to Bees

Risk Rating	Chemicals	Remarks
High risk to bees foraging even 10 hr after spraying	Carbaryl, chlorpyrifos, diazinon, dimethoate, omethoate, methomyl, fenthion, methamidophos, methidathion, monochrotophos	These should never be sprayed on flowering crops, especially if bees are active and the crop requires pollination.
Moderate risk, with some losses expected 10 hr after spraying	Acephate, demeton-s-methyl	
Some risk with low chance of losing bees 3–5 hr after spraying	Endosulfan, dicofol, pirimicarb, petroleum oils, most pyrethroid chemicals, trichlorfon	There is little risk of losing bees if these chemicals are sprayed in the evening when foraging has ceased.
No risk even if sprayed over foraging bees	*Bacillus thuringiensis*, propargite, oxythioquinox	

TABLE 16.3
Pesticides most toxic to bees[1] (LD_{50} 0.001–1.99 µg/bee).

Application of these pesticides on blooming crops or weeds may cause severe damage to bees. Even after 10 hr of spray, these pesticides are still very toxic to bees.

Acephate Fenthion*
Aldicarb G (applied four weeks before bloom) Heptachlor*
Aldrin* Isobenzan
Aminocarb Lean arsenate
Azinphosethyl Malathion D*
Azinphosmethyl Malathion ULV
Benzene hexachloride* Methamidophos
Bomyl Methidathion
Calcium arsenate Methiocarb
Carbanolate Methomyl D
Carbaryl ULV (over 0.4 kg/ha) Methyl-carbophenothion
Carbofuran F Mehyl parathion*
Carbophenothion D Mevinphos
Chlorpyrifos* Mexacarbate
Chlorthion Monocrotophos
Colep Naled D
Crotoxyphos Omethoate
Diazinon* Paraoxon
Dicapthon Parathion*
Dichlorvos* Permethrin
Dicrotophos Phenthoate
Dieldrin Phosmet
Dimethoate* Phosphamidon*
Dinitrocresol Phoxim
Dinoseb Propoxur
EPN Pydrin (over 0.1 kg/ha)
Ethyl-methyl Guthion Pyramat
Famphur Resmethrin
Fenamiphos Sulfotep (p)
Fenitrothion* Sulprofos
Fensulfothion Thionazin

* There is an Indian standard for one or more formulations of these pesticides (India Standards Institution 1979).

1 For minimal hazard to honeybees, the insecticides below should NOT be applied on blooming crops or weeds. Their residual toxicity is usually high even after 10 hr (Table 16.4).

TABLE 16.4
Pesticides Very Toxic to Bees[1]

Malathion EC*
Naled WP
Phorate EC
Pydrin (0.1 kg/ha or less)

* There is an Indian standard for one or more formulations of this pesticide (Indian Standards Institution 1979).

1 For minimal hazard to honeybees, the following insecticides should be applied only during late evening. Their residual toxicity is usually low within 8 hr (Table 16.5).

TABLE 16.5
Pesticides Very Toxic to Bees[1]

Amidithion Kroneton
Aramite D Larvin
Binapacryl* Leptophos
Carbaryl ULV (0.5 kg/ha or less) Lethane 384 special
Carbophenothion Malathion MA
Chlordane* Malonoben
Chlorfenvinphos Menazon
Chlorpyrifos or chlorpyriphos MA Methomyl S, SP
Coumaphos Methoxychlor
DDD Methyl-demeton
Disulfoton EC Propoxur MA
Endosulfan* Propyl thiopyrophosphate
Endrin* Rotenone D
Ethion Sabadilla
*ethyl-DDD Solvigran
Fenchlorphos Strobane
Fenthion G or MA Temephos*
Fonofos TEPP
Formetanate Tetrachlorvinphos
Garrathion Tetram

(Continued)

TABLE 16.5 (*Continued*)
Pesticides Very Toxic to Bees[1]

DDT* Naled EC	Heptachlor G Thanite (isobornyl thiocyanate)
Demeton Nissol	Isobornyl thiocyanate Thioquinox
Dichlorfention or dichlorfenthion Oil sprays (superior type)	Isodrin Toxaphene*
Dichlorvos MA Oxamyl	Isolan Tranid
Dieldrin G Oxydemeton-methyl*	Isopropyl parathion Trichlorfon* Malathion EC*
Dilan Phorate G	Naled WP
Dimetilan Phosalone*	Phorate EC
Dinobuton Phostex	Pydrin (0.1 kg/ha or less)
Dioxathion Pirimicarb	

* There is an Indian standard for one or more formulations of this pesticide (Indian Standards Institution 1979).

[1] For minimal hazard to honeybees, the following insecticides should be applied only during late evening. Their residual toxicity is usually low within 8 hr (Table 16.6).

TABLE 16.6
Pesticides Least Toxic to Bees[1] (LD_{50} 2.0–10 μg/bee)

The following pesticides should be applied in the late evening, when bees are not foraging in the field. Beehives should not be directly exposed to these insecticides. For minimal hazard to bees, dose, the timing, and methods of application are very important.

Allethrin Fenazaflor	Chlorfensulphide Propargite
Amitraz Fenbutatin-oxide	Chlorobenzilate Propoxur G
Bromopropylate Fenson	Chloropropylate Pyrethrum*
Butoxy thiocyanodiethyl ether Fensulfothion G	Cryolite Quinomethionate
Carbaryl G Genite 923 or Genitol 923	Cyhexatin Rotenone EC
Carbofuran G Hydroprene	Dicofol* Ryania
Chlorbenside Lime-sulfur or lime-sulfur*	Dienochlor Schradan
Chlordecone Malathion G	Diflubenzuron Sodium fluosilicate baits
Chlordimeform Mirex G	Dikar Sulphenone
Chlorfenethol Nicotine sulphate	Dinex sulfur or sulfur*
Chlorfenson Oxythane	Dinocap Tetradifon
	Disulfoton G

* There is an Indian standard for one or more formulations of this pesticide (Indian Standards Institution 1979).

[1] The following insecticides can be applied at any time with reasonable safety to honeybees.

Their toxicity is usually low with direct application (Table 16.7).

16.8 PROTECTION OF BEES

About 27 pesticides are likely to be banned by the central government (Table 16.8). India has always been particular about the use of pesticides in agriculture. The country has an Insecticides Act since 1968, which has undergone amendments as time passed. The government had completely banned 12 pesticides, and 6 were phased out (https://krishijagran.com/agriculture-world/government-likely-to-ban-27-pesticides-check-list-here/).

Several management practices are known that may help protect bees from pesticide-caused mortality.

TABLE 16.7
Pesticides Non-Toxic (LD_{50} above 11µg/bee)

These can be used around bees with a minimum injury. They should be applied during late evening, night, or early morning. These products can be used on fields or near hives with minimum damage to bees; in fact, a few of the listed acaricides can be used to control bee mites within the hive.

A. INSECTICIDES AND ACARICIDES

Acaraben (chloro-benzilate)	Ethodan	Omite
	Fundal	OMPA (schradan)
Allethrin	Galecron (chlorophenamidine)	Ovotran (ovey)
Aramite		Phostex
Bacillus thuringiensis	Heliothis virus	Phrethrin
	Kelthane (dicofol)	Rhothane (TDC)
Cryolite	Kepone	Rotenone
Delnav (dioxathion)	Methoxychlor	Ryania
Dessin	Mitox (chlorbenside)	Sabadilla
Dilan	Morestan	Sulphenone
Dylox (trichlorfon)	Nemagon	Tedion (tetradifon)
Eradex	Neotran nicotine	Toxaphene

B. FUNGICIDES

Arasan (thiram)	Cyprex (dodine)	Manzate (maneb)
Benlate (benomyl)	Dexon	Mylone
Bordeaux mixture	Dichlone	Parzate (nabam)
Copper oxychloride sulfate	Difolatan	Phaltan (folpet)
	Dithane M-45	Polyram
Copper sulfate (monahydrate)	(Folcid)	Sulfur
	Glyoxide (glyodin)	Thynon (dithianon)
Cuprous oxide	Karathane (dinocap)	Zerlate (ziram)

C. HERBICIDES

Amitrol	Eptam (EPTC)	Picloram
Ammate (ammonium sulfamate)	Folex (merphos)	Planavin
	Herbisan (EXD)	Princep (simazine)
Atrazine	Hyvar (bromacil)	Randox (CDAA)
Banvel (dicamba)	Igran (terbutryne)	Sinbar (terbacil)
Betanal (phenmedipham)	IPC	Stem F-34 (propanil)
	Karmex (diuron)	
Caparol (promytryne)	MCPA	TOK (nitrofen)
Casoron (dichlobenil)	Milogard (propazine)	Trysben (2,3,6-TBA)
	Monuron	Vegedex (CDEX)
Dalapon	NPA	2,4-D
DEF	Paraquat	2,4-DB
Diquat		2,4,5-T

TABLE 16.8
Pesticides That Are to Be Banned

S. No.	Pesticide
1.	Acephate
2.	Atrazine
3.	Benfuracarb
4.	Butachlor
5.	Captan
6.	Carbendazim
7.	Carbofuran
8.	Chlorpyriphos
9.	2,4-D
10.	Deltamethrin
11.	Dicofol
12.	Dimethoate
13.	Dinocap
14.	Diuron
15.	Malathion
16.	Mancozeb
17.	Methomyl
18.	Monocrotophos
19.	Oxyfluorfen
20.	Pendimethalin
21.	Quinalphos
22.	Sulfosulfuron
23.	Thiodicarb
24.	Thiophanat emethyl
25.	Thiram
26.	Zineb
27.	Ziram

16.9 HOW BEES CAN BE PROTECTED FROM PESTICIDE POISONING

Do not apply any insecticide unless the crop is so heavily infested and the insect in question has reached the economic threshold value (ET-value). If necessary, then:

Read the pesticide label carefully. Read the pesticide label carefully and clearly notify the formulation.

Use less-toxic compounds. Choose insecticide with low toxicity.

16.10 TYPES OF FORMULATION AND THEIR TOXICITY TO BEES

Use less-toxic formulations. Mostly dust, wettable/water-dispersible liquid, and granular formulations of insecticides are used to control insect/pests of various crops. Select the safest formulation of the pesticide that is available for the intended use. "Drifting" of the pesticide from the target pest and/or crop to areas frequented by bees should be minimized.

Dust formulations. These formulations, if used on crop, result in maximum toxicity to bees and other pollinators as compared to other formulation. Dusts almost always drift more than other formulations and are generally more dangerous to bees than are sprays or granular application

WP/WDP Formulations. These formulations are less toxic to bees as compared to dust formulation, but definitely more toxic in comparison to other formulations of insecticides. **Liquid formulations** of course liquid formulations are less toxic to bees as compared to dust and wettable or water-dispersible powder formulations, but they are less toxic as compared to granular formulations.

Granular formulations. Among the insecticidal formulations which are commonly used to control pests of various crops, it is the granular formulation which is least toxic to honeybees visiting flowers to collect nectar and pollen. Further, ground applications are always safer than aerial applications.

Microencapsulated pesticides. This pesticide will not kill the adult bees that collected it, but it will kill the immature stages of the bees and the young adult nurse bees that feed the brood. Bees have little protection from this product.

The mode of pesticide application. Aerial applications are generally more dangerous. Apply pesticide directly over beehive.

Do not treat crops in bloom. Whenever possible, do not treat crops in bloom. If treatments are needed during bloom, choose a short residual material. Make applications during the evening, when bees are not active.

Minimize spray drift. Honeybee hives should not be placed next to fields or orchards that are likely to be treated with pesticides toxic to bees.

Use pesticides only when needed. Vegetable, fruit, and seed crop yields can be adversely affected by reducing the population of pollinating insects and beneficial insect predators. Use pesticides only if necessary, and select those with low toxicity.

Apply pesticide when bees are not flying. Pesticides hazardous to honeybees must be applied to blooming plants when bees are not working, preferably in the early evening. Evening application allows time for these chemicals to partially or totally decompose during the night.

Do not contaminate water. Never contaminate standing water with pesticides or drain spray tank contents onto the ground, creating puddles.

Identify attractive blooms. Bees also visit flowers of some weeds; hence, before spraying, remove such weeds from fields, which might attract bees. In many instances, bees have been killed even though the crop being sprayed was not in bloom. Many times, these attractive blooms can be mowed or otherwise removed, although mowing can result in destroying other beneficial insect habitat or force destructive insects into the crop being cultivated.

Notify beekeepers. Notify beekeeper before making any treatment well in advance so that beekeepers can protect or move their colonies.

Disposing of unused pesticides. Follow proper precautions in disposing of unused pesticides and pesticide containers. Be particularly careful not to contaminate with pesticides, as the water may be collected by bees and result in bee kills.

Inform presence of apiary. A beekeeper needs to inform nearby farmers for the presence of apiary so that farmers can inform the beekeeper if any treatment is to be made.

Use integrated pest management (IPM) to reduce pesticide hazard. Whenever possible, it is wise to use such integrated pest management practices, viz, recognition of existing biological controls and evaluating their value, before applying any pesticides; using pesticides that are less harmful to beneficial organisms; timing pesticide applications so they are less-harmful to beneficial organisms. Before applying pesticides, fields should be examined for the activity of natural enemies. If large numbers are present, pesticide use may be deferred or avoided.

Use of repellents. Chemical repellents have been studied for many years. The repellent is added to the pesticide before field application and is intended to discourage bees from visiting plants until the pesticide becomes relatively nontoxic.

Disposing of unused pesticides. Follow proper precautions in disposing of unused pesticides and pesticide containers. Be particularly careful not to contaminate water with pesticides, as the water may be collected by bees and result in bee kills. Pesticide dusts and small granules should not be left open or carelessly thrown anywhere, because bees are likely to collect such dusts during acute pollen dearth period.

Inform presence of apiary. If your bees are located in any area where pesticides are commonly used, then identify yourself as a beekeeper to any neighbors who may use pesticides. Explain to growers the importance of your honeybees in the pollination of crops being grown in nearby fields so that they may consider the value of bees in pollination before applying any pesticides that may kill the pollinating insects. Do not place apiaries in areas that are used to grow crops that require heavy and frequent usage of pesticides.

Proper space. While confining the bees in the hive, sufficient and proper space for all the bees including foragers should be made available.

Proper ventilation. Proper ventilation must be provided at the top or sides of the hive. Dot only through the entrance, because this may get blocked by deed bees. Ventilation screen should have as large a mesh as possible.

Shade. Shade is usually provided by the use of local materials. It must not hinder the flow of air past the hives.

Covering hives. Covering the hives with wet clothes or gunny bags is very useful because evaporation of water helps reduce the rise in colony temperature. If pesticides are applied from the air, then cover the whole of the hive.

Water availability. Provide water inside the hives for bees. The bees take it and spread out in the hive, where it evaporates and thus reduces the temperature.

Minimize period of confinement. The confinement must, of course, continue as long as the pesticide near the hive retains unacceptable toxicity, and its duration can satisfactorily be reduced by the application of less-persistent insecticide.

Adey et al. (1986) has given the following guide to minimize bee toxicity:

Quick guide to minimize bee toxicity:

↓

Has the infestation reached the economic injury level?

↓

Is there any effective non-chemical method for pest control?

↓

Chose the pesticide carefully.

↓

Read the label in detail.

↓

Are any plants in bloom in/around the crop to be treated?

↓	↓
Yes	No
Are bees foraging in/around the crop to treated?	
↓	↓
Yes	No
↓	↓
Can the bees be moved prior to treatment?	Treat
↓	↓

Yes No

↓

Move bees, treat, use non-persistent pesticides.

Apply when bees are not foraging.

In conclusion, there is a need to reduce treatment of pesticides on the crops dependent on bees for pollination. Catarino et al. (2019) studied the combined effects of pesticides, insect pollination, and soil quality on oilseed rape (*Brassica napus* L.) yield and gross margin and found that yield and gross margins are greater (15–40%) in fields with higher pollinator abundance than in fields with reduced pollinator abundance. This effect is, however, strongly reduced by pesticide use. Greater yields may be achieved by either increasing agrochemicals or increasing bee abundance, but crop economic returns were only increased by the latter, because pesticides did not increase yields, while their costs reduced gross margins. Sihag (1988) reported that in order to maintain or increase productivity, bee pollination and pest management should be given equal importance as factors in agricultural production. Bees can be saved either by providing them with functional protection through the use of repellents (Mahavir 1982) or by using non-toxic or moderately toxic pesticides or by adopting an integrated pollination–pest management practice, for which an understanding of foraging behavior and use of this knowledge are essential.

REFERENCES

Abrol, D. P. and Kumar, R. 2000. Impact of pesticides on the brood rearing activity and foragers of honeybee, *Apis mellifera* L. visiting *Brassica campestris* L var. Toria. *Indian Bee J.* 62, 19–27.

Abrol, D. P. and Kumar, R. 2001. Toxic and morphogenic effects of pesticides on honeybee *Apis mellifera* L. *Brood. Journal of Apiculture* 16(2), 95–112.

Adey, M., Walker, P., and Walker, P. T. 1986. *Pest Control Safe for Bees: A Manual and Directory for the Tropics and Subtropics.* International Bee Research Association (IBRA), London, p. 224.

Atkins, E. L. 1975. Injury to honey bees by poisoning. In: *The Hive and the Honey Bee.* Dadant and Sons, Hamilton, pp. 663–696.

Bai, A. R. K. and Reddy, C. C. 1977. Laboratory toxicity of some insecticides to *Apis cerana indica. J. Apic. Res.* 16(3), 161–162.

Bhattacharya, P., Samui, R. C., Ghosh, M. R., Dasgupta, S. K., and Roy, A. 1982. Sunflower seed yield as influenced by pollination and insect pests. Proceedings of Tenth International Sunflower Conference, March 14–18 1982, Sulfers Paradise, Australia, pp. 132–134.

Bisht, D. D., Naim, M., and Mehrotra, K. N. 1980. Studies on the role of honeybees in rapeseed production. Proceedings of 2nd International Conference of Apiculture in Tropical Climate. Indian Agricultural Research Institute, New Delhi, pp. 491–496.

Cantwell, G. E., Jay, E. G., Pearman, G., Jr., and Thompson, J. V. 1972. Control of the greater wax moth Galleria mellonella (L.) in comb honey with carbon dioxide. Parts I and II. *Am. Bee J.* 112(9), 302–303, 342–344.

Caron, D. M. 2000. Pests of honey bees. MidAtlantic Apicultural Research and Extension Consortium, Publication 4.3, University of Delaware, Newark, Delaware, USA.

Catarino, R., Bretagnolle, V., Perrot, T., Vialloux, F., and Gaba, S. 2019. Bee pollination outperforms pesticides for oilseed crop production and profitability. *Proc. R. Soc. B* 286, 20191550. https://doi.org/10.1098/rspb.2019.1550

Cherian, M. C. and Mahadevan, V. 1946. Preliminary trials with DDT and gammaxene against Indian bee, Apis indica, Indian. *Bee Journal*, 8 (9–10), 128–130.

Church, R. M. and Williams, R. R. 1978. Fungicides toxicity to apple pollen in the another. *J. Horticultural Sci.* 53, 91–94.

Crane, E. and Walker, P. 1983. *The Impact of Pest Management on Bees and Pollination.* International Bee Research Association, England, p. 107.

Cunningham, S. A., FitzGibben, F., and Heard, T. A. 2002. The future of pollinators for Australian agriculture. *Agriculture* 53(8), 893–900.

Davis, A. R. and Shuel, R. W. 1985. Invisible damage to honeybees colonies from pesticides. *Can. Beekeeping* 12, 60–61.

Davis, A. R. and Shuel, R. W. 1988. Distribution of carbofuran and dimethoate in flowers and their secretion in nectar as related to nectary vascular supply. *Can. J. Bot.* 66, 1248–1255.

Dhuley, J. N., Majumdar, S. M., Mudumdar, A. M., and Naik, S. R. 1996. Laboratory studies on contact toxicity of aureofongin in *Apis cerana indica* Fabr. *Indian Bee J.* 58(4), 207–208.

Eaton, G. W. 1963. Germination of apple pollen as influenced by captan sprays. *Proc. Am. Soc. Hortic. Sci.* 83, 101–106.

Fell, R. D., Rajotte, E. C., and Yoder, K. S. 1983. Effect of fungicides sprays during apple bloom on pollen viability and honeybee foraging. *Environ. Entomol.* 12, 1572–1575.

Gupta, B. N. 1994. India. In: *Non-Wood Forest Products in Asia*. Food and Agriculture Organization of the United Nations, Rome.

Hamilton, D. W., Schwartz, P. H., and Townshend, B. G. 1970. Capture of bumble bees and honey bees in traps baited with lures to attract Japanese beetles. *J. Econ. Entomol.* 63, 1442–1445.

Henry, M., Béguin, M., Requier, F., Rollin, O., Odoux, J. F., Aupinel, P., Aptel, J., Tchamitchian, S., and Decourtye, A. 2012. A common pesticide decreases foraging success and survival in honey bees. *Science* 336, 348–350.

Hoskins, W. M. and Harrison, A. S. 1934. The buffering power of the contents of the ventriculus of the honeybees and its effects upon the toxicity of arsenic. *J. Econ. Entomol.* 28, 924–942.

Indian Standards Institution. 1979. Institute of Pesticides Formulation Technology (IPFT) Bureau of Indian Standards Manak Bhavan, 9 Bahadur Shah Zafar Marg, New Delhi 110002.

Johansen, C. A. 1977. Pesticides and pollinators. *Annu. Rev. Entomol.* 22, 177–192.

Johansen, C. A. and Daniel, F. M. 1990. *Pollinator Protection: A Bee and Pesticide Handbook*, Illustrated edition. Wicwas Press, New Delhi, India.

Kapil, R. P. 1970. Role of honeybees in agriculture. *Indian Bee J.* 32, 53–577.

Khan, R. B. and Dethe, M. D. 2004. Median lethal time of new pesticides to foragers of honeybee. *Pestology* 28(1), 28–29.

Kingsmill, G. F. 1917. Sweetened poison mixtures. Proceedings Ontario. Bee Keepers Association. *Annual Report*, 47, 61–62.

Kremen, C., Williams, N. M., and Tharp, R. W. 2002. Crop pollination from native bees at risk from agricultural diversification. *Proc. Nat. Acad. Sci. USA* 99(26), 16812–16816.

Kuriachen, P. M., Thomas, V., and Dave, Y. 1992. Taxonomic and phylogenetic significance of fruit walls in Asclepiadaceae. *Feeds Repertorium* 103, 179–193.

Levin, M. D. 1970. The effects of pesticides on beekeeping in the United States. *Am. Bee J.* 110, 8–9.

Mackenzie, K. E. and Winston, M. L. 1989. Effect of sub lethal exposure to diazinon on longevity and temporal division of labour in the honeybee (Hymenoptera: Apidae). *J. Econ. Entomol.* 82(1), 75–82.

Mahavir, G. 1982. Olfactory physiology of *Apis florea* F. with reference to repellents. PhD thesis. Haryana AgriculturalUniversity, Hisar, India.

Mayer, D. F. 2003. Bee kills from pesticides and the expert witness. *Am. Bee J.* 143(7), 563.

Mayer, D. F. and Johansen, C. A. 1988. How to reduce bee poisoning. Washington State Cooperative Extensions Leaflet No. 15, WREP, p. 12.

Melnichenko, A. N. 1976. Role of insect pollinators in increasing yields of agricultural plants. In: Kozin, R. B. (ed.) *Pollination of Entomophilous Agricultural Crops by Bees*. Amerind Publishing Company Private Limited, New Delhi, p. 331.

Moffett, J. O., Morton, H. L., and Macdonald, R. H. 1972. Toxicity of some herbicidal sprays to honey bees. *J. Econ. Entomol.* 65, 32–36.

Naumann, M. and Isman, M. B. 1996. Toxicity of a neem (*Azadirachta indica* A. juss) insecticides to larval honeybees. *Am. Bee J.* 96, 518–520.

Parkash, R. and Kumaraswami, J. 1984. Toxicity some insecticides of the Indian bee, *Apis cerana* Feb. *Indian Bee J.* 46, 15–17.

Pimental, D., Andow, D., Dyson-Hudson, R., Gallahan, D., Jacobson, S., Irish, M., Kroop, S., Moss, A., Schreiner, I., Shepard, M., Thompson, T., and Vinzant, B. 1980. Environment and pesticides: A preliminary assessment. *Oikos* 34(2), 126–140.

Poehling, H. M. 1989. Selective application strategies for insecticides in agricultural crops. In: *Pesticides and Non Target Invertebrates*. Intercept, Wimborne, pp. 151–175.

Ramakrishna, C., Rajmohan, N., and Subramaniam, T. R. 1974. Effect of insecticidal sprays on the pollination of sunflower. *Madras Agric. J.* 61(6), 175–177.

Rana, B. S. and Goyal, N. P. 1991. Field toxicity of methyl demeton and dimethoate to the foragers of honeybee (*Apis cerana indica* F.) visitors to *Brassica chinensis* L. *Indian Bee J.* 53(1–4), 73–77.

Russell, D., Meyer, D., and Bukowski, J. 1998. Potential impact of micro encapsulated pesticides on New Jersey apiaries. *Am. Bee J.* 38(3), 207–210.

Sedivy, J. 1970. The influence of toxaphene insecticides on the pollen of lucerne. *Ochr. Rost.* 43(3), 187–190 [in Czech., English summary. aa-805/71]

Sharma, P. L. 1993. Effects of insecticides and insect pollinators activity on seed yield of cauliflower. PhD thesis. YS Parmar University of Horticulture and Forestry, Solan, pp. 189 + xviii.

Sihag, R. C. 1988. Effect of pesticides and bee pollination on seed yield of some crops in India. *Journal of Apicultural Research* 27(1), 49–54.

Singh, W. J. K., Singh, R., Hameed, S. F., and Singh, B. 1989. Field toxicity of some insecticides to *Apis cerana indica* Fabr. *Indian Bee J.* 51(4), 137.

Smirle, M. J. 1990. The influence of dextofying enzymes in insecticide tolerance in honeybee (Hymenoptera: Apidae). *J. Econ. Entomol.* 3(3), 715.

Sorthia, B. K. and Chari, M. S. 1985. Toxicity of some insecticides to honeybees, *Apis florea* F. and *Apis mellifera* L. *J. Entomol. Res.* 92(2), 195–197.

Stark, D. J. D., Jepson, P. C., and Mayer, D. F. 1995. Limitation to use of topical toxicity data for production of pesticide side effects in the field. *J. Econ. Entomol.* 88(5), 1081–1088.

Sundararaju, D. 2003. Occurrence of bee fauna and extent of pollination in insecticide sprayed ecosystem of cashew. *J. Palynol.* 39, 121–125.

Thompson, G. M. 1881. Paris green. *Am. Bee J.* 17(1), 85.

Troop, J. 1918. Blossom spraying and bee poisoning. *J. Econ. Entomol.* 11, 433.

Zhong, H. E., Mark, L., Sheve, P., and Cate, B. 2004. Minimizing the import if the mosquito adulterine noted or honeybees, *Apis mellifera* (Hymenoptera: Apidae): Aerial ultra volume application using a high-pressure nozzle system. *J. Econ. Entomol.* 97(1), 1–7.

Section 2

Pulses

17 Pulses: An Introduction

Agriculture is a critical sector of Indian economy, contributing about 15% of national gross domestic product, and more importantly, about half of India's population is wholly or significantly dependent on agriculture and allied activities for their livelihood. But agriculture in India happens to face several constraints and challenges, especially in the areas of food security, natural resource management, and application of advanced farm technology to harness crop yield potential, better farm profitability, minimal adverse environmental impact, better soil health, and food self-reliance. Food security has been a major area of concern for agricultural scientists and planners in India since long. At present, India produces over 328 million tons of food grains every year, with an increase of more than sixfold since independence. But increased efforts to produce more food have resulted in an increase in tremendous shift in cropping systems toward cereal-based cropping systems. This has marginalized the pulses, resulting in quantitative as well as qualitative degradation of productive base, land, and farm resources. The important pulses cultivated include (Table 17.1)

Pulses, which have been consumed for over 10,000 years, are widely recognized as one of the most commonly consumed foods worldwide. Their global cultivation allows for a diverse range of pulses, making them crucial both from an economic and a nutritional standpoint. Pulses are rich in protein, fiber, vitamins (such as iron, zinc, folate, and magnesium), and minerals, providing a substantial contribution to our nutrient intake. Incorporating just half a cup of beans or peas into our daily diet can significantly enhance its quality by boosting the consumption of these beneficial nutrients. Moreover, pulses contain phytochemicals like saponins and tannins, which possess antioxidant and anti-carcinogenic properties, suggesting that pulses may offer notable anti-cancer effects. Consuming pulses also has a positive impact on serum lipid profiles and various other risk factors associated with cardiovascular disease. These factors include blood pressure, platelet activity, and inflammation. Pulses are rich in fiber and have a low glycemic index, which makes them especially advantageous for individuals with diabetes, as they help maintain healthy blood glucose and insulin levels. Recent research exploring the influence of pulse components on HIV and consumption patterns among aging populations suggests that pulses may have additional health benefits. In conclusion, incorporating pulses into the diet is a nutritious way to meet dietary guidelines and is linked to a decreased risk of multiple chronic diseases. However, long-term randomized controlled trials are necessary to establish the direct effects of pulses on these diseases.

In India, pulses hold a distinctive position due to their protein content, often being referred to as "the poor man's meat." India has the highest number of malnourished individuals globally, and given the prevalence of vegetarianism in the country, the future of pulses holds great importance for its sizable impoverished population. Additionally, pulses have suffered from the overshadowing of cereals in agricultural practices over the past few decades, highlighting the involvement of agricultural policy in both the nutritional problems and the potential remedies, as we aim to demonstrate. Pulses hold a distinct position in India, serving as an essential food source for the impoverished population while also offering environmental advantages. Given the significant number of poor and undernourished individuals in the country, it is crucial for India's pulses sector to make a swift recovery. Pulses play a vital role in providing affordable protein for both human and animal consumption in India. Moreover, their recognized value lies in their ability to enhance and revive soil fertility in dry regions.

Pulses, often referred to as precious gems of Indian agriculture, hold a significant role in ensuring food and nutritional security within India. The matter of food security has long been a prominent concern for agricultural scientists and planners in the country. Being predominantly vegetarian in dietary practices, India heavily relies on plant-based sources to fulfill its daily protein needs.

DOI: 10.1201/9781032656724-19

TABLE 17.1
List of Pulses

English Name	Local Name	Scientific Name
Pigeon pea, red gram	Arhar, tur	*Cajanus cajan*
Chickpea, Bengal gram, garbanzo bean	Chana	*Cicer arietinum*
Lentil	Masoor	*Lens culinaris*
Green gram, mung bean	Moong	*Vigna radiate*
Black matpe, black gram, black matpe bean	Urad, udid, urad bean	*Vigna mungo*
Pea	Matar	*Pisum sativum var. arvense*
Grass pea	Khesari	*Lathyrus sativus*
Yellow pea	Matar	*Lathyrus aphaca*
Lablab bean	Matar	*Lablab purpureus*
Moth bean	Moth	*Vigna aconitifolia*
Horse gram, madras gram	Kulti	*Dolichos uniflorus*
Broad bean, faba bean		*Vicia faba*
Cowpea		*Vigna unguiculata*

Pulses, being a valuable source of proteins, vitamins, and minerals, are commonly known as both the "poor man's meat" and the "rich man's vegetable," playing a vital role in bolstering the nation's nutritional well-being.

Pulses provide significant nutritional and health benefits and are known to reduce several non-communicable diseases, such as colon cancer and cardiovascular diseases. Pulses can be grown on a wide range of soil and climatic conditions and play an important role in crop rotation, mixed and intercropping, maintaining soil fertility through biological nitrogen fixation (30–150 kg ha-1) with the help of *Rhizobium* bacteria found in their root nodules, release of soil-bound phosphorus, and thus contribute significantly to sustainability of the farming systems as well as are a consistent source of income and employment to small and marginal farmers; thus, it holds a premier position in the world agriculture. The United Nations declared 2016 as "International Year of Pulses", with the objectives of increasing production and consumption of pulses by 10% by 2020 and creating awareness about the benefits of pulses by utilizing social media. Pulses constitute an essential part of the Indian diet for nutritional security and environmental sustainability. They are important food crops due to their high protein content (20 to 25%), carbohydrates (55 to 60%), calcium, and iron also. Currently, we are in the midway of self-sustaining in pulse production as we are the leading country in production, consumption, and import as well in the entire world. By the end of 2050, we will be able to sustain our production and turn from net importer to net exporter of pulses if everything goes as per plan. Another unique feature is its source of livelihood and still not a commercial business. Improved varieties of high-quality seeds have become the essential factor in increasing pulse production in India, particularly in rainfed environments, which account for over 80% of the total pulse-growing area. India holds the title of being the largest producer and consumer of pulses, contributing approximately 25–28% to the global share (IIPR 2011) and fulfilling 34% of its food requirements. Thus, there is a challenge for agricultural scientists, extension workers, planners and the farming community to enhance and sustain pulses productivity and diversify their cropping systems with pulses to meet national pulses requirement. In view of this, India has to develop and adopt more efficient crop production technologies along with favorable policies to encourage farmers to bring more area under pulses. The advancement in pulses production technologies has further opened a new hope to increase the production and productivity of the pulses in different irrigated and rainfed agro ecosystems of our country. We have to focus on bridging the yield gaps besides area expansion and diversified production systems well equipped with appropriate pulse production technology already generated by research institutions in order to break yield plateau through recent technological advancements.

India continues to predominantly adhere to a vegetarian dietary pattern, relying heavily on plant-based sources to fulfill its daily protein needs. India is poised to become a frontrunner globally in both pulse production and consumption. However, despite India's being a leading importer of pulses, the production of pulse and legume crops has remained stagnant over the years. As a result, there is a growing disparity between demand and supply, with imports alone accounting for approximately 20% of the total pulse requirements. Besides fixing atmospheric nitrogen in an easily accessible form for the next crop, legumes also provide some nitrogen transfer to associated non-legume intercrops. This contributes to the sustainability of the production system by enhancing soil properties through physical, chemical, and biological improvements, known as a rotation effect.

The seed replacement rate is still (<30%), which is lower than cereals, especially wheat and rice. This chapter addresses the bottleneck and focuses on the way forward for improving pulses production in India, vertically and horizontally as well.

Pulses are an important source of proteins, vitamins, and minerals and are popularly known as "poor man's meat" and "rich man's vegetable," contributing significantly to the nutritional security of the country. Currently, we are in the midway of self-sustaining in pulses production as we are the world leader in production, consumption, and import as well. The production of pulses across India was estimated to be around 27 million metric tons in financial year 2022. This was an overall significant increase in production from the financial year 2002 across the South Asian country.

17.1 PULSES SCENARIO IN INDIA

India is the largest producer and consumer of pulses in the world, accounting for about 29% of the world area and 19% of the world's production. Even more importantly, India is also the largest importer and processor of pulses in the world. Production of pulses in 2008–09 was 14.66 million tons, with an average yield of 655 kg/ha. In the year 2013–14, India produced a record of 19.5 MT of total pulses. In 2022–23, pulse production is expected to be around 27 million metric tons. Pulse productivity, which was 441 kg/ha in 1950, increased up to 689 kg/ha during 2011, and 888 kg/ha during 2022–23. However, there are still severe production constraints, especially abiotic stresses, abrupt climatic changes, emergence of new species/strains of insect pests and diseases, and increasing deficiency of secondary and micronutrients in the soil. This requires a two-pronged proactive strategy, improving per unit productivity and reducing cost of production. The yield levels of pulses have remained low and stagnant, also area and total production. Some of the major factors affecting the production of pulses include the use of rainfed marginal land, susceptibility to pest and disease attacks, weather aberrations, lack of genetic breakthrough, and diversion of pulse area to more remunerative crops as and when irrigation facilities become available. Policies formulated and implemented during different plans have hardly been helpful.

17.2 UNFAVORABLE WEATHER CONDITIONS

Poor soil and agro-climatic conditions not only compel late sowing of legumes, leading to reduced length of growing period, but also necessitates sustaining cold injuries at the early vegetative phase, which freezes all biological activities for prolonged period. A sudden rises in temperature after that not only induces forced maturity but simultaneously also invites several biotic stress, viz, diseases and insects pests (Ali et al. 2012; Reddy 2009; Singh and Singh 2008). Traditionally, *rabi* pulses sowing was delayed up to the last week of November, and sometimes, under extreme circumstances, it goes up to the first fortnight of December, obviously due to reasons already explained. However, optimum sowing time of lentil is first fortnight of October (Singh et al. 2011; Ramakrishna et al. 2000). However, few winter legumes, including lentil, are also grown as a paira crop in eastern India, which helps in timely planting of the crop even before the paddy has been harvested (Singh and Singh 1995).

17.3 ABNORMAL SOIL CONDITIONS

In general, pulses crop prefer neutral soil reactions and are very sensitive to acidic, saline, and alkaline soil conditions. Indian soils, especially northwestern soils, have a high pH, contrary to the eastern and northeastern parts, which are characterized as acidic soils. Due to these soil conditions, deficiency of micronutrient is pronounced up to acute shortage level (Singh et al. 2013d).).

17.4 AGRONOMIC CONSTRAINTS

Major agronomic limitations include incorrect timing of sowing, low seed rate, flawed sowing technique, insufficient irrigation, inadequate weed control, and the practice of sowing without appropriate management under unfavorable conditions (Ramakrishna et al. 2000; Reddy 2009). As a result of delayed planting, the lentil crop experiences adverse effects from early exposure to severe cold, which hamper its growth and development for a significant duration. Consequently, plants have a shorter timeframe to complete their life cycle, which often leads to forced early maturity (Ramakrishna et al. 2000)

In Eastern India, lentils are normally planted as a medium-duration crop, taking around 130–150 days to complete their life cycle. However, when planted late, lentils are compelled to complete their life cycle within a shorter timeframe of approximately 105 ± 5 days (Joshi 1998; Ramakrishna et al. 2000; Reddy 2009; Singh and Singh 2008; Singh et al. 2012). Rabi pulses, particularly lentils and chickpeas, sown late in the season go through three distinct phases and undergo significant phenological modifications. As a result, lentil crops experience slow growth during the early seedling phase as their energy is primarily allocated to initial establishment (Singh et al. 2002; Singh et al. 2012).

However, during the mid-phase, there is minimal growth and development observed, which poses a significant threat to achieving the full yield potential due to the occurrence of cold injuries. This phase holds great importance in establishing a source for channeling energy in later stages. In the final and crucial phase, lentils face heat injuries that lead to an early onset of the reproductive phase. Consequently, there are disparities in the distribution of resources and inputs, heightened biotic stress, and accelerated maturation as a consequence (Joshi 1998; Dixit et al. 2009; Reid et al. 2011; Singh and Bhatt 2013). A previous study indicated that the allocation of land to pulse cultivation is mostly predetermined, but as the irrigated area expands, pulses are shifted to rainfed areas, and their area is replaced by cereals or cash crops (Singh et al. 1995). In India, the irrigated area dedicated to pulses constituted only 12%, while for wheat and paddy, it accounted for more than 60% of the total area (Reddy and Reddy 2010).

17.5 INPUT QUALITY AND AVAILABILITY-RELATED CONSTRAINTS

The problem of ensuring timely access to high-quality chemical fertilizers persists in many regions where pulses are grown. One major obstacle in several states/regions is the insufficient availability of affordable sources of sulfur, such as gypsum or pyrites, which hinders progress (Singh et al. 2013d). Legumes have lower nutrient requirements compared to cereals, primarily due to biological nitrogen fixation and relatively lower productivity levels. However, legume crops still exhibit positive responses to higher doses of fertilizer nutrients beyond the commonly applied or recommended levels (Singh et al. 2013e). While pesticides, including herbicides, are generally available in most states, concerns persist regarding their efficacy and environmental friendliness, despite the presence of well-designed regulatory measures (Singh et al. 2012c).

17.6 VARIETAL CONSTRAINTS

Immediate attention is required for varietal constraints, such as the absence of high-yielding varieties, low harvest index, vulnerability to diseases and insect pests, flower dropping, insufficient availability of short-duration varieties, intermediate growth habits, limited responsiveness to inputs, and inconsistent performance (Singh et al. 2013e; Ramakrishna et al. 2000).

17.7 PESTS AND DISEASES

Legume crops face various challenges, such as insect pests and seed-borne diseases, which can have a devastating impact on the crop if not effectively controlled. Fusarium wilt is a common problem in regions where legumes are grown, and pests can cause significant damage to stored legume grains. Generally, legumes are relatively free from pests under normal conditions, when appropriate crop rotation is implemented. However, significant insect and disease pests for legumes include pod borer, aphids, and wilt (*Fusarium lentis*), as mentioned in studies by Singh et al. (2013b, 2013g).

17.8 BLUE BULL TROUBLE

Blue bulls pose a significant threat to legumes in the Indo-Gangetic Plains, leading to farmers in Uttar Pradesh, Bihar, Madhya Pradesh, Rajasthan, and Chhattisgarh avoiding cultivation in potentially suitable areas. Due to the lack of an effective strategy within the country, the widespread menace persists without a viable solution.

17.9 TECHNOLOGICAL CONSTRAINTS

Legumes are cultivated in diverse agro-climatic conditions, such as varying soil types, rainfall patterns, and thermal regimes across the country. This necessitates the implementation of production techniques tailored to specific regions, including the cultivation of crop varieties possessing traits that are suitable for prevailing biotic and abiotic stresses. Moreover, the use of biological fertilizers and pesticides should be based on strains isolated from regions sharing similar agro-climatic conditions to ensure their effectiveness. Unfortunately, our research and development program in pulses has not yet adequately acknowledged and addressed this issue. It is essential to adopt production technologies for legume crops that are specific to the soil type and region, while also considering appropriate tillage and seeding devices. (Singh et al. 2012a)

17.10 INFRASTRUCTURAL CONSTRAINTS

Rainfall received during maturity of *kharif* pulses causes losses in yields and grain quality when farmers usually do not have *pakka* and covered threshing floor. Farmers also lack awareness and means for safe storage of grain/seed of pulses. Many areas are approachable only during fair weather. Warehousing facilities are either inadequate or inaccessible (Anonymous 2013; Singh et al. 2013a).

17.11 CREDIT, MARKETING, AND POLICY CONSTRAINTS

Farmers engaged in cultivation of legumes are mostly small and marginal. A majority are in areas with poor banking infrastructure. They have poor resource base and lack risk-bearing capacity. They therefore either lack access to credit or turn defaulters. Delivery of credit to such farmers is also not hassle-free. There is lack of marketing network in remote areas. Procurement of produce by a dedicated agency is virtually non-existent or ineffective. Although every state has a system in place to regulate the quality of inputs, it requires further enhancement to ensure effectiveness. It is necessary to streamline the delivery of improved technology, inputs, and credits through suitable policy interventions. Additionally, the benefits of crop insurance should be extended to farmers cultivating pulses, as recommended by Anonymous (2013) and Singh et al. (2013a).

17.12 REALIZING POTENTIAL PRODUCTIVITY IN PULSES AND WAY FORWARD

India possesses the second-largest cultivable land area globally, along with a diverse range of agro-climatic conditions, 270 days of sunshine, an annual rainfall of 1,120 mm, an extensive network of rivers, a 7,000 km long coastline, remarkable biodiversity, and a plentiful labor force (Singh and Kumar 2009; Singh, et al. 2015). However, pulses, being susceptible to abiotic stresses that hinder optimal nutrient absorption, are often cultivated using low doses of fertilizers. Moreover, nutrient application is imbalanced and rarely based on soil test results. Nevertheless, the wide range of agro-climatic conditions, favorable temperature patterns that support year-round cropping, and generally sufficient rainfall all indicate a significant untapped potential for enhancing pulse productivity and expanding the cultivation of pulses. There is substantial scientific evidence highlighting a significant disparity between the yield of pulses achieved by farmers and the yield observed in frontline demonstration plots. There is need to promote high-yielding varieties along with a selection of suitable varieties for different agro-climatic conditions. Further use of disease- and pest-resistant varieties is highly advocated under the present climate change scenario (Anonymous 2013). Methods and techniques which can be potentially used for improving pulses production can be classified into three groups, that is, (a) vertical approach, (b) horizontal approach, and (c) policy intervention.

17.12.1 Vertical Approach

In this, possible methods and technique are discussed which help achieve more production without expanding crop area. Here, focus on improving productivity of crop per unit area. The following vertical approach can be efficiently and effectively utilized for increasing pulse productivity.

17.12.1.1 Promotion of Sequential Cropping, Intercropping, and Utera Cultivation of Pulses

Agricultural research stations have developed several promising intercropping systems for pulses. Table 18.1 provides information on the recommended intercropping methods for different pulse-producing states. In rainfed states such as Gujarat, Madhya Pradesh, Chhattisgarh, Maharashtra, Karnataka, and Andhra Pradesh, farmers are already familiar with some of these systems, as they have been traditionally practicing them.

The approach to be followed by the rainfed states should include (a) identification of districts and promising intercropping systems for each agro-climatic zone and setting of area coverage targets; (b) conduction of field demonstrations on intercropping with farmer's active participation and comparing returns with sole cropping system; (c) ensuring availability of seed of pulse varieties recommended for intercropping; (d) demonstration of suitable seeding devices (animal-drawn and tractor-drawn) for simultaneous planting of main and intercrop components; (e) seed minikits of pulses may be given to farmers opting for intercropping only; and (f) Singh et al. (2013e) and Anonymous (2013) suggest that district-level Krishi Vigyan Kendras (KVKs) should play a crucial role in training farmers and conducting field demonstrations for production technology.

17.12.1.2 Better Extension for Adoption of Improved Pulse Production Technique

Improved and better pulse production technique to be disseminated have to be not only region-/agro-climatic zone–specific but should also match the resource base of the farming community (Singh et al. 2014a). Innovative ways of institution building that aggregates the produce of scattered legumes farmers and links them up with the businesses for better quality of inputs and for efficient marketing of the produce need to be found. Similarly, extension strategy to be followed should take into account the prevailing socio-economic status of farmers (Singh et al. 2014b).

17.12.1.3 Mechanization in Pulses

Legumes are grown in different agro-climatic regimes, and soil conditions can vary. In many soils, mechanization is essential to raise productivity. Adoption of deep ploughing, ridge planting, line sowing, inter-culture operations, etc., besides mechanization, contributes to timeliness of operations, reduces cost of production, and improves resource (water, energy, and inputs) use efficiency (Singh et al. 2014a; Patel et al. 2014). Considering the small holding of the farmers, custom hiring of the machines is the only viable option for increasing the reach of farm mechanization. An instance illustrating the positive approach of the Madhya Pradesh government is the "Haldhar" program, which serves as a commendable practice. This initiative offers farmers a subsidy of Rs. 2000/per hectare for conducting deep ploughing on their lands (Anonymous 2013; Singh et al. 2013a).

17.12.1.4 Post-Harvest Handling of Grains for Reducing Losses

Mechanical threshing needs to be promoted by providing incentives for purchase of threshers. Procurement of pulses grains by government-authorized organizations will considerably reduce the need for storage at farmer level. Small dal mills should be popularized and promoted through various incentives. The establishment of dal mills in rural areas/districts with significant land dedicated to pulses should be encouraged, similar to the establishment of sugar mills. Private companies should be encouraged to participate in the processing, packaging, and marketing of pulses. Public sector procurement agencies face limitations in terms of funds and expertise in this field (Anonymous 2013; Singh et al. 2013a).

17.12.1.5 Expansion of Irrigation Using Resource Conservation Technologies

Pulses crops are invariably grown under moisture stress, which leads to sub-optimal productivity levels. Scientific scheduling of irrigation, an estimate of quantity of water to be applied, and deployment of water-saving irrigation methods can lead to enhanced yield, higher water and nutrient use efficiency, and larger area coverage under irrigation. The utilization of sprinkler irrigation holds tremendous promise in terms of conserving irrigation water and increasing the extent of land covered by irrigation. This approach has become increasingly favored in various districts facing water scarcity. Drip irrigation attention of the policymakers, administrators, social workers, as it has assumed social, economic, and ecological dimensions. Fertigation holds promise for widely spaced crops like pigeon pea. These devices can expand irrigation area by 30–50% (Anonymous 2013; Singh et al. 2013a).

17.12.1.6 Managing Blue Bull Effectively

The damage caused to pulses crops by blue bull has been on the rise in the extent and magnitude. The issue has reached a critical point where there has been a significant decrease in the cultivation of pulses, especially summer mung beans, in Punjab, Haryana, Rajasthan, Uttar Pradesh, and Gujarat. Despite the problem persisting for many years, no effective socio-economic measures have been developed and put into practice. Pulse growers continue to suffer heavy economic losses. The issue is very serious and warrants attention (Anonymous 2013).

17.12.2 Horizontal Approach

17.12.2.1 Efficient Utilization of Rice Fallow Lands and Replacement of Low-Productivity Crops With Pulses

The estimated remaining uncropped area after the kharif rice season is approximately 11.65 million hectares. This area is predominantly rainfed and is located in the states of Bihar, Madhya Pradesh, Chhattisgarh, Orissa, Eastern Uttar Pradesh, West Bengal, and Jharkhand. Around 25% of this area has the potential to support rabi pulses cultivation after rice, depending on the soil type and depth. As a result, an additional 3 to 4 million hectares could be dedicated to growing rabi pulses. Assuming an average productivity of 600 kg/ha, this area has the capacity to produce 1.8 to 2.4 million tons of pulses. It is important to incentivize farmers and implement region-specific extension strategies to encourage pulse cultivation in the identified districts. Local research stations should provide the necessary technological support, including suitable short-duration varieties, recommended nutrient application rates, and other agronomic practices.

SAUs/KVKs may be mandated to conduct field demonstrations on pulses in rice fallow lands and train field staff and farmers participating in demonstrations. Approximately 500,000 ha of upland rice, 450,000 ha of millets, and 300,000 ha of barley, mustard, and wheat can be allocated for rabi pulses. It is recommended to replace mustard, barley, and wheat with lentils and chickpeas as rabi pulse crops. If possible, harvested rainwater should be used for rabi crop establishment (Anonymous 2013; Singh et al. 2013a).

17.12.3 Policy Intervention

To address the issue of stock limits, it is necessary to contact the relevant Food and Supply Department of the state government responsible for implementing the pulses control order. Each state has its own specific license requirements and stock limits for pulses. It is worth noting that Madhya Pradesh is the only state that does not impose a stock limit on domestic pulses. While Gujarat has eliminated the need for a control order license, it still enforces a stock limit. As a result of these factors, major trading companies abstain from dealing with domestic pulses.

Acquiring an EC license by applying through the appropriate department in each state is a challenging and time-consuming process. Some states, like Delhi, have experienced delays in issuing pulses dealer licenses for more than a year. Additionally, all channel partners are required to apply for licenses as well. Although the official time limits generally range from one to two months, the actual time taken is often much longer. It is necessary to develop new marketing strategies to address the deficiencies in specific pulses crops. As an illustration, the Ministry of Consumer Affairs has launched a vigorous promotional campaign for yellow dal, as reported by Anonymous (2013) and Singh et al. (2013a).

17.12.3.1 Pollination of Pulses Crops

Legume plants are vital to world agriculture as they help improve livestock and the soil. Large amounts of legume seeds are required annually for planting, especially when intervals between crop rotations become shorter. Improved varieties with higher yields are always difficult to obtain so that good pollination of existing varieties is essential. Poor weather, inadequate irrigation, insect pests, diseases, and inefficient harvesting have caused lower yields. Nevertheless, while insect pests often lead to significant decreases, the primary reason for insufficient pollination remains the most prevalent. The specific pollination needs of legume crops differ, depending on whether they are capable of self-fertilization or self-sterility. For self-pollinating variations such as peas, soybeans, peanuts, snap beans, and subterranean clover, bees contribute minimally. However, bees do provide some benefits for lespedeza, lima beans, and vetches. Conversely, non-self-pollinating varieties rely on external agents like bees for successful pollination.

Included in the assortment are strawberry clover, lotus, crimson clover, and white sweet clover. In the case of self-sterile varieties that necessitate cross-pollination, insects, typically bees, are necessary. Examples of such varieties are yellow sweet clover, white Dutch clover, alsike clover, alfalfa, and red clover. Legume crops have the potential to attract pollinators, particularly Hymenoptera, due to the presence of both pollen and nectar. They can yield substantial honey crops. Some examples of these crops are lima beans, Dutch clover, alsike clover, purple and hairy vetch, yellow sweet clover, and white sweet clover. Ladino clover, crimson clover, and red clover are crops that are primarily attractive for their pollen but less so for their nectar. Alfalfa, on the other hand, is highly attractive for its nectar but less so for its pollen. Snap beans, soybeans, peas, lotus, Hungarian vetch, subterranean clover, and peanuts are crops that exhibit only slight attractiveness to pollinators.

Certain important legume varieties present specific challenges. Take alfalfa, for instance, where the blossoms require tripping to facilitate pollination. The crucial cross-pollination process relies heavily on alkali bees, leaf-cutting bees, and bumblebees. Although honeybees are typically abundant, they are comparatively ineffective as pollinators in comparison to alkali bees. The issue with honeybees is their preference for nectar over pollen when visiting alfalfa. Their frequent pursuit of nectar leads them to avoid the tripping mechanism, resulting in only approximately 1% of their visits leading to successful tripping. However, honeybees can be highly efficient when deliberately seeking pollen. Achieving high yields often necessitates deploying a greater number of bees in the field than is required for honey production, which can be financially burdensome for beekeepers.

Wild bees face several issues, such as their strong preference for alfalfa as their primary pollen source. These bees exhibit swift movements, collecting pollen at a rate twice as fast as honeybees. However, their presence is not widespread enough in most fields to ensure comprehensive pollination, and their reliability is compromised due to fluctuations in their population from one season to another. In order to overcome these challenges, it is beneficial for farmers to collaborate with neighboring farmers in their region, particularly when seeking beekeeping services. These services often come at a cost due to lower honey yields.

Except for sweet clovers, which attract wasps and bees, bees are the sole dependable pollinators for commercial legume crops, with numerous species potentially involved. Despite their usual high efficiency, their population numbers are often insufficient for satisfactory pollination. Wild bees are the most suitable pollinators for crops like red clover and alfalfa. However, certain bumblebee species can have a negative impact on the pollination of vetch and red clover due to their habit of cutting into the corolla tube bases. This behavior can adversely affect red clover seed production in specific regions.

India's pulse demand will persist at a significant level due to population growth and higher disposable income. Pulse crops, to different extents, rely on open pollination, which necessitates the transfer of pollen between flowers. While mechanisms like wind can accomplish pollen transfer and cross-pollination, the most efficient approach involves utilizing insects to transport pollen from one flower to another (Abrol et al. 2012). Bee pollination particularly benefits crops such as tur, urad, mung beans, guar, pea, and cowpea.

Benefit has been reported up to 1,100% in beans and 600% in kidney beans. Like soybean, tur and others are self-pollinated crops, but yield increases by bee pollination have been observed.

In case of faba bean, there was 71.7% increase in seed yield. Additionally, 86.1% increase in yield with plots caged with bees versus a plot where bees are excluded, and 77.4% increase in yield (seed per hectare) in caged plots with bees compared to caged plots without bees. Beans visited by honeybees produced 10.3 pods per plant and 162 g seed per 20 plants compared to beans isolated from bees, which gave 7.3 pods per plant and 25 g seed per 20 plants (Germany) (Singh et al. 2012b, Singh et al. 2012c).

In England, the productivity of enclosed plots amounted to 2,328 kg/hectare, whereas plots with bees confined in cages yielded 4,332 kg/ha. The outcomes for plots with one side of the cage exposed were 4,320 kg/ha and 4,416 kg/ha for open plots. In Ukraine, the inclusion of honeybees led to a 12.1% growth in bean pod production. The percentage of faba bean pollen collected from beehives ranged from 12 to 88% in Czechoslovakia. In France, when flowers in open plots were visited by honeybees eight times or more, the seed yields were at least eight times higher compared to plots where honeybees were kept out. The rate of flower abortion was higher when honeybees were absent compared to when they were present.

Insects that had access to *Vicia faba* flowers (mainly honeybees) produced on average 30% more pods, 13% more seeds/pod, 47% more seeds/m^2, and 23% higher seed yield/m^2. Bee pollination benefits field crops. In case of pigeon pea (*Cajanus cajan*), experiment carried on seed yield revealed that in caged plots with *A. mellifera*, caged plots without bees, and open plots, number of pods per plant ranged from 280 to 326, 284.2 to 317.2, and 934.0 to 1128.4, respectively (Brar et al. 1992). In a similar study, Pando et al. (2011) reported that pigeon pea (*Cajanus cajan*) shows pollination deficit even under normal circumstances, considering that flowers visited by *Chalicodoma cincta cincta* had higher yields compared with those under unlimited access by all visitors. The fruiting rate, the number of seeds per pod, and the percentage of normal seeds of unprotected inflorescences were significantly higher than those of inflorescences protected from insects. Foraging by *Chalicodoma cincta cincta* led to a noteworthy increase in the fruiting rate, with a rise of 19.65%. Additionally, there was a 24.33% increase in the number of seeds per pod and an 11.52% increase in the percentage of normal seeds. To enhance pod and seed production, it is advisable to conserve *Chalicodoma. c. cincta* nests near *C. cajan* fields. Clearly, conducting comprehensive investigations on the pollination needs of pulses is necessary to address the issue of pulse scarcity in the country.

17.13 CONCLUSION

India must accelerate its progress in reducing hunger, malnutrition, and food insecurity. The food policy framework needs to be reconsidered at this point in time. Best agronomic practices (BAP) and their different components showed potential to excel under change climate condition; there is need to adopt all the components of advocated technology as a unit, not to choose few of them at will, which leads to several complications, soil health hazards, and poor response of technology. These steps may be taken on a priority basis for improving pulse productivity: (1) encouraging accelerated adoption of present technology for bridging the yield gap; (2) institutional support to boost seed replacement rate and quality production as well; (3) strengthening of lifesaving irrigation in pulse-growing pockets; (4) guaranteeing availability of critical inputs, viz, seed fertilizer, pesticides; (5) gradual mechanization for pulse production; (6) public–private partnership for sustaining chain and to minimize post-harvest losses; and (7) policy support for value chain for pulse.

REFERENCES

Abrol, D. P., Shankar, U., Chatterjee, D., and Ramamurthy, V. V. 2012. Exploratory studies on diversity of bees with special emphasis on non-*Apis* pollinators in some natural and agricultural plants of Jammu division, India. *Current Science*, 103(7), 780–783.

Ali, M. and Gupta, S. 2012. Carrying capacity of Indian agriculture: Pulse crops. *Cur. Sci.* 102(6), 874–881.

Anonymous. 2011. Agricultural Statistics At a Glance 2011, Department of Economics and Statistics, Department of Agriculture and Cooperation, Ministry of Agriculture, Govt. of India.

Anonymous. 2013. Report of Expert Group on Pulses. Department of Agriculture & Cooperation Government of India, Ministry of Agriculture. 139.

Brar, H. S., Jhajj, H. S., and Gatoria, G. S. 1992. Abundance and activity of the bee visits to pigeonpea [*Cajanus cajan* (L.) Millsp.] and role of *Apis mellifera* (L.) in its pollination. *Indian Bee J.* 54, 76–80.

Dixit, G. P., Katiyar, P. K., Singh, B. B., and Kumar, S. 2009. Lentil varieties in India. AICRP on MULLaRP. Indian Institute of Pulses Research, Kanpur.

IIPR. 2011. Annual Report 2011–12, Indian Institute of Pulses Research, Indian Council of Agricultural Research, Kanpur, India.

Joshi, P. K. 1998. Performance of grain legumes in the Indo-Gangetic plain in residual effects of legumes in rice and wheat cropping systems of the Indo-Gangetic plain. In: Kumarrao, J. V. D. K., Johansen, C. and Rego, T. J. (eds.) *Legumes in Rice and Wheat Cropping Systems of the Indo-Gangetic Plain.* ICRISAT, Patancheru, Andhra Pradesh, pp. 3–13.

Pando, J. B., Fohouo, F. N. T. and Tamesse, J. L. 2011. Pollination and yield responses of pigeon pea (Cajanus cajan L. Mill sp.) to the foraging activity of Chalicodoma cincta cincta (Hymenoptera: Megachilidae) in Yaoundé (Cameroon). *Journal of Animal & Plant Sciences* 11(1), 1346–1357.

Patel, A. K., Singh, D., Baghel, K. S., and Pandey, A. K. 2014. Enhancing water productivity to improve Chickpea production in Bansagar command area of Madhya Pradesh. *Journal of AgriSearch* 1(1), 19–21.

Ramakrishna, A., Gowda, C. L. L., and Johansen, C. 2000. Management factors affecting legumes production in the Indo-Gangetic Plain. In: Johansen, C., Duxbury, J. M., Virmani, S. M. and Gowda, C. L. L. (eds.) *Legumes in Rice and Wheat Cropping Systems of the Indo-Gangetic Plain-Constraints and Opportunities.* ICRISAT, Patancheru, Andhra Pradesh. [Journal of AgriSearch, Vol. 2, No. 2] Pulses Production scenario in India 83, pp. 156–165.

Reddy, A. A. 2009. Pulses production technology: Status and way forward. *Economic & Political Weekly* 44(52), 73–80.

Reddy, A. A. and Reddy, G. P. 2010. Supply side constrains in production of pulses in India: A case study of Lentil. *Agricultural Economics Research Revie.* 23, January–June, 129–136.

Reid, D. E., Ferguson, B. J., Hayashi, S., Lin, Y. H., and Gresshoff, P. M. 2011. Molecular mechanisms controlling legume autoregulation of nodulation. *Annals of Botany* 108, 789–795.

Singh, A. K. and Bhatt, B. P. 2013. Effects of foliar application of zinc on growth and seed yield of late-sown lentil. *Indian J. Agril. Sci.* 83(6), 622–626.

Singh, A. K., Bhatt, B. P., Singh, K. M., Kumar, A., Manibhushan, K. U., Chandra, N., and Bharati, R. C. 2013b. Dynamics of powdery mildew (*Erysiphetrifolii*) disease of lentil influenced by sulphur and zinc nutrition. *Plant Pathology Journal* 12(2), 71–77.

Singh, A. K., Bhatt, B. P., Sundaram, P. K., Chndra, N., Bharati, R. C., and Patel, S. K. 2012a. Faba bean (*Viciafaba* L.) phenology and performance in response to its seed size class and planting depth. *Int. J. of Agril.& Stat. Sci.* 8 (1), 97–109.

Singh, A. K., Bhatt, B. P., Sundaram, P. K., Gupta, A. K., and Singh, D. 2013c. Planting geometry to optimize growth and productivity faba bean (*Viciafaba* L.) and soil fertility. *J. Environ. Biol.* 34(1), 117–122.

Singh, A. K., Bhatt, B. P., Upadhya, A., Singh, B. K., Kumar, S., Sundaram, P. K., Chndra, N., and Bharati, R. C. 2012b. Improvement of faba bean (*Viciafaba L.*) yield and quality through biotechnological approach: A review. *African Journal Biotechnology* 11(87), 15264–15271.

Singh, A. K. and Kumar, P. 2009. Nutrient management in rainfed dryland agro ecosystem in the impending climate change scenario. *Agril. Situ. India.* 66(5), 265–270.

Singh, A. K., Kumar, P. and Chandra, N. 2013e. Studies on yield production of mung bean (*Vigna radiate*) sown at different dates. *J. Environ. Biol.* 34: 1007–1011.

Singh, A. K., Manibhushan, Bhatt, B. P., Singh, K. M., and Upadhyaya, A. 2013a. An analysis of oilseeds and pulses scenario in Eastern India during 2050–51. *Journal of Agril. Sci.* 5(1), 241–249.

Singh, A. K., Meena, M. K., Bharati, R. C., and Gade, R. M. 2013d. Effect of sulphur and zinc management on yield, nutrient uptake, changes in soil fertility and economics in rice (*Oryzasativa*)—lentil (*Lens culinaris*) cropping system. *Indian J. Agril. Sci.* 83(3), 344–348.

Singh, A. K., Singh, K. A., Bharati, R. C., and Chadra, N. 2013e. Response of intercrops and nutrient management on the performance of tobacco based intercropping system and assessment of system sustainability. *Bangladesh J. Bot.* 42(2), 343–348.

Singh, A. K. and Singh, N. P. 2008. Yield and uptake of primary nutrients by large seeded varieties of lentil under varying seed rates in normal and late sown conditions. *Journal of Food Legumes* 20(2), 187–189.

Singh, A. K., Singh, S. S., Prakash, V., Kumar, S., and Dwivedi, S. K. 2015. Pulses production in India: Present status, bottleneck and way forward. *Journal of Agrisearch* 2(2), 75–83.

Singh, A. K., Singh, D., Singh, A. K., Gade, R. M., and Sangle, U. R. 2012c. Good Agronomic Practices (GAP)—An efficient and eco-friendly tool for sustainable management of plant diseases under changing climate scenario. *J. Plant Disease Sci.* 7(1), 1–8.

Singh, A. V., Prasad, B., and Shah, S. 2011. Influence of phosphate solubilizing bacteria for enhancement of plant growth and seed yield in lentil. *Journal of Crop and Weed* 7(1), 1–4.

Singh, D., Kumar, A., Singh, A. K., and Tripathi, H. S. 2013g. Severity of chickpea wilt in north Bihar and nutritional studies on *fusariumoxysporumf.sp. Ciceri J. Plant Disease Sci.* 8(2), 137–140.

Singh, D., Patel, A. K., Baghel, S. K., Singh, M. S., Singh, A., and Singh, A. K. 2014a. Impact of front line demonstration on the yield and economics of Chickpea (*Cicerarietinum*L.) in Sidhi District of Madhya Pradesh. *Journal of AgriSearch* 1(1), 22–25.

Singh, K. M. and Singh, R. K. P. 1995. An economic analysis of Lentil cultivation in N-E alluvial plains of Bihar. *Economic Affairs* 40(3), 157–163.
Singh, P., Singh, R., Kumar, K., and Singh, D. K. 2012. Estimates of genetic parameters in diverse genotypes of lentil. *Journal of Food Legumes* 25(1), 66–69.
Singh, S. K., Varma, S. C., and Singh, R. P. 2002. Integrated nutrient management in rice and its residual effect on lentil. *Indian J. Agric. Res.* 36(4), 286–289.
Singh, S. S., Singh, A. K., and Sundaram, P. K. 2014b. Agrotechnological options for upscaling agricultural productivity in eastern indo gangetic plains under impending climate change situations: A review. *Journal of Agrisearch* 1(2), 55–65.

18 Leguminosae

18.1 LEGUMINOUS PULSES

Leguminous vegetables have been cultivated for more than 6,000 years in different parts of the world. In countries like India, where majority of the population are vegetarians, leguminous vegetables serve as the major source of protein in the diet. Most leguminous vegetables are rich in phosphorus, calcium, iron, and a number of essential vitamins. Leguminous vegetables are used as fresh pod, immature seed, and mature dry seed, and majority of these vegetables are also canned and frozen for the market. Some of the legume seeds are sprouted and are popular in many Asian countries, where many legumes are also processed.

Pulses have been grown since millennia and have been a vital ingredient of the human diet. Legumes are one of the most important components of good-quality dietary proteins. They are also a good source of vitamins, such as folate, and dietary minerals, like calcium, iron, magnesium, and zinc (Heiser 1990). Legumes, including vegetable legumes, contain antioxidants and other beneficial compounds that contribute to lowering the risks associated with diseases like cancer. As a result, legumes, recognized as valuable commercial crops, are cultivated for both domestic and export markets, offering significant benefits to human health.

The productivity of legume crops, especially pulses, has not increased markedly in the last 30 years. This is a serious concern in light of the exponential growth of the human population, because legume crops provide essential dietary components that are not present in cereals. The demand for food is likely to increase by about 50% by 2050. This target will have to be achieved in the backdrop of decrease in cultivable land, increased water deficit, and increased uncertainty of climatic conditions. With the prediction of 2–4°C rise in temperature in the next 100 years, conditions will become further challenging to grow crops successfully.

Legume plants have the ability to enhance soil fertility by harnessing atmospheric nitrogen through a symbiotic relationship with nitrogen-fixing bacteria like *Rhizobium* or *Bradyrhizobium*. This sets leguminous crops apart from other vegetables, as they can access and utilize atmospheric nitrogen, resulting in improved soil quality. The nitrogen fixed in the root nodules is not only available to the plant but also enrich the soil, in varying amounts, when the plants complete their life cycle.

Large amounts of legume seed are required annually for planting, especially when intervals between crop rotations become shorter. Improved varieties with higher yields are always difficult to obtain so that good pollination of existing varieties is essential. Poor weather, inadequate irrigation, insect pests, diseases, and inefficient harvesting have caused the lower yields. Vegetable legumes are also highly susceptible to insect pests and diseases, which cause up to 80% yield losses or more (Afun et al. 1991; Dreyer et al. 1994; Ulrichs and Mewis 2004). Moreover, the significant decrease in crop yields is predominantly attributed to insufficient pollination, making it crucial to adopt sustainable pest management approaches. These strategies present a viable and secure alternative to the excessive and improper use of pesticides in the cultivation of food legumes in tropical regions of Asia and Africa. By implementing these methods, we not only safeguard pollinators and other beneficial insects but also ensure the attainment of optimal yields.

18.2 POLLINATION OF LEGUMINOUS VEGETABLES

The pollination needs of legume crops differ based on whether they are self-fertile or self-sterile. For self-pollinating varieties like peas, soybeans, peanuts, snap beans, and subterranean clover, bees have minimal importance. However, bees do have some value for lespedeza, lima beans,

DOI: 10.1201/9781032656724-20

and vetches. In the case of non-self-pollinating varieties, an external agent like bees is necessary. Legume crops can be highly appealing to pollinators, particularly Hymenoptera, as they offer both pollen and nectar. They can yield substantial honey production and serve as excellent sources of honey and pollen for bees. A typical legume flower consists of sepals, five petals (including a standard, two wings, and two keels), stamens, and carpels. Additionally, supplementary floral structures like stipule extrafloral nectaries may also contribute to attracting pollinators (Westerkamp and Weber 1999).

Legumes attract a wide range of social, eusocial, and solitary bees, including commercial, feral, and wild species (Free 1993; Delaplane and Mayer 2000). In sustainable agriculture, legumes play a crucial role in providing foraging areas and nesting sites for both solitary and social bees. By offering floral resources within the crop, legumes have the potential to bolster beneficial insect pollinator populations and improve the environmental function of these plants.

Consequently, we face a situation where the development of pollinator-friendly cultivars is needed. Such a situation may require a re-thinking of crop breeding strategies and objectives. The role of bee pollinators as agents of pollination and of hybridization needs to be considered. Pollinators help increase seed set in self-pollination but more importantly enhance cross-pollination (Richards 2001). Pollinators as agents of cross-pollination can contribute to breeding strategies. Utilizing crops that possess desirable floral attractiveness and rewards can be employed to promote the preservation of pollinators. This, in turn, can potentially result in enhanced seed production, improved crop yields, and heightened resilience against both biotic and abiotic stresses. The improved focus on crop pollinator relationships has both ecological and breeding justifications, as highlighted by Westerkamp and Gottsberger (2000) and Aizen et al. (2008).

Optimizing crop pollinator relationship would be a key to the establishment of breeding strategies that increase yield and its stability by using social and solitary bees, thereby facilitating the development of "pollinator-friendly" cultivars. Farmers have the potential to receive extra income through conservation payments while also contributing to the preservation of biodiversity, which is a remarkable aspect of this breeding strategy (Morandin and Winston 2006). And the farmer would still continue to get high yield and yield stability. Important leguminous pulses include cowpea (*Vigna unguiculata* [L.] Walp.), chickpea (*Cicer arietinum*), pigeon pea (*Cajanus cajan* [L.] Millsp.), beans, and soybeans (*Glycine max* [L.] Merr.).

18.3 BEANS

18.3.1 Scarlet Runner Bean (*Phaseolus coccineus* L.)

Phaseolus coccineus L., family Leguminosae, is the scarlet runner bean, or runner bean, native to Mexico or Central America (Graham 1941). This bean is of major importance in parts of Europe, and of minor importance in the United States. The showy blossoms are borne on long racemes. They are attractive to various species of bees for both nectar and pollen. Darwin (1874) showed that the blossom is attractive to insects, ensuring pollination. This is brought about by a pair of nectar holes at the base of the staminal tube, almost covered by a flap on one stamen. When the bee forces this flap aside, it rubs across the staminal column and transfers pollen from anthers to stigma.

18.4 POLLINATION REQUIREMENTS

Darwin (1889) covered scarlet runner beans to exclude pollinating insects and obtained only one-eighth and one-third as many beans as were obtained from open plants. In a more elaborate experiment, Free (1966) caged scarlet runner beans with and without honeybees to determine the need for insect pollination. Plants caged with bees set 1,479 g of pods per plant, compared with 180 g per plant where bees were excluded. Other factors varied correspondingly: pods per plant, 96:11; grams of seed per plant, 356:50; number of seeds, 206:30; and seeds per pod, 3.9 in bee cages compared to

2.7 in cages where bees were excluded. Only the mean weight per seed was unchanged. The pods in the cages without bees could not be harvested until later in the season than those in the bee cages. Tedoradze (1959) obtained similar benefits from bees. Mommers (1971) studied the pollination of runner beans under glass and reported that bees had no effect on production of beans. He, however, concluded that the results obtained by Free (1966) may be due to varietal differences.

18.5 POLLINATORS

Henslow (1878) reported that scarlet runner beans were pollinated by bumblebees. Free and Racey (1968) reported that honeybees were equally effective as bumblebees in pollinating scarlet runner beans cultivated in greenhouses. However, Free's findings from the same year indicated that individual bumblebees worked at a quicker pace and visited a greater number of flowers per plant compared to honeybees. Additionally, bumblebees often engage in nectar theft by piercing a hole at the flower's base, thereby bypassing their role in pollination. Subsequently, honeybees visit this hole, affecting no pollination (Jany 1950).

18.6 POLLINATION RECOMMENDATIONS AND PRACTICES

Blackwall (1964) reported that adequate pollination is a must for early yields and that lack of insect pollination is the most important single cause for low yield of beans. She further recommended that where lack of pollination is a regular problem, honeybee colonies must be placed to supplement the local pollinator population. Beekeepers must supply small, rapidly expanding colonies, which have a high need for pollen. Free (1966) stated that because honeybees are clearly effective in pollinating scarlet runner beans, their use should increase yields where pollination is a limiting factor. However, he did not make a recommendation to growers on the use of honeybees as pollinators of this crop. The evidence indicates, however, that if the grower wants to obtain maximum yields and harvest the seeds as early as possible, he should have a high pollinator population in his field, particularly during early flowering.

18.6.1 The Lima Bean (*Phaseolus lunatus* L.)

18.6.1.1 Flowering

The white- or cream-colored, legume-type flower of the lima bean is borne at the end of a pedicel on a 2–4 in long raceme. The keel of the legume-type flower is elongated into a twisting beak. Inside the keel are the style, stigma, and the pollen-bearing anthers (Figure 18.1). Anthers drop pollen on the portion of style below the stigma. When pressure is applied to the flower, as from a visiting bee, the stigma and a portion of the style project out of the keel; they retract when pressure is removed. Pollen from the pollen-dusted style rubs onto the end of the keel and contacts the stigma as it retracts. Anthesis occurs between 0700 and 0800 h. The flower never closes, but the corolla is shed after a few days. Cross-pollination is possible if the stigma is pollinated by a bee before it retracts. Not all flowers on a raceme set fruit. The problem may be high temperature, low humidity, or poor pollination (McGregor 1976). The lima bean flower produces nectar and pollen, and bees will visit the plant if no richer forages are available.

18.6.1.2 Lima Bean Pollination Requirements

Both self-pollination and cross-pollination can occur in lima bean. Non-caged plants exhibit an increase in the number of pods, beans per pod, and total bean weight when compared to plants that are caged to prevent bee access. However, there are contradictory findings from other studies on this matter. Crossing promotes hybrid vigor in lima bean, but this sometimes happens without the help of bees. Production of hybrid seed increases in the presence of bees. According to McGregor (1976), cross-pollination is generally not essential, but it is likely to be beneficial.

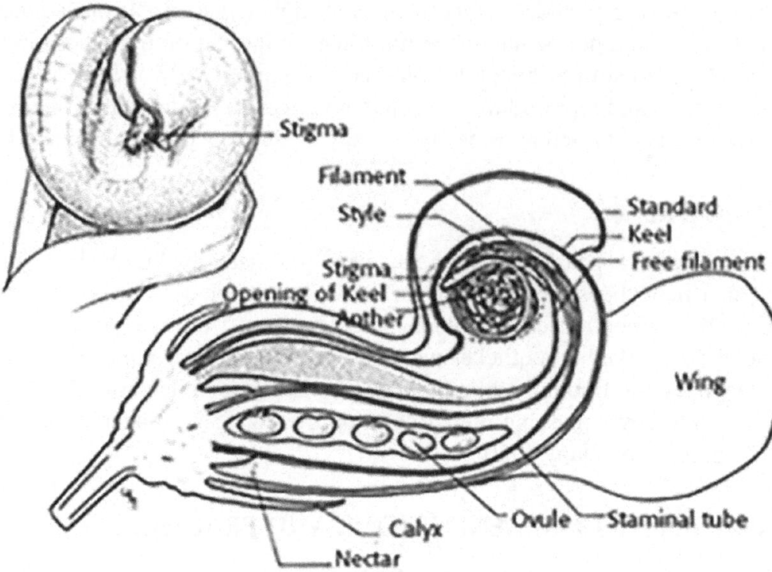

FIGURE 18.1 Flower of lima bean, *Phaseolus lunatus.*

Amos (1943) attributed a 30% increase in yield due to the presence of bees. He stated that 30 colonies were within 1 mi of the test plots, but the acreage of beans in bloom was not given. There was increase in numbers of pods, beans per pod, and total weight of beans, but he gave no indication as to the pollinator density in the field. Allard (1954) caged plants with and without bees to obtain a measurement of bee activity in terms of hybrid beans produced. He achieved a hybridization rate of 2.38% in plants confined within cages that were visited by bees, whereas a rate of 1.06% was observed in cages where bees were kept out. These observations indicate that the lima bean, like other beans, is capable of self-pollination, but cross-pollination can and does occur. The extent of cross-pollination and its significance are influenced by the abundance and types of insect pollinators or the level of insect activity in relation to the crop. According to Boswell (1961), it was suggested that insects do not contribute significantly to the pollination of lima beans, but other studies have presented differing viewpoints on this matter.

18.6.1.3 Lima Bean Pollinators

Honeybees and bumblebees readily visit lima bean and promote cross-pollination. A few studies have credited bees with higher yields (McGregor 1976). Thrips are numerous in lima bean flowers and may pollinate the flowers; they may even cross-pollinate if crowding forces them to leave one flower and go to another (Free 1993). There are few or no research-based recommendations for bee densities in lima bean. However, there is evidence that the presence of bees helps in improved yield. Supplying one honeybee hive per acre (2.5/ha) is suggested for optimizing cross-pollination and yields. Additional benefit may be realized by conserving local bumblebee populations, by leaving field margins undisturbed and by planting bee pasture plants in order to offset natural dearth periods.

Mackie and Smith (1935) and Allard (1954) gave thrips the credit for the cross-pollination of lima beans. Jones and Rosa (1928) reported that honeybees, bumblebees, and other insects visit the flowers and cause cross-pollination. Vansell and Reinhardt (1948) conducted a comparative study on production from caged and open plots as well as in areas characterized by varying levels of general bee activity during the flowering period. They noted that 34.4% of the flowers set in the area where bee activity was high and continuous, but only 5 to 22% set where insecticides were applied. Because of the insecticide applications, most of the colonies were moved away during part of the flowering period, resulting in low pollinator activity. Initially, one bee was recorded per yard of row prior to the start of colony evacuation. However, after the evacuation began, the number decreased significantly to one bee per 33 yards of row. Despite this reduced bee activity, they observed a higher yield in terms of both bean quantity and weight from open plants compared to caged plants. Wester and Jorgensen (1950) found that bumblebees and honeybees were primarily responsible for cross-pollination of lima beans in the field. Anderson (1959) found that lima bean plants enclosed in cages with bees exhibited a greater bean production and larger bean size during the early blooming period, in contrast to cages where bees were not present. He also reported an abundant visitation of bees to lima bean flowers

18.6.1.4 Pollination Recommendations and Practices

Despite the relatively positive findings regarding the significant impact of insect pollination on the enhanced yield of lima beans by Amos (1943) and Vansell and Reinhardt (1948), Drayner (1956) holds a contrasting view on the contribution of cross-pollination to increased production. The recommended number of colonies for lima bean pollination remains uncertain, emphasizing the need for further studies to provide conclusive recommendations. In the meantime, the limited available data suggests that growers could greatly benefit by strategically placing multiple colonies within and around their lima bean fields during the flowering period and implementing measures to safeguard these pollinators from harmful pesticides.

18.6.2 Beet (*Beta vulgaris*)

18.6.2.1 Flowering

The small greenish flower of beet (*Beta vulgaris*) occurs on a branched seed stem in clusters of two or three. Although flowers are perfect, self-pollination is infrequent due to the delayed maturation of stigmas compared to anthers. The flowers typically bloom in the morning, with anthers releasing pollen before noon. Stigmas, on the other hand, start to open in the afternoon and reach full openness by days 2–3, when the anthers of the same flower cease pollen production. Once open, stigmas may stay receptive for more than two weeks. The pollen is wind-borne. Many insects visit beet flowers. Honeybees visit beet, but they readily move to more attractive food sources.

18.6.2.2 Pollinators

Recorded pollinators include various species, such as honeybees (*Apis mellifera* Lin., *Apis cerana* Fab., *Apis dorsata* Fab.), cowpea borer (*Lampides boeticus* L.), bumblebees (*Bombus* spp.), tabanid flies (*Tabanus* spp.), syrphid flies (*Syrphus* sp.), houseflies (*Musca domestica* Lin.), rice skippers (*Pelopidas mathias* (F.)), castor butterflies (*Ergolis merione* Cram.), cabbage butterflies (*Pieris brassicae* Lin., *P. canidia* Lin.), golden wasps (*Vespa magnifica* (Smith)), and mustard sawflies (*Athalia lugens proxima* (Klug.)).

18.6.3 Broad Bean (*Vicia faba* L.)

Rowlands (1960) recorded an average of 24% set in the open but only 7.3% set in the glass house. He concluded that some plants are highly self-sterile and some cross better than others. Free (1970) stated that the amount and concentration of field bean floral nectar was relatively low, so the flowers are not attractive to nectar-collecting insects. In discussing field bean inflorescence, nectar secretion, and bee visitation, Free (1970) found that bee activity at extrafloral nectaries peaked during midday, while visits to flowers were highest between 1400 and 1600 h, coinciding with the availability of pollen. He further reported that bees were more attracted to pollen than floral nectar.

18.6.3.1 Pollination Requirements

Free (1966) examined the pollination of broad bean and field bean and determined that insect pollination significantly boosted broad bean production but had minimal impact on field bean production. Later, Free (1970) concluded that the pollination requirements of broad bean and field bean were similar, although inadequate pollination limited seed production in broad beans more so than in field beans. Darwin (1889) reported that 17 broad bean plants covered with a net to exclude pollinators produced only 40 seeds, whereas 17 unprotected plants produced 135 seeds. Free (1966) showed that plants caged with bees produced about twice as many seeds as similar plants caged without bees.

18.6.3.2 Pollinators

Bond and Hawkins (1967) and Free (1962) have studied the behavior of bees on field bean. Free (1962) observed that the activity of the honeybees may vary. They may visit only the extrafloral nectaries, they may visit only the holes cut in the bases of the corollas by bumblebees, or they may visit the corolla for nectar and pollen. Only in the latter case are the bees of value to the beans as a pollinating agent. Bond and Hawkins (1967) supplemented one acre of field bean with two colonies of honeybees. They found that bees collected significant quantities of pollen. Further, they reported that bumblebees were mainly responsible for cross-pollination. Wafa and Ibrahim (1960) concluded that the carpenter bee, *Xylocopa aestuans* (L.), was the most important wild pollinating insect in the Gaza region. Free (1959) concluded that better visitation to the bean flowers resulted when the honeybee colonies were moved to the crop after it started to flower. Watts and Marshall (1961) observed a 26% increase in seed production due to the presence of bees. They also highlighted the significance of bees in facilitating early seed development. Comparing plots with bees and open plots, they observed an average of nine pods at the bottom, nine in the middle, and two at the top of the plant. However, in the no-bee cage, only three pods were found at the bottom, five in the middle, and four at the top. Free (1970) concluded that insects with long tongues were the only ones capable of accessing the nectar in bean flowers. Honeybees and short-tongued bumblebees that entered the flowers likely obtained mainly pollen. However, some bumblebee species with short tongues, male carpenter bees (*Xylocopa aestuans*), and certain ants (*Cataglyphis bicolor* (F.)) bite holes in the base of the corolla and "rob" the nectar without contributing to pollination. Honeybees do not make holes, but they will rob nectar from holes made by other insects.

Pollination recommendations and practices. Allen and Scriven (1957) concluded that one colony of honeybees per acre should give adequate coverage of field beans.

Mung bean/green gram, *Phaseolus* spp. The inflorescence is an axillary raceme, with 10 to 25 pale-yellow flowers, 1 1/2 to 2 cm long, and clustered at the top. The shedding of pollen occurs in the afternoon prior to the flower's opening the next morning. The flower wilts later in the same afternoon. Approximately 64% of the flowers open, allowing for potential cross-pollination.

Pollination Requirements. Purseglove mentioned that the flowers achieve complete self-fertility when enclosed in bags and are predominantly self-pollinated. Rheenen (1964) reported that when he interchanged varieties within the row, he observed a crossing rate of 2.8 to 3%.

18.6.4 The Common Green Snap Bean (*Phaseolus vulgaris*)

18.6.4.1 Flowering

The common bean (*Phaseolus vulgaris*) flower has an elongated, twisted keel containing the style and ten stamens. Inside the flower, the anthers drop pollen onto the style in the evening before it opens. The next day, if pressure is applied to the flower, by an insect, for example, the style and stigma protrude from the keel and retract when the pressure is released. The style leaves pollen at the opening of the keel. Cross-pollination is possible if the stigma contacts a pollen-coated bee while it is extended. Otherwise, the stigma will be self-pollinated when it retracts and contacts its own pollen at the opening of the keel (McGregor 1976).

18.6.4.2 Pollination Requirements

Self-pollination is a regular feature with green snap beans, and it normally happens automatically at or before flower opening in the morning. However, it takes 8–9 hr for pollen tubes to grow and fertilize seeds, during which time honeybees and bumblebees can visit the flower and cross-pollinate it. Free (1993) reported that the growth rate of foreign pollen tubes is likely faster than that of the plant's own pollen, increasing the chances of cross-pollination occurring. Tubes of foreign pollen probably grow faster than the plant's own pollen, so crossing is likely when the plant is cross-pollinated (Free 1993). For the grower, there is no yield or other economic advantage to cross-pollination. For the bean breeder, cross-pollination is actually a hazard to maintaining purity of varietal seed. In order to minimize undesired hybridization, it is recommended to create a physical distance of 6–12 ft (1.8–3.7 m), along with a tall and dense barrier between different varieties. For Canadian varieties, the standard practice involves separating "registered" seed stocks by 150 ft (46 m) and "elite" seed stocks by 0.3 mile (0.5 km) (Free 1993).

18.6.4.3 Pollinators

When it comes to pollinators, bumblebees demonstrate a higher preference for green snap beans compared to honeybees (McGregor 1976). Nevertheless, the overall contribution of bees to green snap bean production is likely to be minimal. In *Phaseolus vulgaris* L., the flower of the bean exhibits the typical legume shape, typically whitish, but with possible variations of violet, purple, or red, and measures approximately 1/2 to 3/4 in in length.

The keel is prolonged in a spirally twisted beak. The style follows the spirals of the keel. There are the usual ten stamens, the upper one free, while the other nine are united into a tube enclosing the long ovary and part of the style. The blooms are loosely scattered along a 2– 3 in rachis (stem).

Weinstein (1926) reported that numerous pollen grains germinate within the anther, but it remains uncertain whether such pollen contributes to fertilization. Knuth (1908) stated that although the anthers encompass the style and the pollen is released prior to the flower's opening, the pollen does not reach the stigma until after a triggering event. For instance, when pressure is exerted, such as by a large bee's weight, the style, carrying pollen attached to its brush of hairs, protrudes from the tip of the keel. Upon the release of pressure, the style retracts into the keel, occasionally leaving some pollen exposed. During the initial exposure of the stigma, it is devoid of pollen and may come into contact with a bee covered in pollen during this brief interval. In such cases, cross-pollination can occur. Additionally, as the stigma retracts into the keel, it encounters the pollen released by the anthers, allowing for the possibility of self-pollination.

Pollination requirements. The flowers of most bean cultivars are capable of self-pollination, although cross-pollination can and does occur to varying degrees, depending upon the cultivar and the pollinator population (Hawthorn and Pollard 1954; Mackie and Smith 1935; Barrons 1939). Much of the research on bean pollination has been concerned only with the impact of the cross-pollination on varietal seed contamination, not on quantity or quality of fruit set. Rutger and Beckham (1970) stated that *P. vualgaris* is primarily a selfing species, but studies of controlled insect pollination are needed. Free (1966) reported that it is not possible to predict the pollination needs of various bean species, emphasizing the need for individualized studies.

Thorough studies on the possible benefits of maximum pollinator activity on the various types and cultivars would appear to be highly valuable. Knuth (1908) stated that the anthers surround the style and shed their pollen upon it, but the stigma is never dusted until the keel is depressed. Then the style, with its pollen-covered brush, springs out of the opening. In case the stigma is touched by the bee, coated with pollen from other plants, before being contacted by its own pollen, cross-pollination results. However, Taylor (1919) stated that the flowers are entirely self-fertile, as is abundantly proved by their productiveness when grown in greenhouses where bees are not present. Yet he stated that they can be cross-fertilized by bees, although the percentage of crossing is not great. Mackie and Smith (1935) and Barrons (1939) indicated that cross-pollination may amount to more than 8%. The pollinator population associated with the crossing was never measured. Jones and Rosa (1928) stated that because the stigma and the anthers are enclosed within the keel, insects reach them only with difficulty; however, visitation occurs and crossing results. They considered the amount negligible. They also stated that selfing occurs freely when flowers are enclosed.

Kristofferson (1921) documented crossing rates ranging from 0 to 1.4%, while Emerson (1904) observed crossing rates between 0 and 10%. Free (1966), who conducted research on the "Processor" cv. of *P. vulgaris*, concluded that honeybees were unlikely to have a significant impact on the yield. However, his findings actually indicated that plants in cages with bees produced 21% more seed weight and 6% more pod weight compared to plants without bees. It should be noted that honeybees visit beans infrequently, and beekeepers do not consider this crop to be a significant source of nectar or pollen.

In Kenya, Kasina et al. (2009) found that cross-pollinated common beans resulted to increased seed set and nutrition improvement of the dry harvested seeds. Yet since French bean marketing quality is based on individual pod weight as well as uniformity in pod shape, commercial yield might be affected by insect pollination. *Phaseolus vulgaris* (Bliss 1980) does not exhibit parthenocarpy or apomixis. The uniformity of pods is influenced by the consistency of seed set, meaning, that when all ovules are fertilized, the pods grow uniformly, resulting in high-quality products for the producer. Inadequate pollination, on the other hand, leads to fewer developing ovules, causing curved pods that are unmarketable. While French beans are known to be self-fertile (White and Izquierdo 1991), the presence of large and robust bees like carpenter bees and bumblebees can potentially enhance both auto- and allo-pollen pollination by tripping flowers. Free (1966) found that the runner bean (*Phaseolus multiflorus*) plants, isolated from insects, produced fewer pods or seeds than those visited by honeybees, but excluding bees made little difference to the yield of dwarf beans (*P. vulgaris*).

Pollinators. Twenty-four insect species visit *P. vulgaris* flowers. *X. olivacea* was the most frequent visitor, and they intensely and exclusively foraged nectar. Foraging speed of *X. olivacea* was measured at 9.94 flowers per minute. The foraging behavior of *X. olivacea* had a significant impact on the fruiting rate, resulting in a notable increase of 48.43% and 78.18% in 2009 and 2010, respectively. Moreover, the number of seeds per pod showed a respective increase of 19.38% and 18.58%, while the number of normal seeds per pod increased by 15.67% and 38.25% in the same years. In contrast, undisturbed bean flowers did not produce any viable seeds, while flowers manipulated by wing petals yielded a substantial number of seeds, as observed by researchers (Darwin 1858).

Bumblebees were indicated to be the most effective pollinator of beans when visiting blossoms to collect pollen and nectar in South Africa (Palmer 1967). Genus *Bombus* was associated with pollination activity in common beans in North America (McGregor 1976). *P. vulgaris* flowers were reported to produce fewer seeds per pod in the absence of efficient pollinators in the United States of America (Ibarra-Perez et al. 1999).

A recent study conducted in Kenya has found that the most abundant insect species visiting *P. vulgaris* flowers are *Apis mellifera*, followed by *X. calens* (currently *Xylocopa olivacea*) and *X. inconstans* (Kasina et al. 2009). *Xylocopa olivacea* was the primary visitor of *P. vulgaris* flowers during the observation period. In other regions, bumblebees in South Africa (Palmer 1967) and *A. mellifera* in Western Kenya (Kasina et al. 2009) have also been reported as the main visitors of this crop. The peak activity of *X. olivacea* on *P. vulgaris* flowers occurred between 1000 and 1300 h, which coincided with the period of highest nectar availability on the flowers. However, the activity decreased after 1600 to 1700 h, possibly due to the declining temperature in the experimental field. While foragers preferred warm and sunny days for optimal floral activity (Kasper et al. 2008), higher temperatures positively influenced insect activity on the flower. Similarly, rainfall has been documented as an environmental factor that can disrupt the activity of floral insects (McGregor 1976). The high abundance of X. olivacea foragers on 1,000 flowers and the strong, positive correlation observed between the number of *P. vulgaris* flowers in bloom and the number of visits by *X. olivacea* indicate the attractiveness of *P. vulgaris* nectar to this bee species. Indeed, it has been demonstrated that weather conditions during bloom can influence the abundance and foraging behavior of pollinator insects (Bramel et al. 2004; Julianna and Rufus 2010). Out of the 24 insect species observed visiting *P. vulgaris* flowers, *X. olivacea* was the most abundant, accounting for 28.7% of the visitors, followed by *Coryna* sp. (17.92%), *Chalicodoma* rufipes (15.24%), and *Ceratina* sp. (7.65%).

During nectar collection from each flower, *X. olivacea* foragers frequently come into contact with the stigma. They can also carry pollen on their hairs, legs, and mouth parts, transferring it from one flower to the stigma of another flower on the same plant (geitonogamy), to the same flower (autogamy), or to a flower on a different plant (xenogamy). The pollination efficiency of *X. olivacea* was found to make a significant contribution to *P. vulgaris* production, surpassing that of all other insects visiting the exposed flowers. The weight of *X. olivacea* played a positive role during nectar collection, as the bees' shaking action on the flowers facilitated the release of pollen from the anthers for optimal contact with the stigma.

Our results confirmed those of Mensah and Kudom (2011), who revealed that the development of fruits from *L. aegyptiaca* flowers that have received a single visit of *X. olivacea* produced a mean weight of 428.7 g, which was 1.5 times heavier than fruits from flowers visited by *A. mellifera* (286.76 g). Vanderborght and Rasmont (1987) also observed this occurrence in *X. bariwal*, an efficient pollinator of *P. coccineus*. Comparing the productivity of pods and seeds between unlimited insect visits and bagged flowers revealed that insect visits effectively increased cross-pollination. Therefore, it can be inferred that *P. vulgaris* black seed outlets, being highly rich in nectar, derive advantages from insect pollination, with *X. olivacea* being particularly significant.

The comparison of pods and seeds set of unprotected flowers with that of flowers visited exclusively by *X. olivacea* underscores the value of this bee in increasing pods and seed yields as well as seed quality. The installation of *X. olivacea* nests at the proximity of *P. vulgaris* small black seed fields should be recommended for the increase of pods and seed yields of this valuable crop.

Mommers (1971) studied pollination of beans cultivated in a greenhouse and found that the development of *P. vulgaris* is not influenced by bees. Diwan and Salvi (1965) reported that beans attract *Apis dorsata*, *A. florea*, and *Trigona* spp., while *A. cerana* ignored the flowers. Studies have revealed that bumblebees and honeybees are the most frequent visitors to bean flowers in the United States. Darwin (1858) and Palmer (1967) suggested the beneficial role of bees in relation to beans, although the specific type of bean they referred to remains unclear.

The data indicate that, if pollinating insects are of benefit, the bumblebee is the best pollinator. Honeybees are apparently next in importance. They visit the blossom for both pollen and nectar. The precise impact on the flower or the significance for various *P. vulgaris* cultivars remains uncertain.

18.7 DEPENDENCE OF CROPS ON BEE POLLINATION

To evaluate the impact of bee pollination on crop yield, both bagged and un-bagged flowers were employed. The bagged flowers were considered to undergo self-pollination without bee involvement, whereas it was highly likely that the un-bagged flowers were visited by bees. A notable disparity was observed in the number of seeds per pod of legumes (pulses) obtained from the bagged and un-bagged flowers.

However, this difference was relatively lower for beans (25%) and green grams (27%) compared to the other legumes, while cowpeas showed the highest difference (43%). The results show that the mean number of seeds was lower than the maximum attainable in the un-bagged flowers, implying that bee pollination of the legumes in the study area was not sufficient.

18.8 IMPACT ON YIELD

Kingha et al. (2012) evaluated the impact of the carpenter bee (*Xylocopa olivacea*) on pod and seed sets of *Phaseolus vulgaris* during 2009 and 2010 cropping seasons and found that the foraging activity resulted in a significant increment in fruiting rate by 48.43 and 78.18%, the number of seeds per pod by 19.38 and 18.58%, and the normal seeds per pod by 15.67 and 38.25%, respectively, in 2009 and 2010. Therefore, it is advisable to promote the preservation of *X. olivacea* nests near *P. vulgaris* crop fields as a means of enhancing pod and seed production in the area. Higher productivity of pods and seeds in unlimited visits when compared with bagged flowers showed that insect visits were effective in increasing cross-pollination. Earlier, Webster et al. (1982), Wells et al. (1988), and Ibarra-Perez et al. (1997) revealed that *P. vulgaris* flowers set little pods in the absence of insect pollinators. Similar experiments in England (Free 1966) and in Brazil (Free 1993) have shown that pollination by insects was not always needed. Darwin (1876) showed that self-pollination of *P. vulgaris* flowers produced as many pods and seeds as exposed plants. Thus, pollination requirements may differ between plant varieties (Table 18.5).

Anthophilous insects, such as bees, are widely recognized for their role in enhancing the fruit and seed yields of various plant species by providing pollination services (Keller and Waller 2002; Fluri and Frick 2005; Sabbahi et al. 2005; Klein et al. 2007; Tchuenguem Fohouo et al. 2009a). *Phaseolus vulgaris* is an annual plant (Graham and Ranalli 1997) with pink flowers but can vary from white to purple, depending on the different varieties (Debouck 1991), and produces nectar/pollen which attract insects (Ibarra-Perez et al. 1999). *Phaseolus vulgaris* is cultivated as vegetable and can be consumed raw or cooked; its pods are sold fresh (green beans) or transformed into flour, while the stems and leaves are used to feed livestock (Debouck 1991). The plant is allogam/autogam (Ibarra-Perez et al. 1999), and its cross-pollination is ensured by insects (Mackie and Smith 1935; Barrons 1939; Wells et al. 1988; Ibarra-Perez et al. 1997). Previous researches have shown no set of seeds from undisturbed bean flowers, whereas flowers manipulated by wing petals have a great number of seeds (Darwin 1858). Bumblebees have been reported as the most effective pollinator of beans when visiting blossoms to collect pollen and nectar in South Africa (Palmer 1967) and North America (McGregor 1976). *Phaseolus vulgaris* flowers were reported to produce fewer seeds per pod in the absence of efficient pollinators in the United States of America (Ibarra-Perez et al. 1999). A recent study carried out in Kenya has discovered that *Apis mellifera* is the insect species with the highest population visiting *P. vulgaris* flowers, followed by *X. calens* (currently known as *Xylocopa olivacea*) and *X. inconstans* (Kasina et al. 2009) (Table 18.2).

Carpenter bees visit and pollinate different cultivars of *P. vulgaris* grown for dry seeds in Cameroon (Kingha et al. 2012) and in Puerto Rico (Bliss 1980). Ibarra-Pérez et al. (1999) in the

TABLE 18.1
***Phaseolus vulgaris* Mean Yields under Different Pollination Treatments**

Treatments	Years	Flowers	Pod	Fruiting Rate	Seeds/Pod mean	Seeds/Pod Sd	Total Seeds	Normal Seeds	% Normal Seeds
Ff (unlimited visits)	2009	120	99	82.5	6.33	1.25	627	572	91.22
Pf (bagged flowers)	2009	120	32	26.67	5.16	1.76	165	129	78.18
Ff (unlimited visits)	2010	120	114	95	6.39	1.22	728	678	93.13
Pf (bagged flowers)	2010	120	24	20	5.17	1.74	124	70	56.45
Fva (*X. olivacea* flowers)	2009	29	15	51.72	6.4	1.18	96	89	92.71
Fva (*X. olivacea* flowers)	2010	144	132	91.67	6.4	1.34	851	778	91.42

Note: Ff: free flower; Pf: protected flowers; Fva: flowers visited exclusively by *X. olivacea*.
Source: Kingha et al. (2012).

TABLE 18.2
Diversity of Floral Insects on *P. vulgaris* Flowers in 2009 and 2010, Number and Percentage of Visits of Different Insects

	Insects		2009		2010	
Order	**Family**	**Genus, Species, Sub-Species**	**n1**	**P1 (%)**	**n2**	**P2 (%)**
	Apidae	*Xylocopa olivacea* (nectar)	61	34.46	36	22.93
		Xylocopa sp. (nectar)	5	2.82	3	1.91
		Apis mellifera adansonii (nectar)	14	7.91	3	1.91
		Amegila sp. 1 (nectar)	2	1.13	13	8.28
		Amegila sp. 2 (nectar)	-	-	2	1.27
		Braunsapis sp. (nectar)	11	6.21	-	-
		Ceratina sp. 1 (nectar+ pollen)	-	-	24	15.29
		Ceratina sp. 2 (nectar)	-	-	11	7.01
Hymenoptera						
	Halictidae	*Lasioglossum* sp. (nectar)	1	0.56	-	-
	Megachilidae	*Chalicodoma rufipes* (nectar)	28	15.82	23	14.65
		Megachile sp. 1 (nectar)	1	0.56	-	-
		Megachile sp. 2 (nectar)	-	-	1	0.64
		Megachile sp. 3 (nectar)	-	-	1	0.64
		Megachile sp. 4 (nectar)	-	-	5	3.18
		Megachile sp. 5 (nectar)	-	-	2	1.27
	Formicidae	*Camponotus flavomarginatus* (nectar)	2	1.13	-	-
		Camponotus sp. (nectar)	1	0.56	-	-
	Total Hymenoptera		126	71.19	124	78.98
Lepidoptera	Pieridae	*Eurema* sp. 1 (nectar)	1	0.56	8	5.1
		Eurema sp. 2 (nectar)	2	1.13	-	-
	Lycaenidae					
		(sp. 1) (nectar)	-	-	2	1.27
		(sp. 2) (nectar)	-	-	4	2.55
	Hesperiidae	*Lambrix* sp. (nectar)	2	1.13	-	-
	Total Lepidoptera		5	2.82	14	8.92
Diptera	Syrphidae	*Episyrphus* sp. (nectar)	4	2.26	-	-
Coleoptera						
	Meloidae	*Coryna* sp. (eat flowers)	42	23.73	19	12.1
	Total		177	100	157	100

Note: n1: number of visits on 120 flowers in 11 days; n2: number of visits on 120 flowers in 09 days; p1 and p2: percentages of visits, p1 = (n1), p2= (n2) *X. olivacea* visits for two years: $\chi 2 = 5.37$; $t = 2.32$; $p < 10–2$; sp: unidentified species.
Source: Kingha et al. (2012).

USA discovered that bumblebees (*Bombus* spp.) were the most efficient pollinators of *P. vulgaris*, and their visits to the flowers had a significant positive impact on seed production in two specific varieties. While honeybees are abundant visitors to French bean flowers in Kenya (Kasina et al. 2009c), exposing bean plants to honeybee pollination did not lead to higher yields of dry seeds in the United Kingdom or the United States (Free 1966; McGregor 1976; Delaplane and Mayer 2000). However, in Kenya, Kasina et al. (2009b) found that cross-pollination of common beans resulted in increased seed set and improved nutritional content of the harvested dry seeds.

Earlier studies have revealed *X. olivacea* foraging on *V. unguiculata* (Pando et al. 2011a, 2011b), *Phaseolus vulgaris* (Kingha et al. 2012), and *Phaseolus coccineus* (Fameni et al. 2012). *X. olivacea* foraged *V. unguiculata* flowers for nectar and pollen. *X. olivacea* has been reported as the main floral visitors of this crop (Pando et al. 2013) in Yaounde. This rank could be due to the relative abundance of *X. olivacea* in different study areas, though the abundance and diversity of floral insects to a plant vary depending on the region (Roubik 2000).

It is known that the anthophilous insects of a plant vary over time (Enfawal et al. 1976; Tchuenguem 2005). The peak of *X. olivacea* activity on flowers in the morning corresponds to the period of high availability of pollen and nectar on *V. unguiculata* flowers. The decreased in the activity of *X. olivacea* between 1100 h could be related to elevated temperature and the closure of flowers in the experimental field. Although foragers preferred warmth or sunrays for maximum floral activity (Kasper et al. 2008), high temperature negatively affects insect activity on foraged flowers. Similarly, rainfall has been documented as an environmental factor that can disrupt the floral insect activity (McGregor 1976). The abundance of *X. olivacea* foragers on 1,000 flowers, as well as the positive and highly significant correlation between the number of *V. unguiculata* flowers at bloom and the number of *X. olivacea* visits, indicates the attractiveness of *V. unguiculata* nectar and pollen with respect to this bee. In fact, weather conditions during bloom were revealed to affect the abundance and foraging of pollinator insects as *Apis mellifera Ligustica* and *Bombus impatiens* in Michigan (Julianna and Rufus 2010).

During each of the two flowering periods of the host plant, *X. olivacea* intensely and regularly harvested nectar and pollen. This could be linked to the needs of individuals at the flowering period. The disruptions of visits by other insects reduced the duration of certain *X. olivacea*. This has obliged some bees to visit more flowers during a foraging trip in order to maximize their pollen loads. Similar observations were made for *Chalicodoma cincta cincta* (Hymenoptera: Megachilidae) foraging on *Cajanus cajan* (Fabaceae) flowers (Pando et al. 2011b), *Xylocopa olivacea* workers foraging on *Phaseolus vulgaris* of small black seeds flowers (Kingha et al. 2012), and *Xylocopa calens* foraging on flowers of *P. coccineus* (Pando et al. 2011a). *X. olivacea* foragers had a high affinity with respect to *V. unguiculata* when compared to the neighboring plant species, indicating their faithfulness to this fabaceae, a phenomenon known as "floral constancy" (Louveaux 1984; Backhaus 1993; Basualdo et al. 2000). Flower constancy is an important aspect in the management of pollination. For this research, it indicates that *X. olivacea* can provide benefits to pollination management of *V. unguiculata*.

During the collection of nectar and pollen on each flower, *X. olivacea* individuals regularly come into contact with the stigma. They were also able to carry pollen with their hairs, legs, and mouth accessories from a flower of one plant to the stigma of another flower of the same plant (geitonogamy) and to the same flower (autogamy). The foragers might thus have influenced self-pollination and cross-pollination as early as reported (Moffett et al. 1975). The significant contribution of *X. olivacea* in pod and seed yields of *V. unguiculata* lines with the findings of Pando et al. (2013). This high productivity of pod and seeds in unlimited visits when compared with bagged flowers showed that insect visits were effective in increasing cross-pollination. Earlier, Pando et al. (2013) in Ngaoundere revealed that *V. unguiculata* and *P. vulgaris* flowers set little pods in the absence of insect pollinators. The contribution of *X. olivacea* to *V. unguiculata* production through its pollination efficiency was significantly higher than that of all insects visiting the exposed flowers. The weight of *X. olivacea* might have played a positive role during nectar and pollen collection. In fact, *X. olivacea* shook flowers when landing on plants, facilitating the liberation of pollen by the anthers for optimal occupation of the stigma. Similar experiments in England (Free 1966) and in Brazil (Free 1993) have shown that pollination by insects was not always needed. Thus, pollination requirements might differ between regions or countries.

18.9 SOYBEAN

18.9.1 *Glycine max* (L.) Merr., Family Leguminosae

Soybeans comprise the most important oilseed crop grown in the United States. Approximately 41 million acres of soybeans were grown, with a farm value of $2.6 billion. The crop is widely grown

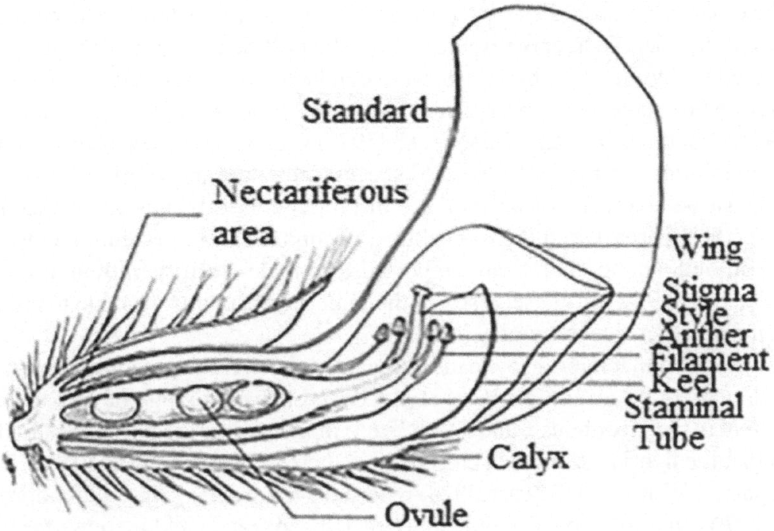

FIGURE 18.2 Longitudinal section of soybean flower x25.

in all areas except New England and the West. The primary reason for growing the crop is for the beans, which are processed to obtain oil and a high-protein meal.

The soybean is a bushy, upright, annual summer legume, 1–4 ft tall, with dense, alternate, and usually trifoliate oval leaves 2–4 in across. There may be from 1 to 35 purple or white florets, three-eighths of an inch long, on each short raceme or flower cluster. A single plant may bear as many as 800 florets but may set only 13 to 57% (Shaik and Probst 1958). The floret has the characteristics and shape of many other legume flowers—a large standard petal, two small wing petals, and a keel petal that encloses the staminal column (Figure 18.2). The calyx is relatively large in proportion to the flower or even to the calyx of other legumes. Each floret is capable of producing a bean pod. Southern-grown cultivars stop growing when flowering begins. Flowering usually continues for four to six weeks. There may be one-half million florets per acre. There are no extrafloral nectaries (Jaycox 1970).

Soybean flowers attract relatively few bees (Blickenstaff and Huggans 1962); however, bees do visit the flowers for both pollen and nectar. The anthers dehisce before the flower opens so that the stigma is in contact with and receptive to the pollen on the anthers. Pollination and fertilization are usually accomplished before the flower opens. Beekeepers in some areas frequently report honey crops from soybeans (Hambleton 1936; Jaycox 1970; Pellett 1947). Beekeepers in other areas report that bees never visit the flowers, or if they do, no honey is obtained.

18.10 POLLINATION REQUIREMENTS

The soybean is regarded as self-fertile and does not rely on insect pollination for its reproductive success (Morse and Cartter 1937; Rubis 1970). Although there is no experimental evidence to support them, some soybean growers in Arkansas have indicated that bees increase production of beans, and they encourage the presence of apiaries near their fields. Tests with plants caged to exclude bees have shown no decrease in production over exposed plots (Woodhouse and Taylor 1913; Milum 1940; Piper and Morse 1910). Culter (1934) achieved approximately 5% cross-pollination in open plots surrounded by 36 in cheesecloth fences and an adjacent apiary. Gordienko (1960) conducted experiments using caged honeybees and two different soybean cultivars. By providing scented syrup to attract bees to the flowers, he obtained hybridization rates of 29% for one cultivar and 44% for the other. Several studies have been carried out to assess cross-pollination levels in various locations, but the specific agents responsible for the pollination were often not identified. This could be attributed to the fact that these tests were conducted by agronomists who did not consider themselves qualified to make such observations (Caviness 1966, 1970; Caviness et al. 1970; Garber and Odland 1926; Woodworth 1922). Weber et al. (1970) stated that for a self-fertilizing species like soybeans, two conditions need to be fulfilled to successfully produce hybrid seeds.

First, there must be heterosis (or an increase in production, or some other beneficial effect, of the cross over the best parent), and there must be an economical, large-scale method of producing the hybrids. Brim and Young (1971) have reported that male sterility, although not cytoplasmic, has been found and that 99% of the seed set on the male sterile plants was due to cross-pollination; however, the total seed set was extremely low.

Veatch (1930) showed that hybrid soybeans out-yielded the average parent variety up to 95.9% and the higher parent variety up to 71.5%. Weiss et al. (1947) showed a range of 14.2 to 71.3% heterosis over the best parent in the greenhouse, and 5.9 to 38.6% increase over the best parent in the field. Weiss (1949) showed that the hybrid exhibited a considerable increase ranging from 19.6 to 117% compared to the superior parent. Wentz and Stewart (1924) reported even greater gains, with improvements ranging from 60 to 397%. In a study by Weber et al. (1970), the average yield increase for all experiments involving hybrids surpassed the best parent by 13.4%. It is worth noting that these yield enhancements for hybrids were observed using a relatively small number of plants that were widely spaced.

Because of the problems in making a sufficient number of hand-pollinations to satisfactorily evaluate F_1s in a normal-type planting, essentially no such studies have been conducted.

The significant increases reported are believed to be partially due to the additional space provided for the plants. According to Bradner (1969), all the necessary breeding components for hybrid soybean seed production have been described, except for the identification of a suitable pollen vector. However, Bradner expressed optimism about the potential role of honeybees in this regard. Piper and Morse (1923) observed that soybean flowers received significant attention from bees. They further noted that three of their varieties, while in full bloom in Jackson, Tennessee, emitted a strong and pleasant fragrance reminiscent of lilacs. Although they did not specifically indicate whether these varieties were more attractive to bees than others, this information suggests that cultivars could be selected based on their aroma or attractiveness. Incorporating such traits into a hybrid seed program may effectively attract pollinating insects.

Because of the potential value of hybrid soybeans, the breeders may monitor for traits that show attractiveness to bees.

18.11 POLLINATORS

Soybeans have received little attention regarding the presence of pollinating insects. In a study by Jaycox (1970), it was reported that there were only seven honeybees per 100 yards of row, but no information was given about the number of honeybee colonies or the acreage of soybeans in the area. Blickenstaff and Huggans (1962) collected 25,346 individual insects on soybeans using a sweep net and found only one honeybee, three leaf-cutting bees, and nine halictids. However, this method is not ideal for accurately determining pollinator populations in dense foliage like soybeans. Previous research by Culter (1934) and Gordienko (1960) demonstrated that honeybee populations can thrive on soybeans. In a study by Beard and Knowles (1971), it was observed that the rate of crossing varied in different locations in California, with Five Points having a 10% crossing rate, 1 to 5 % at Davis, and 0 to 7% at Shafter.

Surprisingly, no research has been conducted on the potential benefits of concentrating gregarious types of wild bees on soybeans, even though these insects might prove to be more efficient pollinators than honeybees. While Woodworth's 1932 work suggested that thrips and honeybees might play a role in pollen transfer, Caviness (1970) revealed that thrips are ineffective as soybean pollinators, whereas honeybees accounted for 7.7% of the crossing observed.

18.12 POLLINATION RECOMMENDATIONS AND PRACTICES

Although there are currently no specific guidelines regarding the utilization of bees in pollinating soybeans, the topic is being examined due to the growing fascination with hybrid soybeans and the potential for incorporating pollinating insects in their production.

18.13 *VICIA* SPECIES

18.13.1 Vetch

18.13.1.1 *Vicia* spp., Family Leguminosae

Inadequate pollination poses a significant limitation to the achievable yields of faba bean (*Vicia faba*) crops. Faba beans are a broad bean originating from the Middle East. They are an important human food crop in China, the Middle East, the Mediterranean, and in African countries. Faba beans have been used extensively in Australia for livestock nutrition, but an emerging export market may, in the future, be more important.

The inflorescence of the vetch plant typically consists of a raceme with a varying number of solitary flowers, ranging in color from light-blue to dark-purple or lavender. Similar to other legume flowers, the reproductive column is enclosed within the keel petal. When the column remains closed, self-pollination is the only possible outcome, provided that the style and pollen are compatible. However, if the column is released or triggered, allowing the stigma to come into contact with pollen from other flowers, cross-pollination can take place. Certain species of vetch can benefit from triggering even in the absence of cross-pollination.

Some species benefit by tripping even if cross-pollination does not occur. Repeated visitation to the flower by pollinating insects also increases productiveness in some species. The flower must be tripped if bees are to collect pollen from it.

Nectar is usually secreted inside the corolla, sometimes on the outside of the base of the corolla. In many species, including at least *V. sativa*, *angustifolia*, and *pannonica*, nectar is also secreted in a nectary located on the leaf stipule (Knuth 1908; Herman 1960). Normally, when a bee visits the vetch flower for pollen or floral nectar, it settles on the wing petals, then forces its proboscis or "tongue" down the corolla. The pressure depresses the keel petal and forces the hairy style and pollen-laden anthers out. The pollen is thus shed onto the stigma, and some of it rubs off onto the bee, which then carries it to other flowers, resulting in cross-pollination. However, Weaver (1956b) stated that bees sometimes collect nectar from the side of the corolla without depressing the keel petal. Common and hairy vetches are an important source of excellent-quality honey, and the production is usually sufficient for the storage of surplus quantities for the beekeeper.

Hairy vetch exhibits inconsistency in both its nectar production and appeal to pollinating insects, as stated by Alex et al. (1950). Common vetch, on the other hand, generates a dense stipular nectar that is more appealing to bees than its floral nectar, as noted by Sculler in 1956. Hungarian vetch is renowned for providing abundant nectar for an extended duration, surpassing other cultivated plants (Schoth 1923).

18.13.2 Pollination Requirements

There is limited knowledge regarding the pollination needs of various species and cultivars of Vicia. McKee and Pieters (1937) reported that vetches are predominantly self-pollinated, with rare instances of cross-pollination. However, it is now evident that certain species benefit significantly from bee pollination, leading to increased seed production. Alex et al. (1950), Bieberdorf (1952, 1954), Coe (1949), Mlyniec (1962), Mlyniec and Wojtowski (1962), Pritsch (1966), von Schelhorn (1942), Thomas (1950), Weaver (1954, 1956a, 1956b, 1957), and Wojtowski and Mlyniec (1964) have demonstrated the significant impact of bee visits on the yield of hairy vetch (*V. villosa*). Svetsugo and Kobayashi (1952) stated that fertilization of *V. villosa* is unachievable without insect intervention. Although it is unclear if some researchers considered the smooth-stemmed *V. dasycarpa* as a subspecies of *V. villosa*, it appears that both species share similar pollination requirements. Von Schelhorn (1946) concluded that Carniolan honeybees were the most effective pollinators of hairy vetch in his area. Todd (1951) listed both hairy and purple vetch among the plants that exhibited increased seed production due to insect pollination. He also noted that the vetch's brush mechanism for tripping required multiple insect visits for thorough pollination.

According to his findings, he classified Hungarian and narrow-leaved vetches as predominantly self-pollinated plants. Scullen (1956) observed that Hungarian vetch and the Willamette strain of common vetch are rarely visited by bees, suggesting that pollination is likely carried out by bumblebees or other native bee species. Scullen did not discuss the significance of insect pollination or whether floral visitation could be increased with a higher population of bees compared to those used by beekeepers for honey production. McKee and Schoth (1925) stated that common vetch is self-pollinated. However, Schoth (USDA 1942) later provided a list of vetch species that benefit from insect pollination, including *V. villosa*, *V. benghalensis*, *V. pannonica*, and *V. sativa*. In the revised edition (USDA 1946), only *V. villosa* and *V. pannonica* were identified as vetches that benefit from bee pollination.

Knuth (1908) stated that *V. sativa* anthers dehisce in the bud so that automatic self-pollination is inevitable. However, he noted that the flowers were visited by pollinating insects. Further, on the same page, he stated that the flower mechanism of *V. angustifolia* agrees with that of *V. sativa*, which would indicate that this species is capable of producing self-pollinated seeds. These flowers were visited by honeybees and other pollinating insects. In summary, therefore, the data indicate that hairy vetch (*V. villosa*) is greatly benefited by insect pollination, and that this probably applies equally to woolly-pod (smooth) vetch (*V. dasycarpa*). Purple, common, Hungarian, and narrow-leaved vetches may be benefited by insect visitation either in increased seed production during the current year, or the ability to produce at the maximum potential in future years, although literature is scant in this area.

The pollination requirement of all the species of vetch grown in the United States is badly in need of clarification. Like many other of our agricultural crops and cultivars, the problem offers an excellent opportunity for plant scientists to conduct experiments likely to yield data of considerable economic importance to our agriculture.

Knuth (1908) reported that *V. sativa* anthers release pollen while still in the bud, making self-pollination inevitable. However, he observed that these flowers were also visited by pollinating insects. Additionally, on the same page, he mentioned that *V. angustifolia* exhibits a flower mechanism similar to *V. sativa*, suggesting its ability to produce self-pollinated seeds. Honeybees and other pollinating insects were observed visiting these flowers. Overall, the data indicate that insect pollination greatly benefits hairy vetch (*V. villosa*) and likely applies to woolly-pod (smooth) vetch (*V. dasycarpa*) as well. Purple, common, Hungarian, and narrow-leaved vetches may also benefit from insect visitation, leading to increased seed production in the current year, or the potential to maximize production in future years, although literature on this topic is limited. The pollination requirements of all vetch species cultivated in the United States require further clarification. Similar to other agricultural crops and cultivars, this issue presents an excellent opportunity for plant scientists to conduct experiments that could yield economically important data for agriculture.

18.14 POLLINATORS

Honeybees are the primary pollinators of hairy and smooth vetch. Von Schelhorn (1946), Weaver (1956a, 1956b, 1957, 1965) and other researchers have provided substantial evidence to support this fact. However, limited information is available regarding the floral visitors of other species.

According to Scullen's (1956) observations, bumblebees actively interacted with hairy vetch and played a role in its pollination. He also suggested that bumblebees or other native bees were likely involved in pollinating Hungarian vetch, while the flowers of the "Willamette" cv. of common vetch were seldom visited by them. In the United States, honeybees are commonly utilized in large-scale vetch seed production as a standard practice.

Alex et al. (1950) observed that fields with no colonies of bees within 2 mi produced 410 lb seed per acre, and fields with 0.1 to 1.0 colonies per acre within a mile produced 713 lb/acre, whereas fields with 1.5 to 3.0 colonies per acre within one-half mile produced 1,277 lb/acre. From these and other observations, they concluded that the bee saturation point had not been determined for hairy vetch.

Al-Ghzawi et al. (2009) found six different species belonging to six genera and four families visiting the flowers of *Vicia villosa* spp. *dasycarpa*. The most frequent visitor species were *A. mellifera* and *Anthophora albigena* of family Apidae. Both made up more than 68.3% of the total number of the collected bees. With the exception of *Polistes dominulus* (Vispidae), which ranked as the third most abundant visitor, the remaining flower visitors were present in very small numbers.

18.15 POLLINATION RECOMMENDATIONS AND PRACTICES

There have been few suggestions regarding the pollination of vetch. In their study, Alex et al. (1950) conservatively concluded that having more than one colony of honeybees per acre increases the production of hairy vetch seeds. Weaver (1954) stated that several colonies per acre were necessary for optimal pollination of hairy vetch, but using more than one colony per acre might not be cost-effective. Later, Weaver (1956a) estimated that a single strong colony could sufficiently pollinate 3–5 ac. The findings indicate that, due to the need for repeated visits to individual flowers and the fact that vetches are not always highly attractive to bees, a relatively high bee population on the crop is desirable. Therefore, the previous recommendation of several colonies per acre appears to be more practical than relying on one colony for several acres. Notably, Drayner's results (1956) were particularly interesting, revealing that self-fertility in vetches was progressively lost through inbreeding but could be restored through hybridization. This underscores the overlooked importance of bee pollination for vetch crops, which are generally considered self-fertile or capable of self-fertilization.

The findings indicated that employing pollinating insects would be financially beneficial and advisable for crops like vetches, which can self-pollinate.

Yield instability is a common problem in domestic *Vicia* species cultivated under semi-arid conditions. As in many other crops, the number of both flowers and ovules that are formed almost always greatly exceeds the resulting number of pods and seeds that are produced. According to McGregor (1976), the reproductive system of *Vicia* plants is self-pollinated, with some extent of cross-pollination. In general, there are low fruit and yield production by *V. villosa* under natural conditions in Jordan. Such patterns of low fruit set within inflorescences could be attributed to insufficient pollen receipt (Lee 1988; Berry and Calvo 1991; Brunet and Charlesworth 1995), to competition among ovaries of the same inflorescence for a limited source of nutrients (Bawa and Webb 1984; Guitián 1994; Medrano et al. 2000), and finally, it may be attributed to intrinsic limitations on the reproductive success of flowers in different positions within the inflorescence (Lee 1988; Berry and Calvo 1991; Diggle 1995).

Variations in fruit set among flowers within inflorescences are influenced not only by pollination success but also by post-pollination processes, such as variation in the quality of pollen arriving to flowers, pollen germination activity, pollen tube growth, and development of fertilized seeds (Lee 1988). Fruit set is also influenced by the time of flowering during the different months of the year. Completion of reproductive events during the warm period of the season may be advantageous to assure the dispersal of pollens from anthers and their arrival to the stigmas to increase seed production. Bees play an important role in crop pollination, and it can be concluded that the role of wild bees and honeybees must be enhanced to ensure more visitation rates to *V. villosa* flowers.

In addition to its significance in fruit or seed production, pollination plays a crucial role in improving both the quantity and quality of yields (McGregor 1976; Free 1993). Certain crops also experience advantages in terms of consistent ripening, resulting in decreased yield losses during cultivation. Cross-pollination has been found to enhance plant vigor as well, as observed in the case of broad beans (*Vicia faba* L.), which rely on flower tripping to produce viable seeds (Stoddard and Bond 1987).

Apart from the importance of pollination in fruit or seed set, the process enhances higher yields of better quality (McGregor 1976; Free 1993). Some crops benefit also in terms of uniform ripening, which reduces yield losses in the field. Plant vigor has also been shown to be enhanced by cross-pollination, for example, in broad beans (*Vicia faba* L.), which requires flower tripping to produce viable seeds (Stoddard and Bond 1987).

The varying pollination requirements of the flowering crops and their varieties make research needs more complex. For example, self-pollination is spontaneous in a flower. The flower structures are arranged such that it is easy for the pollen to touch stigma without the need of an external agent. Physiologically, the male and female parts also mature at the same time. Although self-pollinated flowers may not require an external agent, sometimes the presence of agents such as bees enhances self-pollination or enables cross-pollination that may be vital to the offspring performance. As an example, broad bean (*Vicia faba*) seedlings resulting from bee-pollinated flowers do not require bee visits to set good seeds, but those from unvisited (self-pollinated) flowers do not set seed (Stoddard and Bond 1987). Hybridization (cross-pollination) in such cases reduces inbreeding depression (Dryner 1956, 59; Bond and Fyfe 1962). Other than self-pollination, which requires bees to enhance yield, there are other forms that require bees entirely for pollen transfer. These are geitonogamy and cross-pollination. Geitonogamy is the transfer of pollen from anthers to stigma of different flowers, but in the same plant. It is also referred to as neighbor pollination. In "true" cross-pollination, pollen source and pollen receiver flowers occur in different plants but of the same species.

18.16 PODS AND SEED SET

Al-Ghzawi et al. (2009) studied the role of bee visitors on seed set and production of *Vicia villosa* spp. *dasycarpa* grown under semi-arid conditions of Irbid, Jordan. They found that the percentage of pod set of the uncovered plants averaged 14% out of the total florets on the plants, which was significantly higher than the covered plants (2%) (Table 18.3).

These results indicated that the percentage of flower abscission was high and averaged more than 86%. Plant covering significantly reduced seed yield by reducing seed and pod number per plant and

TABLE 18.3
Number of Florets, Pods, and Seeds per Inflorescence, Seed per Pod, and Percentage of Formed Pods and Seeds for *Vicia villosa* spp. *dasycarpa* Grown under Uncovered and Covered Conditions during Flowering Stage of Development

	No./Inflorescence				Percentage of Total Florets	
Treatments	**Florets**	**Pods**	**Seeds**	**Seed/Pod**	**Pods**	**Seeds**
Uncovered	18a	2.4a	8.8a	3.8a	14.1a	51.3a
Covered	17a	0.3b	1.1b	0.3b	1.7b	6.4b

Source: a,b: $P < 0.05$.

TABLE 18.4
Seed Yield and Number, Pod Number, Seed Number per Pod, and Seed Weight of *Vicia villosa* spp. *dasycarpa* Plants Grown under Uncovered and Covered Conditions during Flowering Stage of Development

	Seed Yield	Seed No.	Pod No.	Seed No.	Seed Weight
Treatments	(g.plant-1)	(no. plant-1)		(no. pod-1)	(mg.seed-1)
Uncovered	18.6a	528a	176a	2.9a	43a
Covered	0.23b	5b	4b	1.4b	35a

Note: a,b: P<0.05.

seed number per pod but had no effect on individual seed weight (Table 18.4). In conclusion, exclusion of bees from visiting during flowering of *V. villosa* spp. *dasycarpa* resulted in reduced seed set, seed yield, and various components of yield.

It is clearly evident that plant covering significantly reduced seed yield by reducing all yield components, except the individual seed weight. The covered plants had significantly lower seed and pod numbers per plant and seed number per pod. Individual seed weight was not significantly different between the covered and uncovered plants, which might be due to filling the remaining seeds on plants to large size after the covering treatment was released. These results indicated that *Vicia villosa* spp. *dasycarpa* is very sensitive to covering at this critical stage of development, which negatively affected seed yield and yield components due to lack of pollinator access to plants.

REFERENCES

Afun, J. V. K., Jackai, L. E. N., and Hodgson, C. J. 1991. Calendar and monitored insecticide application for the control of cowpea pests. *Crop Protection* 10(5), 363–370.

Aizen, M. A., Garibaldi, L. A., Cunningham, S. A., and Klein, A. M. 2008. Long-term global trends in crop yield and production reveal no current pollination shortage but increasing pollinator dependency. *Curr. Biol.* 18, 1572–1575.

Alex, A. H., Thomas, F. L., and Warne, B. 1950. Importance of bees in vetch seed production. *Tex. Agr. Expt. Sta. Prog. Rpt.* 1306, 3 pp.

Al-Ghzawi, A., Samarah, N., Zaitoun, S., and Alqudah, A. 2009. Impact of bee pollinators on seed set and yield of *Vicia villosa* spp. *dasycarpa* (*Leguminosae*) grown under semiarid conditions. *Ital. J. Anim. Sci.* 8, 65–74.

Allard, R. W. 1954. Natural hybridization in Lima beans in California. *Amer. Soc. Hort. Sci. Proc.* 64, 410–416.

Allen, H. and Scriven, W. A. 1957. Notes on the pollination of field beans both in the open and in cages. *Brit. Bee Jour.* 85, 288–290.

Amos, J. M. 1943. A measurement of the value of bees in the pollination of Lima beans. *Amer. Bee Jour.* 83, 240–241.

Anderson, E. J. 1959. Two part pollination program carried out on university farms. *Penn. Beekeeper* 34(3), 4–8.

Bawa, K. S. and Webb, C. J. 1984. Flower, fruit and seed abortion in tropical forest trees: Implications for the evolution of paternal and maternal reproductive patterns. *American Journal of Botany* 71(5), 736–751.

Backhaus, W. 1993. Colour vision and colour choice behaviour of the honey bee. *Apidologie* 24, 309–331.

Barrons, K. C. 1939. Natural crossing in beans at different degrees of isolation. *Am. Soc. Hort. Sci. Proc.* 36, 637–640.

Basualdo, M., Bedascarrasbure, E., and DeJong, D. 2000. Africanized honey bees (Hymenoptera: Apidae) have a greater fidelity to sunflowers than European bees. *Journal of Economic Entomology* 93, 304–307.

Beard, B. H. and Knowles, P.F. 1971. Frequency of cross pollination of soybeans after seed irradiation. *Crop Sci.* 11, 681–684.

Berry, P. E. and Calvo, R. N. 1991. Pollinator limitation and position dependent fruit set in the high Andean orchid *Myrosmodes cochleare* (*Orchidaceae*). *Plant Systematics and Evolution* 174, 93–101.

Bieberdorf, G. A. 1954. Honey bees as vetch pollinating agents. Okla. Agr. Expt. Sta., Mimeo. Cir. M-250, 3 pp.

Bieberdorf, G. A. 1952. Honey bees and vetch pollination. *Okla. Acad. Sci. Proc.* 33, 36–37.

Blackwall, F. L. C. 1964. Runner beans: Adequate pollination is a must for early yields. *Grower* 61, 1272–1273.

Blickenstaff, C. C. and Huggans, L. J. 1962. Soybean insects and related arthropods in Missouri. *Mo. Agr. Expt. Sta. Bul.* 803, 51 pp.

Bliss, F. A. 1980. Common bean. In: Fehr, W. R. and Hadley, H. H. (eds.) *Hybridization of Crop Plants*. American Society of Agronomy & Crop Science Society of America, Madison, WI, pp. 273–284.

Bond, D. A. and Fyfe, J. L. 1962. Breeding field beans. *Plant Breed. Inst. (England) report*. 1960–1961, pp. 4–26.

Bond, D. A. and Hawkins, R. P. 1967. Behavior of bees visiting male-sterile field beans (Vicia FABA). *Jour. Agr. Sci. [England]* 68, 243–247.

Boswell, V. R. 1961. Flowering habits and production of seeds. *U.S. Dept. Agr. Yearbook*, 57–64.

Bradner, N. R. 1969. Hybrid soybeans: Fiction or fact? *Soybean Digest* 29(7), 16–17. 338.

Bramel, J., Kiran, S., Reddy, J., Ford-Lloyd, B., and Chancra, S. 2004. Degree and distribution of pigeon pea landrace morphological diversity in traditional cropping systems in Andhra Pradesh. In: Bramel, P. J. (ed.) *Assessing the Risk of Losses of Biodiversity in Traditional Cropping Systems: A Case Study of Pigeon Pea in Andhra Pradesh*. International Crop Research Institute for the semi-arid Tropics. Pantacheru Andhra Pradesh, India, pp. 1–45.

Brim, C. A. and Young, M. F. 1971. Inheritance of male-sterile character in soybeans. *Crop Sci.* 11, 564–566.

Brunet, J. and Charlesworth, D. 1995. Floral Sex Allocation in Sequentially Blooming Plants. *Evolution* 49, 70–79, https://doi.org/10.1111/j.1558-5646.1995.tb05959.x

Caviness, C. E. 1966. Estimates of natural cross-pollination in Jackson soybeans in Arkansas. *Crop Sci.* 6(2), 211.

Caviness, C. E. 1970. Cross-pollination in the soybean. In The Indispensable Pollinators, Ark. *Agr. Ext. Serv. Misc. Pub.* 127, 33–36.

Caviness, C. E., Walters, H. J., and Johnson, D. L. 1970. A partially male sterile strain of soybean. *Crop Sci.* 10(1), 107–108.

Coe, L. E. 1949. Pollination of hairy vetch in Oklahoma: The nearer the bees the better the harvest. *Amer. Bee Jour.* 89, 558–559.

Culter, G. H. 1934. A simple method for making soybean hybrids. *Amer. Soc. Agron. Jour.* 26, 252–254.

Darwin, C. 1889. *The Effects of Cross and Self Fertilization in the Vegetable Kingdom*. Appleton, New York.

Darwin, C. R. 1876. *The Effects of Cross and Self Fertilisation in the Vegetable Kingdom*. John Murray, London.

Darwin, F. 1858. On the agency of bees in the fertilization of papilionaceous flowers and on the crossing of kidney beans. *Gard. Chron.* 46, 828–829.

Darwin, F. 1874. Bees visiting flowers. *Nature* 9, 189–190.

David, B. V. 2001. *Elements of Economic Entomology*, Revised and Enlarged edn. Popular Book Depot, Chennai, India. 590 p.

Debouck, D. 1991. Systematics and morphology. In: Van Schoonhoven, A. and Voyset, O. (eds.) *Common Beans: Research for Crop Improvement*. Cali, Colombie, pp. 55–118.

Delaplane, K. S., Mayer, D. F. 2000. *Crop Pollination by Bees*. CABI Publishing, Wallingford, UK and New York.

Diggle, P. K. 1995. Architectural effects and the interpretation of patterns of fruit and seed development. *Annual Review of Ecology and Systematics* 26(1), 531–552.

Diwan, V. V. and Salvi, S. R. 1965. Some interesting behavioural features of *Apis dorsata* Fab. *Indian Bee J.* 27(1), 52.

Drayner, J. M. 1956. Regulation of outbreeding in field beans. *Nature* 177, 489–490.

Dreyer, H., Baumgärtner, J., and Tamò, M. 1994. Seed-damaging field pests of cowpea (*Vigna unguiculata* L. Walp.) in Benin: Occurrence and pest status. *International Journal of Pest Management* 40(3), 252–260, DOI: 10.1080/09670879409371893

Dryner, J. M. 1956. Regulation of out breeding in field beans (*Vicia faba*). *Nature* 177(4506), 489–490.

Emerson, R. A. 1904. Heredity in bean hybrids. (*Phaseolus vulgaris*). *17th Ann. Rep. Nebr. Agric. Exp. Stat.* 34–61.

Enfawal, M. A., Bishr, M. A., and Hassoub, E. K. 1976. Natural cross pollination in Egyptian Cotton (*Gossypium barbadense* L.). *Journal of Agricultural Sciences* 86, 205–209.

Fameni, S. T., Tchuenguem Fohouo, F. N., and Brueckner, D. 2012. Pollination efficiency of *Apis mellifera adansonii* (Hymenoptera: Apidae) on *Callistemon rigidus* (Myrtaceae) flowers at Dang (Ngaoundéré, Cameroon). *International Journal of Tropical Insect Science* 32(1), 2–11.

Fluri, P. and Frick, R. 2005. Apiculture in Sweden: State and perspectives. *Rev. Suisse. Agric.* 37(2), 81–86.

Free, J. B. 1959. The effect of moving colonies of honeybees to new sites on their subsequent foraging behavior. *Jour. Agr. Sci.* [England] 53, 1–9.

Free, J. B. 1962. The behavior of honeybees visiting fiebd beans (Vicia faba). *Jour. Anim. Ecol.* 31, 497–502.

Free, J. B. 1966. The Pollination of the Beans *Phaseolus Multiflorus* and *Phaseolus Vulgaris* by Honeybees. *Journal of Apicultural Research* 5(2), 87–91.

Free, J. B. 1970. *Insect Pollination of Crops*. Academic Press, London and New York, p. 544.

Free, J. B. 1993. *Insect Pollination of Crops*, 2nd edn. Academic Press, London.

Free, J. B. and Racey, P. A. 1968. The pollination of runner beans (phaseolus multiflorus) in a glasshouse. *Jour. Apic. Res.* 7, 67–69.

Garber, R. J. and Odland, T. E. 1926. Natural crossing in soybeans. *Am. Soc. Agron. J.* 18, 967–970.

Gordienko, V. 1960. Sexual hybrids of soya beans obtained by directed bee pollination. In Mel'nichenko, A. N., [Pollination of Agricultural Plants by Bees.] Izd-vo Minist. Sel'sko Khoz. SSSR. [In Russian.], Pp. 400–407.

Graham, E. H. 1941. *Legumes for Erosion Control and Wild Life*. U.S. Department of Agriculture Miscellaneous Publication No. 412, Ohio State University, p. 153.

Graham, P. and Ranalli, P. 1997. Common *Bean* (Phaseolus vulgaris L.). *Field Crops Research* 53, 131–146.

Guitián, J. 1994. J Selective fruit abortion in Prunus mahaleb (Rosaceae). *American Journal of Botany* 81, 1555–1558.

Hambleton, J. I. 1936. Soybean for pollen and nectar. *Gleanings Bee Cult.* 64, 431.

Hawthorn, L. R. and Pollard, L. H. 1954. *Vegetable and Flower Seed Production*. McGraw-Hill, New York, p. 626.

Heiser, C. B. 1990. *Seed to Civilization: The Story of Food*. Harvard University Press, Cambridge. MA, 228 pp.

Henslow, G. 1878. The fertilization of the scarlet runner bean by bumble-bees. *Gard. Chron.* n.s., 10, 561.

Herman, F. J. 1960. Vetches in the united statesñnative, naturalized and cultivated. *U.S. Dept. Agr., Agr. Handb.* 168, 84 pp.

Jany, E. 1950. [The piercing of scarlet runner flowers by bumble bees.] *Ztschr. f. Angew. Ent.* 32(2), 172–183. [In German.] AA-45/54.

Jones, H. A. and Rosa, J. T. 1928. *Truck Crop Plants*. McGraw-H Book Co., New York.

Ibarra-Pérez, F. J., Barnhart, D., Ehdaie, B., Knio, K. M., and Waines, J. G. 1999. Effects of insect tripping on seed yield of common bean. *Crop Sci.* 39, 425–433.

Ibarra-Perez, F. J., Ehdaie, B. and Waines, J. G. 1997. Estimation of outcrossing rate in common bean. *Crop Sci.* 37, 60–65.

Jaycox, E. R. 1970. Ecological relationshtps between honey bees and soybeans. *Amer. Bee Jour.* 110(8), 306–307, (9), 343–345, (10), 383–385.

Julianna, K. T. and Rufus, I. 2010. Weather during bloom affects pollination and yield of high bush blueberry. *Journal of Economic Entomology* 103, 557–562.

Kasina, J. M., Hagen, M., Kraemer, M., Nderitu, J., Martius, C., and Wittmann, D. 2009. Bee pollination enhances crop yield and fruit quality in kakamega, western Kenya. *East African Agricultural and Forestry Journal* 75(1), 1–11.

Kasina, M., Kraemer, M., Martiusand, C., and Wittmann, D. 2009c. Diversity and activity density of bees visiting crop flowers in Kakamega, western Kenya. *Journal of Apicultural Research* 48, 134–139.

Kasper, M. L., Reeson, A. F., Mackay, D. A., and Austin, A. D. 2008. Environmental factors influencing daily foraging activity of *Vespula germanica* (Hymenoptera, Vespidae) in Mediterranean Australia. *Insectes Sociaux* 55, 288–296.

Keller, L. F. and Waller, D. M. 2002. Inbreeding effects in wild populations. *Trends Ecol. Evol.* 17, 230–241.

Kingha, B. M. T., Tchuenguem Fohouo, F. N., Ngakou, A., and Brückne, D. 2012. Foraging and pollination activities of Xylocopa olivacea (Hymenoptera, Apidae) on Phaseolus vulgaris (Fabaceae) flowers at Dang (Ngaoundere-Cameroon). *Journal of Agricultural Extension and Rural Development* 4(6), 330–339.

Klein, A.-M., Vaissière, B. E., Cane, J. H., Steffan-Dewenter, I., Cunningham, S. A., Kremen, C., and Tscharntke, T. 2007. Importance of pollinators in changing landscapes for world crops. *Proceedings of the Royal Society B: Biological Sciences* 274, 303–313.

Knuth, P. 1908. *Achillea millefolium* L. In: Knuth, P. (ed.) *Handbook of Flower Pollination*. Oxford Clarendon Press, London, UK, pp. 611–616.

Kristofferson, K. B. 1921. Spontaneous crossing in the garden bean. *Phaseolus vulgaris Hereditas* 2, 395–400.

Lee, T. D. 1988. Patterns of fruit and seed production. In: Lovett Doust, L. and Lovett Doust, L. (eds.) *Plant Reproductive Ecology*. Oxford University Press, New York, pp. 179–202.

Louveaux, J. 1984. L'abeille domestique dans ses relations avec les plantes cultivées. In: Pesson, P. and Louveaux, J. (eds.) *"Pollinisation et productions végétales"*. INRA, Paris, pp. 527–555.

Mackie, W. and Smith, F. L. 1935. Evidence of field hybridization in beans. *J. Am. Soc. Agron.* 27, 903–909.

McGregor, S. E. 1976. Insect pollination of cultivated crop plants. *Agricultural Research Service. United States Department of Agriculture, Agric. Handb, Washington, USA*. 496, 411.

Mckee, R. and Pieters, A. J. 1937. Miscellaneous forage and cover crop legumes. *U.S. Dept. Agr. Yearbook*, 999–1031.

Mckee, R. and Schoth, H. A. 1925. Common vetch and its varieties. *U.S. Dept. Agr. Bul.* 1289, 20 pp.

Medrano, M., Guitian, P., and Guitián, J. 2000. Patterns of fruit and seed set within inflorescences of Pancratium maritimum (Amaryllidaceae): Nonuniform pollination, resource limitation, or architectural effects? *American Journal of Botany* 87, 493–501.

Mensah, B. A. and Kudom, A. A. 2011. Foraging dynamics and pollination efficiency of Apis mellifera and Xylocopa olivacea on Luffa aegyptiaca Mill (Cucurbitaceae) in Southern Ghana. *Journal of Pollination Ecology* 4, 34–38.

Milum, V. G. 1940. Bees and soybeans. *Amer. Bee Jour.* 80, 22.

Mlyniec, W. 1962. The mechanism of pollination and generative reproduction in *Vicia villosa* Roth. Genet. Polon. 3: 285 [In Polish.] *Abstract in Plant Breeding Abs.* 34, 669, 76, 1964.

Mlyniec, W. and Wojtowski, F. 1962. Use of bumble bees for research on flowering biology of hairy vetch. *Ekol. Pol. Ser. B* 8(1), 59–65. [In Polish, English summary.] AA-38/67.

Moffett, J. O., Stith, L. S., Burkhart, C. C., and Shipman, C. W. 1975. Influence of cotton genotypes on floral visits of honey bees. *Crop Science* 15, 782–784.

Mommers, J. 1971. [The pollination of beans phaseolus under glass.] *Maandschr. V. De bijent.* 73(5), 79–80. [in Dutch English summary.]

Morandin, L. A. and Winston, M. L. 2006. *Pollinators* provide economic incentive to preserve natural land in agroecosystems. *Agriculture Ecosystems and Environment* 116, 289–292.

Morse, W. J., and Cartter, J. L. 1937. Improvement in soybeans. *U.S. Dept. Agr. Yearbook*, 1154–1189.

Pellett, F. C. 1947. *American Honey Plants*, 4th edn. Illus, New York, p. 467. (30).

Piper, C. V. and Morse, W. J. 1910. The soybean: History, varieties and field studies. *U.S. Dept. Agr. Burl Plant Indus. Bul.* 197, 84.

Piper, C. V. and Morse, W. J. 1923. *The Soybean*. McGraw-Hill Book Co., Inc., New York, p. 329.

Palmer, R. 1967. Bees and bean pollination. *S. Afr. Bee J.* 39(4), 12–13.

Pando, J. B., Tchuenguem Fohouo, F.-N., and Tamesse, J. L. 2011a. Pollination and yield responses of pigeon pea (Cajanus cajan L. Millsp.) to the foraging activity of Chalicodoma cincta cincta (Hymenoptera: Megachilidae) in Yaounde´ (Cameroon). *Journal of Animal and Plant Science* 11, 1346–1357.

Pando, J. B., Tchuenguem Fohouo, F.-N., and Tamesse, J. L. 2011b. Foraging and pollination behaviour of Xylocopa calens Lepeletier (Hymenoptera: Apidae) on Phaseolus coccineus L. (Fabaceae) flowers at Yaounde´ (Cameroon). *Entomological Research* 41, 185–193.

Pando, J. B., Tchuenguem Fohouo, F. N., and Tamesse, J. L. 2013. Activité de butinage et de pollinisation de *Xylocopa olivacea* Fabricius 1787 (Hymenoptera: Apidae) sur les fleurs de *Vigna unguiculata* (L.) Walp. 1843 (Fabaceae) à Yaoundé-Cameroun. Faunistic entomology 66, 47–59.

Pritsch, G. 1966. Investigations on the significance of honey bees (*Apis mellifera* L.) in assuring seed yield of winter vetches (*Vicia villosa* Roth).] *Arch. f. Geflugelz. u. Kleintierk.* 15(3), 205–211. [In German, English summary.

Rheenen, H. A. Van. 1964. Preliminary study of natural cross-fertilization in mung bean, (Phaseolus aureus ROXB.). *Netherlands Jour. Agr. Sci.* 12(4), 260–262.

Richards, A. J. 2001. Does low biodiversity resulting from modern agricultural practice affect crop pollination and yield? *Ann Bot (Lond)* 88, 165–172.

Rowlands, D. G. 1960. Fertility studies in the field bean (Vicia faba L.). 1. Cross- and self-fertility. *Heredity* 15(2/3), 161–173.

Roubik, D. W. 2000. Pollination system stability in tropical America. *Conservation Biology* 14, 1235–1236.

Rubis, D. D. 1970. Breeding insect pollinated crops. In The lndispensable Pollinators. *Ark. Agr. Ext. Serv. Misc. Pub.* 127, 19–24.

Rutger, J. N. and Beckham, L. S. 1970. Natural hybridization of Phaseolus vulgaris X P. coccineus *L. Amer. Soc. Hort. Sci. Proc.* 95, 659–661.

Sabbahi, R., De OLiveira, D. and Marceau, J. 2005. Influence of honey bee (Hymenoptera: Apidae) density on the production of canola (Crucifera: Brassicaceae). *Journal of Economic Entomology* 2, 367–372.

Schelhorn, M. Von. 1942. [Biological studies on the blossoms of hairy vetch.] *Pflanzenbau* 18(10), 311–320. [In German.]

Schelhorn, M. Von. 1946. [Biology of flowering and seed setting in *Vicia villosa*.] *Zuchter* 17/19, 22–24. [In German.]

Schoth, H. A. 1923. Hungarian vetch in Oregon. *Oreg. Agr. Expt. Sta. Cir.* 46, 4 pp.

Scullen, H. A. 1956. Bees for legume seed production. Circular of information. Agricultural Experiment State Oregon State College Corvallis, p. 15.

Shaik, P. H. Van and Probst, A. H. 1958. Effects of some environmental factors on flower production and reproductive efficiency in soybeans. *Agron. Jour.* 50, 192–197.

Stoddard, F. L. and Bond, D. A. 1987. The pollination requirements of the faba bean. *Bee World* 68, 144–152.

Svetsugo, I. and Kobayashi, S. 1952. [Studies on *Vicia villosa* roth, with special reference to the fertilization and seed production.] *Crop Sci. Soc. Proc.* 21(1/2), 150–151. [In Japanese, English summary.]

Taylor, W. H. 1919. Bees and flower fertilization. *New Zeal. Jour. Agr.* 18, 203.

Tchuenguem, F. F. N., Djonwangwé, D. and Messi, J. and Brückner, D. 2009a. Foraging and pollination activities of *Apis mellifera adansonii* (Hymenoptera: Apidae) at Ngaoundéré (Cameroon). *Cameroon J. Exp. Biol.* 5(1), 1–9.

Tchuenguem Fohouo, F. N. 2005. Activité de butinage et de pollinisation d'*Apis mellifera adansonii* Latreille (Hymenoptera: Apidae, Apinae) sur les fleurs de trois plantes à Ngaoundéré (Cameroun): *Callistemon rigidus* (Myrtaceae), *Sygygium guineense* var. macrocarpum (Myrtaceae) et *Voacanga africana* (Apocynaceae). Thèse du Doctorat d'Etat, Université de Yaoundé I, Yaoundé, Cameroun. 103 p.

Tedoradze, S. G. 1959. [The role of bees in the shape-forming process of [scarlet runner] beans under Georgian conditions.] *Pchelovodstvo* 36(9), 40–42. [In Russian.]

Thomas, F. L. 1950. Vetch seed growers ask for honey bees. *Amer. Bee Jour.* 90, 315.

Todd, F. E. 1951. The community approach to pollination problems. *In Iowa State Apiarist Rpt.*, 1950, 104–108.

Ulrichs, C. and Mewis, I. 2004. Evaluation of the efficacy of Trichogramma evanescens Westwood (Hym., Trichogrammatidae) inundative releases for the control of Maruca vitrata F. (Lep., Pyralidae). *Journal of Applied Entomology* 128, 426–431. http://dx.doi.org/10.1111/j.1439-0418.2004.00867.x

United States Department Of Agriculture. 1942. The dependence of agriculture on the beekeeping industry. A review. *U.S. Dept. Agr. Burl Ent. and Plant Quar.* E-584, 39 pp.

United States Department Of Agriculture. 1946. The dependence of agriculture on the beekeeping industry. A review. *U.S. Dept. Agr. Burl Ent. and Plant Quar.* E-584, revised, 39 pp.

Vanderborght, I. and Rasmont, P. 1987. Utilization of the domestic bees (Apis mellifera L.) and the Xylocope (Xylocopa bariwal MAIDL) (Hymenoptera, Apidae) for pollination in cage of Phaseolus coccineus L. *Bull. Rech. Agron. Gemb.* 22(4), 279–284.

Vansell, G. H. and Reinhardt, J. F. 1948. Do honey bees help pollinate baby lima beans? *Gleanings Bee Cult.* 76, 678–679.

Veatch, C. 1930. Vigor in soybeans as affected by hybridity. *Jour. Amer. Soc. Agron.* 22, 289–310.

Wafa, A. K. and Ibrahim, S. H. 1960. The effect of the honeybee as a pollinating agent on the yield of broad bean. *Bul. Fac. Agr. Cairo Univ.* 205, 36 pp.

Watts, F. H. and Marshall, P. R. 1961. Pollination of field beans. yield response due to bees. *Rpt. Field Expts. Observ. Studies Hort., Natl. Agr. Advisory Serv., E. Midlands Reg.* [England], 76–79.

Weinstein, A. I. 1926. Cytological studies on phaseolus vulgaris. *Amer. Jour. Bot.* 13, 248–263.

Wells, W. C., Isom, W. H., and Waines, J. G. 1988. Outcrossing rates of six common bean lines. *Crop Sci* 28, 177–178.

Wentz, J. B. and Stewart, R. T. 1924. Hybrid vigor in soybeans. *Jour. Amer. Soc. Agron.* 16, 534–540.

Westerkamp, C. and Gottsberger, G. 2000. Diversity pays in crop pollination. *Crop Science* 40(5), 1209–1222.

Westerkamp, C. and Weber, A. 1999. Keel flowers of the Polygalaceae and Fabaceae: A functional comparison. *Botanical Journal of the Linnean Society* 129, 207–221.

White, J. W. and Izquierdo, J. 1991. Physiology of yield potential and stress tolerance. In: van Schoonhoven, A. and Voysest, O. (eds.) *Common Beans*. Research for Cop Improvement, CAB Int CIAT, Colombia, pp. 287–382.

Weaver, N. 1954. Pollination of hairy vetch by honeybees. *Tex. Agr. Expt. Sta. Prog. Rpt.* 1649, 5 pp.

Weaver, N. 1956a. The pollination of hairy vetch by honey bees. *Jour. Econ. Ent.* 49, 666–671.

Weaver, N. 1956b. The foraging behavior of honeybees on hairy vetch. i. foraging methods and learning to forage. *Insectes Soc.* 3, 538–549.

Weaver, N. 1957. The foraging behavior of honeybees on hairy vetch. ii. the foraging area and foraging speed. *Insectes Soc.* 4, 43–57.

Weaver, N. 1965. The foraging behavior of honeybees on hairy vetch. iii. differences in the vetch. *Insectes Soc.* 12, 321–325.

Weber, C. R., Emptg, L. T., and Thorne, J. C. 1970. Heterotic performance and combining ability of two-way f_1 soybean hybrids. *Crop Sci.* 10, 159–160.

Webster, B. D., Ross, R. M., and Evans, T. 1982. Nectar and nectary of *Phaseolus vulgaris*. L. *J. Am. Soc. Hort. Sci.*, 107, 497–503.

Weiss, M. G. 1949. Soybeans. *Adv. in Agron.* 1, 77–157.

Weiss, M. G., Weber, C. R., and Kalton, R. R. 1947. Early generation testing in soybeans. *Jour. Amer. Soc. Agron.* 39, 791–811.

Wester, R. E. and Jorgensen, H. 1950. Emasculation unnecessary in hybridizing Lima beans. *Amer. Soc. Hort. Sci. Proc.* 55, 384–390.

Wojtowski, F. and Mlyniec, W. 1964. [Behaviour of bees during pollination of hairy vetch.] *Roczn. Wyz. Szk. Roln. Poznan.* 19, 197–207. [In Polish, English and Russian summaries.] AA-245167.

Woodhouse, E. J. and Taylor, C. S. 1913. The varieties of soybeans found in bengal, bikar and orissa and their commercial possibilities. *India Dept. Agr. Mem. Bot. Ser.* 5, 103–175.

Woodworth, C. M. 1922. The extent of natural cross-pollination in soybeans. *Jour. Amer. Soc. Agron.* 14, 278–283.

19 Cowpea (*Vigna unguiculata* (L.) Walp.)

19.1 INTRODUCTION

Cowpea (*Vigna unguiculata* (L.) Walp.) is grown in tropical Africa, Asia, and North and South America mostly as a grain, but also as a vegetable and fodder crop. It is favored because of its wide adaptation and tolerance to several stresses. It is an important food source and is estimated to be the major protein source for more than 200 million people in sub-Saharan Africa and is in the top ten fresh vegetables in China. Cowpea (V*igna unguiculata* L. Walp) belongs to the family Fabaceae. It is a grain legume grown in savanna regions of the tropics and subtropics with some 150–190 species. The cultivated cowpea is based on five so-called groups (Cultigroups), namely, *unguiculata*, *sesquipedalis*, *textiles*, *melanophtalmus*, and *biflora* (Westphall 1974; Maréchal et al. 1978; Ng and Marechal 1985). Its value lies in its high protein content (23–29%), with potentials for perhaps 35%, and its ability to fix atmospheric nitrogen, which allows it to grow on and improve poor soil (Steele 1972). Cowpea is called as vegetable meat due to high amount of protein in grain, with better biological value on dry weight basis. On dry weight basis, cowpea grain contains 23.4% protein, 1.8% fat, and 60.3% carbohydrates, and it is a rich source of calcium and iron (Gupta 1988). Apart from this, cowpea forms excellent forage, and it gives a heavy vegetative growth and covers the ground so well that it checks the soil erosion. As a leguminous crop, it fixes about 70–240 kg per hectare of nitrogen per year.

At present, this species is the second most important pulse crop (Maréchal et al. 1978), and in Nigeria, it is generally cultivated for its seed (shelled green or dried), pods, and leaves, which are consumed in fresh form as green vegetables, while snacks and main meal dishes are prepared from the dried grain. All the plant parts used for food are nutritious, making it extremely valuable where many people cannot afford protein foods such as meat and fish. The seed is valued as a nutritional supplement to cereals and an extender of animal proteins (Kay 1979; Coetzee 1995). It is very palatable, highly nutritious, and relatively free of metabolites or other toxins (Kay 1979). The rest of the cowpea plant, after pods are harvested, is also used as a nutritious livestock fodder.

Cowpea is mainly grown in tropical and sub-tropical regions in the world for vegetable and seed purposes and, to a lesser extent, as a fodder crop. It is a most versatile pulse crop because of its smothering nature, drought-tolerant characters, soil-restoring properties, and multi-purpose uses. As a pulse crop, cowpea fits well into most of the cropping systems.

In Africa, cowpea can be cultivated up to 1,800 m altitude but is mainly grown in the lowlands. The center of maximum diversity of cultivated cowpeas and land races is found in West Africa in a region comprising the Sudan savannah zone of Nigeria (at 5 million hectares, Nigeria has the largest area of cowpea cultivation), central Burkina Faso, Ghana, Togo, northern Benin, and the northwestern part of Cameroon (Padulosi and Ng 1997). Substantial cowpea cultivation also occurs in the semi-arid Sahelian zone, which is a transition zone between the Sahara desert in the north and the Sudan savannah zone in the south. The Sahel encompasses northern and central Senegal and southern Mauritania in the west to central Sudan in the east, passing through central Mali, northern Burkina Faso, southern Niger (at 4 million hectares, Niger has the second-largest area of cowpea cultivation), and central Chad. Significant cowpea production also occurs in the Northern Guinea savannah zone and the Forest and Southern Guinea savannah zones of West Africa, the United Republic of Tanzania and Uganda, and some cowpeas are cultivated in central, southern, and northeastern Africa.

DOI: 10.1201/9781032656724-21

The wild relatives of cowpea are widely distributed across sub-Saharan Africa. They occupy a range of habitats to an elevation of 2,600 m. *Vigna monantha* has been found in Somalia in the coastal plain from Hobyo to Bender Bayla.

In Asia, cowpea ("asparagus bean") ranks as one of the top ten fresh vegetables. It is cultivated across a broad geographic range, except for some permanently cold regions. According to the FAO statistics, Myanmar is the main cowpea producer in Asia (FAOSTAT 2015). The People's Republic of China, Korea, Japan, Thailand, and India are among the major asparagus bean–producing countries. The estimated annual cultivation area in Asia in total is 1 million hectares, the People's Republic of China alone making up roughly one-fifth of the world's fresh pods production with over 1.5 million tons (equivalent to an additional 0.2 MMT of dry matter). Compared with the African cowpea, "asparagus bean" is more adapted to cool climates and is less tolerant to very high temperatures.

Cowpea is a legume species usually considered as resistant to droughts. Droughts often occur in the Sahelian zone and Sudan savannah zones (Dancette and Hall 1979). Cowpea has a greater ability to withstand these droughts and to produce significant grain than any other crop grown, including the drought-resistant crops pearl millet, sorghum, and peanut. In addition, cowpea hay is an important source of forage for livestock, which plays a particularly critical role in feeding animals during the dry season in many parts of West Africa (Singh and Tarawali 1997; Tarawali et al. 2002).

In Asia and Brazil, both sole-cropping and intercropping are practiced (Pandey and Ngarm 1985; Watt et al. 1985), while in the United States, generally only sole crops are grown. In India and Brazil, some intercropping of cowpea is still practiced, but the majority of the crop is produced under sole-cropping with inputs. Cowpea production in the United States is entirely mechanized with machinery and agronomic practices adapted from other crops, such as common beans or soybeans. Large growers in Brazil have adopted similar modern farming practices to produce high yields (Freire Filho et al. 2011).

In the People's Republic of China, "asparagus bean," as a vegetable, is usually intercropped with common bean or cucumber. Smallholder farming and hand harvest of the immature fresh pods of asparagus bean still remain the dominant production system in China, as pod quality/appearance, rather than yield, is usually more important.

Cowpea is one of the most important food legume crops in the semi-arid tropics, covering Asia, Africa, Southern Europe, and Central America, and it is native of Central Africa (Singh 2003). It is a warm-weather crop well adapted to many areas of the humid tropics and subtropical climate with a drought-tolerant nature. But it is intolerant to frost and waterlogged condition (Singh 2003). It fixes 30–60 kg N per hectare (Daniel and Patrick 2014). In addition, it is a compatible intercrop in maize, millets, papaya, banana, and sorghum (IITA 2003).

Cowpea plays an important role in nutritional aspects also; an edible green pod of 100 g contains 84.6 g moisture, 4.3 g protein, 0.2 g fat, 0.9 g minerals, 2 g fiber, and 8 g carbohydrates (Sebetha et al. 2010), and its seed is a nutritious component in the human diet, as well as a nutritious feed for livestock (Fatokun 2002). In India, it is grown highest in states like Rajasthan, with an area of 12.31 lakh per hectare, production of 5.73 lakh per hectare, and productivity of 976 kg per hectare, and in Gujrat with an area of 2.15 lakh per hectare, production of 2.10 lakh per hectare, and productivity of 465 kg per hectare (FAO 2015).

Among the pulses, cowpea stands fourth in Karnataka in terms of area and production. In Karnataka, cowpea is grown in an area of 0.88 lakh per hectare, with a production of 0.42 lakh per hectare. The productivity as low as 505 kg per hectare (FAO 2015). In Karnataka, Arka Garima, Arka Suman, and Arka Mangala vegetable cowpea varieties are extensively grown in the districts of Dharwad, Gadag, and Mysore, with an optimal yield of 15–20 quintal per hectare. However, its spread is not much in Bagalkot district as compared to other pulse crop.

Cowpea, *Vigna unguiculata* (L.) Walp. is an annual plant originated from occidental Africa (Graham et al. 1997). It is an economically important grain legume and cash crop with high-quality protein (24%) to supplement the staple food crops in most African communities (Singh et al. 2011).

In the worldwide production of 3 million tons seed yield harvested from 12.5 million hectares, Africa accounts for about 68%, and Cameroon only 0.4% of this production (Singh et al. 1997). Cowpea is a far from negligible item of internal trade, but it is an important component of some dietary staples in Cameroon. There is a real need to increase its production in order to improve on human nutrition. Cowpea yield are generally high (1.5 to 3 $t.ha^{-1}$), but the actual yields on small-scale farm in developing countries is just averagely 0.2 to 0.4 $t.ha^{-1}$ (Murdock et al. 2003).

19.2 TAXONOMY AND PHENOLOGY OF COWPEA

Cowpea (*V. unguiculata*) belongs to the Fabaceae family and the sub-family Faboidae (Marechal and Stainier 1978). It is an annual herb with a strong taproot and several fibrous roots. The leaves are trifoliate and alternate, with the first pair usually simple and opposite. They have long peduncles holding flowers and the seed pod. They are usually green-colored. The petiole is 5–25 cm long, while stems have variations from striate, slightly hairy, or smooth. The plant height varies from 20 cm to 2 m tall (Sheahan 2012), while the colors are variable amongst cultivars from purple, blue, yellow, and pink. Floral arrangement is an intermediate inflorescence usually at distal ends of the peduncles and borne on alternate pairs with one or two.

19.3 GEOGRAPHIC DISTRIBUTION

Cultivated cowpeas are grown as warm-season-adapted annuals in tropical and subtropical zones (Hall 2001) in all countries in sub-Saharan Africa and in Asia, South America, Central America, the Caribbean, the United States, and around the Mediterranean Sea. In regions with subtropical climates, cowpea can only thrive during the summer season, whereas in tropical regions, it can grow well throughout the year. The majority of global cowpea production, accounting for over 95%, occurs in sub-Saharan Africa. As of 2013, approximately 12 million hectares of land worldwide were dedicated to cultivating cowpea (Singh et al. 2002; FAOSTAT 2015). Asia is the second producing region, representing less than 3% of the global production in average over the 1993–2013 period, most of it being cropped in Myanmar (FAOSTAT 2015).

19.4 ORIGIN OF COWPEA

Unlike other economically important crops, there is little information that is known about cowpea's domestication, dispersal, and cultivation (Xiong et al. 2016). It is universally agreed that West Africa is cowpea's center of origin and domestication (Ogunkanmi et al. 2005), although there is no archaeological evidence that exists for its early cultivation. Research with molecular markers have shown that cowpea's domestication could have taken place in East Africa instead, and both theories are currently of same weight (Xiong et al. 2016).

19.5 ECONOMIC IMPORTANCE OF COWPEA

Cowpea is an economically important crop which can be utilized as both a vegetable and a grain. The semi-spreading varieties of cowpea are best utilized as vegetables (Noubissié et al. 2011). In addition, the cowpea seeds utilized as seed vegetables are a relatively cheap and reliable source of protein (Rusike et al. 2013). The seeds of cowpea are a rich source of dietary calories, proteins, vitamins, and minerals (Gonçalves et al. 2016), which complement cereals in countries where cowpea is the main food crop (Phillips et al. 2003). Research has shown that cowpea grains are a rich source of folic acid and vitamins, which help protect unborn babies from neural tube defects (Witthöft and Hefni 2016). Cowpea also plays a key role in improvement of soil fertility by deriving atmospheric nitrogen and fixing 75–150 kg in the soil for the next crop (Simão et al.

2015). The high protein content of cowpea, its adaptation in drought-prone areas, and its role in soil fertility improvement and soil erosion make it an economically important crop in developing countries (Hall 2004).

19.6 DESCRIPTION OF THE PLANT

The cowpea *Vigna unguiculata* (L.) Walp. is an annual herbaceous legume cultivated for its edible seeds or for fodder. Cultivated cowpeas are herbaceous annuals that are either erect, prostrate, or climbing annuals with a taproot, and virtually all are glabrous. The majority of these plants are cultivated primarily for their grain, although a minor portion (approximately 10%) is cultivated in Africa as green leafy vegetables and fodder, or in eastern Asia as fresh pods (Boukar et al. 2015).

Cowpea *V. unguiculata* can grow up to 80 cm and up to 2 m for climbing cultivars. It has a well-developed root system. The germination process of this plant is epigeal, characterized by the initial pair of true leaves being simple and opposite. The following leaves are trifoliate, consisting of oval leaflets measuring 6–15 cm long and 4–11 cm broad, and they grow in an alternate arrangement. The papilionaceous flowers are found on racemose inflorescences situated at the tips of peduncles emerging from leaf axils. These flowers can exhibit colors such as white, yellowish, pale blue, or violet.

Peduncles are stout and grooved and usually much longer than the leaves (2 to 20 cm long). For each inflorescence, flowers are sequentially produced in alternating pairs on thickened nodes at the tip, with cushion-like extrafloral nectaries between each pair of flowers. The flower is of considerable size, typically measuring 2–3 cm in diameter. It features a straight keel, stamens that are diadelphous (consisting of one free stamen and nine fused stamens), a sessile ovary with numerous ovules, and a style that is adorned with a bearded interior and ends in an oblique stigma (Figure 19.1). The pods are found in pairs, forming a V shape, primarily hanging or upright, but occasionally erect. They have a cylindrical shape, ranging from 2 to 6 cm in length and 3 to 12 mm in width, and contain 8 to 20 seeds. The seeds themselves can exhibit colors such as white, pink, brown, or black (Heuzé et al. 2013).

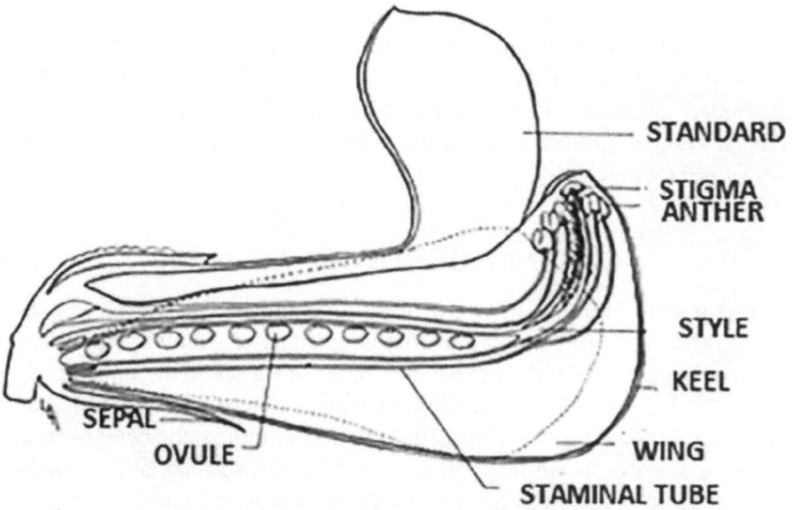

FIGURE 19.1 Flower of cowpea (*Vigna unguiculata* L.).

19.7 HABIT AND FLORAL MORPHOLOGY

The species is an annual herb reaching a height of up to 80 cm with a strong taproot and many lateral roots in the soil surface. Leaves are alternate and trifoliate. The initial set of leaves is characterized by their simplicity and opposite arrangement. Leaf characteristics encompass a wide range of sizes (6–16 × 4–11 cm) and shapes (ranging from lanceolate to ovate), while typically displaying a dark-green hue. The petiole, or leaf stalk, measures approximately 5–25 cm in length. The stems are striate, smooth, or slightly hairy. The flowers are organized in racemose or intermediate clusters at the outer ends of peduncles that can range from 5 to 60 cm in length. These clusters typically contain only two flowers per inflorescence, arranged in alternate pairs. The flowers themselves are easily noticeable, are capable of self-pollination, and emerge on short stalks. The color of the corollas can vary, ranging from white and pale blue to dirty yellow, pink, or purple. The flowers open during early daylight hours and close around midday. Once they have bloomed and opened, they subsequently wither and collapse. The flowers of cowpea possess the ability to self-pollinate and exhibit various colors, such as white, dirty yellow, pink, pale blue, or purple. They are arranged in raceme or intermediate inflorescences, forming alternate pairs. These flowers typically open during the early part of the day and close around midday. After blooming, they eventually wither and collapse. While the activities of pollinating insects can positively impact the number of pod sets and seeds per pod, there are currently no specific recommendations regarding the utilization of pollinating insects for cowpeas (McGregor 1976). *Cowpea* is a widely recognized term in the English language (Tindall 1983; Aveling 1999). It is also referred to internationally as lubia, niebe coupe, or frijol. However, all these names pertain to the same species, *Vigna unguiculata* (L) Walp., which might have been previously identified as *Vigna sinensis* (L.) in older references (Quinn 1999).

The morphological characteristics of the inflorescences of the three varieties are the same, and each inflorescence bears about 4–5 flowers on each variety, and each flower produces nectar and pollen. The flowers are borne in multiple racemes; the flower stalks that arise from leaf axil have about two to three pods per peduncle and about four in some case. The flower head samples collected from each variety showed that there are similarities in the color, shape, and size of the flowers. The flowers have variable colors which range from white, cream, yellow, or light purple to dark purple, and the colors can be in different combinations. In all the varieties, flower opening is initiated between the hours of 0600 and 0630 h and is closed between 1130 and 1200 h. When the weather is hot and dry, the flowers close earlier compared to when the weather is cold and humid. After about one or two days of flower opening and closing, the flower wilts and falls off from the cowpea plant. This signifies that cowpea flowers do better or bloom under a cool and humid condition, but the onset of hot and dry weather initiates flower closing. The stigmas are receptive over a short period of time, and unfertilized flowers drop off within 24 hr after anthesis, and the fertilized ovary may remain attached for 48 hr after anthesis.

19.8 POLLINATION OF COWPEA

Insect pollination is an important ecosystem function and essential input in crop production worldwide. Many plants grown for fruits, vegetables, oilseeds, legumes, and fodder depend on insect pollinators (Richards 2001). Insect pollination contributes to fruits and seed production in 70% of tropical crops (Roubik 1995). Out of 124 leading food crops, 84 crops depend on insect pollination (Klein et al. 2007). Insufficient pollination can therefore be a limiting factor, causing reductions in crop yield, translating to food insecurity, hunger, malnutrition, and elevated levels of poverty. Efforts to increase crop yield call for the employment of approaches that are insect pollinators–friendly to manage them on the farm. It has been shown that managed pollination contributes to fruit and seed formation in vegetable crops (Hogendoorn et al. 2006). Due to a decline in pollinators' population, studies in pollinator diversity in natural and crop ecosystems are important (Kevan et al. 1990). Modern agriculture is increasingly dependent on managed pollinators and less on wild insects living on the periphery of crop fields (Richards 2001).

Insect pollinators are threatened on a global scale and are declining, leading to food insecurity, which is likely to disrupt the ecosystem functions of insect pollinators (Potts et al. 2010). Global declines are attributed to weather alteration (Le-Conte and Navajas 2008), air contamination (Girling et al. 2013), electro-magnetic radiation from the sun (Ferrari 2014), exposure to life pressure (Wray et al. 2011), nanomaterials (Milivojević et al. 2015), and improper use of chemical pesticides with lethal and sub-lethal effects on insect pollinators (Ngowi et al. 2007). When other factors influencing crop yield, such as nutrients, water, microclimate, diseases, or pest status are sub-optimal, the positive effects of pollination can be hidden or reduced (Klein et al. 2007).

19.9 POLLINATION CHARACTERISTICS

In general, cultivated cowpeas have a high level of self-pollination. Based on their work in Texas, Blackhurst and Miller (1980) noted that the pollination process in cultivated cowpeas is complete before the flower opens. However, once they have begun flowering, cultivated cowpeas, wild cowpeas, and wild relatives have the ability to produce flowers every day for several weeks (Gwathmey et al. 1992). Consequently, some opportunities for cross-pollination occur, providing pollinators are present. Outcrossing in limited amount has been observed and quantified in literature. Fatokun and Ng (2007) report it at two locations in Nigeria and one location in Benin, and in one case, pollen travelled up to 31 m between parental plants. The authors concluded that outcrossing occurred at a frequency of less than 1%. In Senegal, outcrossing rates at 2% have been observed. In the southeastern United States, outcrossing of 0–1.4% was observed with six cultivars (Williams and Chambliss 1980). Some non-quantitative observations have also been made. Significant outcrossing has been observed in cowpea fields that are next to wild lands in Botswana. In California, some cowpea cultivars have exhibited a few percent outcrossing in some locations.

Cross-pollination is usually less than 1% but will vary somewhat with the cultivar and, more particularly, with the population of some insects. In several cases, the pollinators are not known, but honeybees (*Apis mellifera*) have been observed around cowpea flowers and thus have been implicated in pollination (Ige et al. 2011). Purseglove (1968) reported that the extrafloral nectaries at the base of the corolla attract ants, flies, and bees, but noted that a heavy insect would be required to depress the wings of the flower and expose the stamens and stigma (tripping). In coastal Kenya and Burkina Faso, several large carpenter bee species (*Xylocopa* spp.) and leaf-cutting bee species (*Megachilidae* spp.) were considered potential cross-pollinators of cowpea (R. S. Pasquet, personal communication), and it was shown that these same leaf-cutting and carpenter bees were the likely pollinators of the wild progenitor of cowpea (Kouam et al. 2012). Casual observations made in California, Texas, and Nigeria indicate that large bumblebees (*Bombus* spp.) may be responsible for the cross-pollination that occurs in cowpeas in these regions.

19.10 POLLINATION MECHANISM

Cowpea is a self-pollinating crop, and usually, pollination occurs in cowpea flower before they open. The enclosure of the pistil and stamen within the keel enhances self-pollination. Pollen grains are transferred from the anther to the stigma, and thus, pollination occurs. However, Cowpea flowers produce nectars which attract insects to them. Various insects such as butterflies, houseflies (*Musca domestica*), and honeybees (*Apis mellifera*) were seen around the cowpea flower. Honeybee was observed to be the most prominent out of the various insects observed. *Apis mellifera* was observed to start visitation to the flower immediately after flower opening was initiated. The bee lands on the wing of the flower, thereby depressing it and allowing the stamens and stigma to be pushed out of the corolla. The bee then adopts a supine feeding posture, with the head positioned toward the nectar chamber, and collects nectar. After nectar feeding, it gradually withdraws from the chamber to its landing position and then flies away onto the next flower or away from the vicinity.

However, during this process of feeding, the hairs on the bee brush the anthers, and pollen adheres to them, and these pollen grains are transferred to another flower. Pollen grains of cowpea are heavy and sticky and could not be readily transferred by wind; therefore, insects are responsible for transfer of pollen grains and, consequently, cross-pollination in cowpea plant. *Apis mellifera* has been previously reported as good pollen collectors and, as such is an effective pollinator (Nepi and Pacini 1993; Luo and Li 1999)

19.11 POLLEN PRODUCTIVITY

Pollen yield pertains to the amount of pollen generated per anther within a flower or inflorescence. When assessing the average pollen measurements from the polar to equatorial regions, it was observed that *Var. Drum* possesses larger pollen dimensions compared to the other two variations. Examination of pollen structure revealed certain resemblances in the exine design across all varieties, characterized by a reticuloid pattern. The pollen grains exhibit a circular form, which could be linked to a structural adjustment that enhances pollination efficiency through insects, as noted by Gemenes (1991) and Edeoga et al. (1996).

19.12 POLLEN DISPERSION

Cultivated cowpea flowers do not experience mechanical pollen dispersion due to specific timing and characteristics. The anthers of these flowers release pollen while the flowers are still closed, specifically during the first half of the night, as observed by Ladeinde and Bliss in 1977. Furthermore, the pollen itself is both sticky and heavy. During the second half of the night, the protective cuticle on the stigmatic surface breaks, allowing the release of a stigmatic exudate. This stage marks the start of self-fertilization. Following this, the flower opens in the early morning and subsequently closes later in the morning.

19.13 SEED VIABILITY

Cultivated varieties of cowpeas generally do not form persistent seed banks in the soil, because their seed coats are usually water-permeable, and the seeds have low dormancy levels (Lush et al. 1980). However, certain land races and cultivated types with smooth seed coats may contain a proportion of hard seeds. In contrast, wild cowpeas and cowpea relatives have seeds with impermeable seed coats, resulting in seed dormancy (Lush and Evans 1980). These hard seeds have the ability to remain viable in the soil for several years, particularly under dry soil conditions.

19.14 INSECT POLLINATORS

Cowpea (*Vigna unguiculata* L. Walp) belongs to the Fabaceae family and is classified as a grain legume. It is predominantly cultivated in the savanna regions of the tropics and subtropics, with a wide range of species numbering around 150–190. The cultivated varieties of cowpea can be categorized into five distinct groups, known as Cultigroups, namely, unguiculata, sesquipedalis, textiles, melanophtalmus, and biflora (Westphall 1974; Maréchal et al. 1978; Ng and Marechal 1985). The significance of cowpea lies in its notable protein content, which ranges from 23% to 29%, and it has the potential to reach up to 35%. Additionally, it possesses the unique ability to fix atmospheric nitrogen, allowing it to thrive in and enhance the quality of nutrient-poor soils (Steele 1972). Currently, cowpea ranks as the second most important pulse crop (Maréchal et al. 1978). Cowpea is a self-pollinating crop, facilitated by the arrangement of its floral parts. However, insects visiting cowpea flowers have been identified as contributors to the transfer of pollen from one cowpea plant to another.

19.15 *VIGNA SINENSIS* SAVI (L.) EX HASSK., FAMILY LEGUMINOSAE

The flowers are arranged in racemose or intermediate inflorescence at the distal ends of 5–60 cm long peduncles. The flowers appear in alternate pairs, typically with only two flowers per cluster. They are easily noticeable, capable of self-pollination, and are held on short stalks. The color of the corollas can range from white, dirty yellow, pink, pale blue, to purple. These flowers bloom in the early morning and close around midday. Once they have bloomed, they wither and collapse. The inflorescence is made up of two to eight papilionaceous flowers, which are whitish, yellowish, or violet in color. These flowers are arranged in pairs and densely clustered at the tip of a slender stalk or stem, measuring from 1–6 in in length. Each flower, measuring 3/4 to 1 in, features a curved style with a bearded inner curve just below the slanted stigma. The anthers are uniformly grouped in two fused clusters around the style. A single ovary is present, containing 8–20 ovules. There are extrafloral nectaries located at the base of the corolla. Although it is not entirely clear, it is highly likely that the flower secretes nectar. Bumblebees and other insects are attracted to these flowers as they gather both nectar and pollen from them.

19.16 POLLINATION REQUIREMENTS

Robbins (1931) reported that the cowpea has the ability to self-fertilize, which is likely the most frequent phenomenon. However, the flowers are often frequented by honeybees or bumblebees, primarily attracted by the additional nectar from the extrafloral source. Cowpea (*V. unguiculata*) is both self- and cross-pollinating, producing nectar that attracts a variety of insects that forage on its flowers (Ige et al. 2011). Despite cowpea being an important food source, there is little information that is known about the contribution of insect pollinators to its yield and quality improvement. Although it is an autogamous species, it has a cross-pollination rate of 10%. However, the mean productivity has declined during the recent years, which is linked to a decrease in or absence of pollinating insects. Vas et al. (1998) found that fruit set in the presence of pollinators is 83% more than in their absence caged flowers, at 77%. However, cross-pollination reduced the number of seeds per pod and fruit set relative to self-pollination. The results reveal that insect pollination has a complimentary role in the yield of cowpea by creating a mixed pollination system where self-pollination dominates.

19.17 POLLINATORS

According to Purseglove (1968), a heavy insect is necessary to press down on the flower's wings, thus revealing the stamens and stigma. The pollen's stickiness and weight indicate that the plant does not rely on wind for pollination (Mackie 1946). The occurrence of cross-pollination appears to be linked to regions abundant in bumblebees.

Mackie and Smith (1935) stated that bumblebees are the primary pollinators. The actual value of frequent visitation by pollinating insects to blossoms of cowpeas has not been determined; however, the study by Warnock and Hagedorn (1954) would indicate that such activity is beneficial in increasing the number of pod set, the number of seeds per pod, or both. The pollinators recorded include European honeybee (*Apis mellifera* Lin.), Asiatic honeybee (*Apis cerana* Fab.), rock bee (*Apis dorsata* Fab.), carpenter bee (*Xylocopa* sp.), rice skipper (*Pelopidas mathias* (F.)), cowpea borer (*Lampides boeticus* L.), bumblebee (*Bombus* spp.), Oriental wasp (*Vespa orientalis* (Lin.), Tabanid fly (*Tabanus* spp.), Syrphid fly (*Eristalis* sp.), housefly (*Musca domestica* Lin.), fruit fly (*Bactrocera* sp.), castor butterfly (*Ergolis merione* Cram.), cabbage butterfly (*Pieris brassicae* Lin., *P. Canidia* Lin.), golden wasp (*Vespa magnifica* (Smith)), Cyntomid fly (*Cyntomis passalis* (F.)), and mustard sawfly (*Athalia lugens proxima* (Klug)).

Nghia and Srivastava (2015) studied the diversity of beneficial insects visiting cowpea flowers. They observed a total of 33 insect species, including 16 insect species, visit on flowers, 13 predators, and 4 parasitoids. Among the insect visitors/pollinators, a total of 16 species belong to three orders, Lepidoptera (8), Hymenoptera (4), and Diptera (4). The various groups of predators belong to five orders, Coleoptera (6), Neuroptera (1), Hemiptera (1), Odonata (3), and Arachnida (2), that were also observed preying upon the insect pests of cowpea crop, such as aphids, thrips, Jassids, and some small flies. Small insects were observed on both the crop and gathered on eggs or larvae. These insects were identified as parasitoids from the order Hymenoptera, specifically four species from the families Eulophidae, Eurtomymidae, and Braconidae. It was observed that these parasitoids targeted lepidopteran larvae and leaf miners for parasitization. The findings of this study could be beneficial in developing a comprehensive strategy to control insect pests and improve the effectiveness of beneficial insects, thereby maximizing crop yield.

Hordzi (2011) reported that the main insects that visited cowpea flowers from the survey and experimental farms were beetles, *Apis mellifera adansoni* (Hymenoptera: Apidae), *Ceratina sp* (Hymenoptera: Apidae; Xylocopinae), *Megachile* sp. (Hymenoptera: Megachilidae; Megachilinae), *Xylocopa calens* (Hymenoptera: Apidae; Xylocopinae), *Xylocopa imitator* (Hymenoptera: Apidae; Xylocopinae), *Brausepis* sp. (Hymenoptera: Halictidae; Halictinae), *Lipotriches* sp. (Hymenoptera: Halictidae; Halictinae), *Melecta* sp. (Apidae; Apinae), and *Amegilla* sp. (Hymenoptera: Apidae; Apinae).

Other insects included thrips (Thripidae), flies (Dipterans), Lepidopterans (butterflies and moths), and *Dysdercus* sp. Thrips and beetles were the commonest and were observed quite a number of times on the stigma and inside the flowers. Beetles and *Ceratina* sp. feed on flowers. *Xylocopa calens* and *Megachile* sp. are probable pollinators of cowpea flowers, as they were frequently observed on the stigma's tip, indicating their active role in pollination. The findings revealed that the differences between the frequencies of occurrence of the insects on flowers were not due to chance. This shows that thrips and beetles are actually the most occurring insects on the flowers during the major season, and *Megachile* sp., as well as thrips, is the most common for the minor season.

Stephanie et al. (2015), when comparing the yields of flowers that were unprotected to those that were isolated and then exposed to *X. olivacea*, observed a significant increase in the fructification index by 67.89% and an increase in the number of seeds per pod by 13.58% due to the presence of this bee. Additionally, the combined effect of insects and Rhizobium resulted in a 62.28% increase in the number of seeds per pod and an 87.15% increase in the percentage of normal seeds. The synergistic activity of insects and *Rhizobium* increased the number of seeds per pod by 62.28% and the percentage of normal seeds by 87.15%. Based on our findings, we recommend the practice of introducing Rhizobium to cowpea plants during sowing, along with establishing nests of *X. olivacea* near the field. This approach has been shown to effectively enhance both the quality and quantity of pods and seeds, promoting sustainable yield improvement for this crop.

Dingha et al. (2021) recorded honeybees, bumblebees, carpenter bees, wasps, and butterflies and moths visiting cowpea flowers; however, their distribution varied significantly among varieties and sampling methods. Cowpea has the potential to enhance resources available for pollinators and has the capacity to enhance pollinator abundance and diversity in various agricultural systems. Various vegetable crop varieties, such as cowpea (*Vigna unguiculata* (L.) Walp.), have been identified as appealing to pollinators. This versatile crop serves as both a vegetable and a dry bean. It is adapted to a wide range of growing conditions and considered one of the best drought-resistant food crops (Folger et al. 2012; Gonçalves et al. 2016), and therefore, it can help counter the negative impact of climate change, which can undermine agricultural production and sustainability (Folger et al. 2012; Gonçalves et al. 2016). The ability to fix atmospheric nitrogen makes cowpea crucial for maintaining soil fertility in a sustainable manner (Quin 1997; Singh and Rachie 1985). Moreover, cowpea possesses various morphological characteristics, such as floral resources, which attract honeybees and other bee species (Fohouo et al. 2009). An increasing body of empirical evidence shows that the management of floral resources in local agricultural fields can mitigate adverse human impact on pollinator diversity and pollination services (Kennedy et al. 2013). Cowpea floral resources, especially nectar, are attractive to pollinators (Fohouo et al. 2009). There is a lack of information regarding the successful utilization of cowpea as an intercrop to enhance pollinator activity and increase crop production. In a study conducted by Wousla et al. (2019), it was observed that the main flower visitors were large bees belonging to the family Megachilidae and the genus Xylocopa (Apidae, Xylocopinae). *Xylocopa caffra* (Linnaeus) was the most prevalent bee species, accounting for 59% of the total bees observed and 58% of the total flowers visited. It was found that pod set was higher in visits by megachilid bees compared to Xylocopa bees. Among the Xylocopa species, there was a strong correlation between their size and efficiency ($r = 0.95$, $p = 0.0477$), indicating that larger insects were more efficient in pollination. Although honeybees, small bees (*Ceratina* and *Nomia* species), and Lepidopterans also visited the flowers, they did not have any significant effect on pollination. The pollination system of cowpea is specialized, with pollinators mainly restricted to Megachilidae and Xylocopa species. The variation in pollen load deposit and pod set highlights that the success of pollination depends on the specific insect species involved and their size. Megachilids were found to be the most efficient pollinators, particularly due to their ability to promote outcrossing more effectively than Xylocopa species.

Musa et al. (2013) Studied the effect of honeybees' (*Apis mellifera adansonii* Latreille) (Hymenoptera: Apidae) visitation on pod and seed set production of *Vigna unguiculata* L. Walp. (Leguminosae). They found there were significant differences ($p < 0.05$) in mean plant height, weight of filled pods, number of empty pods per plant, and seeds per pod exposed to honeybee visitation as compared to plants covered without honeybees (no pollination). Mean weight of pods for plants not covered (open-pollinated) was 3.64% more than plants covered in muslin with honeybees. The mean number of empty pods was higher in the non-pollinated compared to the insect-pollinated. Results showed that pollination activity of *A. mellifera* caused an increase in mean weight of pods per plant and number of seeds per pod of cowpea exposed to open-pollination. The mean number of seeds per pod (11.15) in plants pollinated by *A. mellifera* (open-pollinated) was significantly higher ($p <$ 0.05) than a mean of 9.53 obtained in plants not pollinated, showing a percentage increase of 17.0%. The mean number of empty pods was higher in the non-pollinated as compared to the pollinated cowpea. The installation of hives of *A. mellifera* close to *V. unguiculata* fields could therefore be recommended to improve cowpea pod and seed production in the study area.

Chesney et al. (2010) had earlier reported that cowpea plant height varied between 31.7 and 33.4 cm. Plant height could be influenced by the growing environment, crop variety, and insect visitation. The number of seeds per pod recorded in this experiment ranged from 9.53 to 11.15, which were in variance with a range of 4.84 and 7.99 reported by Vas et al. (1998) in different pollination treatments of cowpea flowers. The results showed that cowpea plant exposed to pollination by honeybees gave an increase in pod formation, leading to high seed production. The smallest mean weight of pods and number of seeds per pod in the absence of honeybees gave the impression that for higher seed production, there is a need for honeybees to pollinate the host plant flowers (Cane 2002; Tchuenguem et al. 2008). *Mellifera* has been previously reported as good pollen collectors and, as such, effective pollinators (Luo and Li 1999).

The daily activity of the domestic bee on flowering plants may depend on the production of either pollen (Stone et al. 1998) or nectar (Suzo et al. 2001; Pouvreau 2004). The floral visits of honeybees were reported to be mostly concerned with pollen (Benachour et al. 2007). The decrease in pod weight and seed yields could be linked to the occurrence of empty pods and infestation by lepidopterous larvae, even though the seeds were treated before planting. Infestation by caterpillars of lepidopterans particularly, *Maruca testulalis* (Geyer), possibly caused abortion of pods (Allen et al. 1996).

The higher number of seeds per pod produced in open-pollinated plants when compared to those plants on a plot 10 m away from other treatments may be due to pollination by honeybees and other insects, respectively. The highest weight of pods per plant was found in open-pollinated plants exposed to honeybees' visitation, suggesting that population density of honeybees may have contributed to variation in seed yield. The number of seeds per pod obtained in plants on a plot 10 m away from other treatments was comparable to plants covered without honeybees (no pollination), which produced more empty pods. Preventing bees from visiting during flowering of *V. villosa* spp. *Dasycarpa* decreased set, seed yield, and yield components (Al-Ghzawi et al. 2009).

Findings were in accordance with the report of Partap and Partap (2002), which indicated that absence of pollinators causes decline in fruit and seed production. The high reduction in the mean number of empty pods per plant and the high mean number of seeds per pod in open-pollinated plants showed that pollination activity of honeybees had a positive impact on seed production of cowpea.

Similar results had been reported on pollination effect of honeybees by Al-Ghzawi et al. (2009). In this study, plants covered without honeybees (no pollination) showed significant reduction in mean weight of pods per plant, which could be attributed to absence of insects to pollinate the plants, and no significant effect on number of seeds per pod was observed.

Yield of various plants can be decreased by half or more in the presence of insufficient pollinating insects (Philippe 1991). The yields of certain fruit, seed, and nut crops decrease by more than 90% without bee pollination (Klein et al. 2007). In this study, the use of pollinators, especially honeybees, was significant in improving the quantity of pods and seeds produced

Dukku and Mukaddas (2019), in their study on diurnal visitation pattern of insect pollinators on cowpea, found that 52.3% of the visitors belonged to the order Hymenoptera (honeybees, wasps, and ants), 26.2% to Lepidoptera (butterflies and moths), and 21.5% to Diptera (flies). Among the hymenopterans, 42.3% were honeybees, 40.9% ants, and 16.8% wasps. In earlier studies, *A. mellifera* has been reported as the most frequent floral visitor of cowpea (Fohouo et al. 2009; Ige et al. 2011; Kwapong et al. 2013). This underscores the importance of this species as a pollinator of this plant.

While self-pollination is common in cowpea, this crop still benefits from the presence of animal pollinators, particularly in genotypes that produce male sterile flowers. Moreover, some genotypes experience hindered self-pollination due to the positioning of the anthers within the corolla, which prevents contact with the emerging stigma (Rachie et al. 1975). As a result, it is estimated that cross-pollination, predominantly carried out by insects (Aouar-Sadli et al. 2008), accounts for approximately 1% to 10% of pollination in cowpea (Blackhurst and Miller 1980; Rachie and Silvester 1977; Williams and Chambliss 1980). Field observations conducted in Ghana (Kwapong et al. 2013), Cameroon (Fohouo et al. 2009), and Nigeria (Ige et al. 2011) support these findings.

A. mellifera was recorded as the most abundant floral visitor to cowpea, and the pollination activity of this bee significantly increased the number of seeds produced, per pod, by 9.31% (Fohouo et al. 2009) and 17% (Musa et al. 2013). The bee visited the flowers continuously from 0700 h to 1200 h, a period that coincided with the dehiscence of the flowers, which was between 0600 h and 1200 h (Ige et al. 2011). Findings suggest that pollination of cowpea is largely dependent on flower visitation for nectar and pollen by honeybees. The role played by honeybees' pollination has impact on pod and seed sets of cowpea plants in the study area. The establishment of cowpea field close to honeybee hive installation could be recommended to increase pod and seed set in cowpea.

19.18 POLLINATION RECOMMENDATIONS AND PRACTICES

Cowpea is a self-pollinating crop, and usually pollination occurs in cowpea flower before they open. The keel structure surrounding the pistil and stamen promotes self-pollination by facilitating the transfer of pollen grains from the anther to the stigma. In addition to this mechanism, cowpea flowers produce nectar to attract insects toward them. Ige et al. (2011) observed various insects such as butterflies, houseflies (*Musca domestica*), and honeybees (*Apis mellifera*) visiting the cowpea flower. They found that honeybee, *Apis mellifera*, observed to be the most prominent out of the various insects observed, started visitation to the flower immediately after flower opening was initiated. The bee lands on the wing of the flower, thereby depressing it and allowing the stamens and stigma to be pushed out of the corolla. The bee then adopts a supine feeding posture, with the head positioned toward the nectar chamber, and collects nectar. After consuming nectar, the bee gradually retreats from the chamber and assumes its landing position before taking flight toward either the next flower or away from the surrounding area. However, during the feeding process, the bee's hairs come into contact with the anthers, causing pollen to stick to them. These pollen grains are then transferred to another flower. The pollen grains of cowpea are heavy and sticky, making them unsuitable for wind dispersal. Consequently, insects play a crucial role in transferring the pollen grains and facilitating cross-pollination in cowpea plants. Previous studies have identified *Apis mellifera* as excellent pollen collectors and effective pollinators (Nepi and Pacini 1993; Luo and Li 1999).

Fernand-Nestor Tchuenguem Fohouo et al. (2009) reported that number and dry weight of seed per pod, the pod length, and the percentage of normal seeds per pod from unprotected flowers were significantly higher than those of flowers protected from insects. The fruitification rates were 62 and 48%, while the percentages of healthy seeds were 97.61 and 76.17%, respectively, in unprotected and protected inflorescences. The installation of *A. mellifera adansonii* could be recommended near cowpea (*Vigna unguiculata*) field to improve its seed and pod production.

REFERENCES

Al-Ghzawi, A., Samarah, N., Zartoun, S., and Alqudah, A. 2009. Impact of bee pollinators on seed set and yield of *Vicia villosa* spp. *dasycarpa* (Leguminosae) grown under semiarid conditions. *Ital. J. Anim. Sci.* 8, 65–74.

Allen, D. J., Ampofo, J. K. O., and Wortman, C. S. 1996. Pests, diseases and nutritional disorders of the common bean in Africa. A field guide. A CIAT/CTA Publication, 132 pp. Centro Internacional de Agricultura Tropical (CIAT): Publications Wageningen, Netherlands.

Aouar-Sadli, M., Louadi, K., and Doumandji, S. 2008. Pollination of the broad bean (*Vicia faba* L.var. *major*) (Fabaceae) by wild bees and honey bees (Hymenoptera: Apoidea) and its impact on the seed production in the Tizi-Ouzou area (Algeria). *Afr. J. Agr. Res.* 3(4), 266–272.

Aveling, T. 1999. Cowpea pathology research. Available at www.ap.ac.za/academic/microbio/plant/pr-colwpea.html

Benachour, K., Louardi, K., and Terzo, M. 2007. Role of wild and domestic bees (Hymenoptera: Apoidea) in the pollination of *Vicia faba* L. var. *major* (Fabaceae) in the Constantine region (Algeria). *Ann. Soc. Entomol. Fr.* 43(2), 213–219.

Blackhurst, H. T. and Miller, J.C. 1980. Cowpea. In: Fehr, W. R. and Hadley, H. H. (eds.) *Hybridization of Crop Plants, American Society of Agronomy – Crop Science Society of America*, Wisconsin, Madison, pp. 327–337.

Boukar, O. et al. 2015. Cowpea. In: De Ron, A. M. (ed.) *Grain Legumes, Series Handbook of Plant Breeding*, Springer-Verlag, New York, pp. 219–250.

Cane, J. H. 2002. Pollinating bees (Hymenoptera: Apiforms) of United States alfalfa compared for rate of pod and seed set. *J. Econ. Entomol.* 95(1), 22–27.

Chesney, P. E. K., Simpson, L. A., Cumberbatch, R. N., Homenauth, O., and Benjamin, F. 2010. Cowpea yield performance in an alleycropping practice on an acid infertile soil at Ebini, Guyana. *Theopen Agric. J.* 4, 80–88.

Dancette, C. and Hall, A. E. 1979. Agroclimatology applied to water management in the Sudanian and Sahelian Zones of Africa. In: Hall, A. E., Cannell, G. H., and Lawton, H. W. (eds) *Agriculture in Semi-Arid Environments. Ecological Studies*, vol. 34. Springer, Berlin, Heidelberg.

Daniel, N. and Patrick, A. 2014. Effects of *Bradyrhizobium japonicum* and phosphorus supplementation on the productivity of legumes. *Int. J. of Plant & Soil Sci.* 3(7), 894–910.

Dingha, B. N., Jackai, L. E., Amoah, B. A., and Akotsen-Mensah, C. 2021. Pollinators on cowpea *Vigna unguiculata*: Implications for intercropping to enhance biodiversity. *Insects* 12, 54.

Dukku, U. H. and Mukaddas, J. P. 2019. Diurnal visitation of flowers of cowpea by insect pollinators with emphasis on the Western honeybee, *Apis mellifera*, in Bauchi, Nigeria. *Science Forum (Journal of Pure and Applied Sciences)* 17, 10–14.

Edeoga, H. O., Ogbebor, N. O., and Amayo, A. D. 1996. Pollen morphology of some Nigeria species of (*Anielrma* R. Br and *Ludwigia* L). *New Bot.* 23, 223–231.

FAO (Food and Agriculture Organisation). 2015. Fertilizer use by crop in Ghana, Rome 39, 225–262.

FAOSTAT. 2015. FAO statistics. Available at http://faostat3.fao.org/home/E (accessed 27 October 2015)

Fatokun, A. C. 2002. Breeding cowpea for resistance to insect pest. *J. Agric. Sci.* 125, 426–521.

Fatokun, C. A. and Ng, Q. 2007. Outcrossing in cowpea. *Journal of Food Agriculture & Environment* 5, 334–338.

Ferrari, T. E. 2014. Magnets, magnetic field fluctuations and geomagnetic disturbances impair the homing ability of honey bees (*Apis mellifera*). *Journal of Apicultural Research* 53(4), 452–465.

Fohouo, F. T., Albert, N., and Kengni, B. S. 2009. Pollination and yield responses of cowpea (*Vigna unguiculata* L. Walp.) to the foraging activity of *Apis mellifera adansonii* (Hymenoptera: Apidae) at Ngaoundéré (Cameroon). *Afr. J. Biotech.* 8(9), 1988–1996.

Folger, P., Cody, B. A., and Carter, N. T. 2012. *Drought in the United States: Causes and Issues for Congress.* Congressional Research Service, Washington, DC, p. 36.

Freire Filho, F. R., Ribeiro, V. Q., Rocha, M. M., Silva, K. J. D., Nogueira, M. S. R., and Rodrigues, E. V. 2011. *Feijão-caupi no Brasil: Produção, melhoramento genético, avanços e desafios.* Embrapa Meio-Norte, Teresina, PI.

Gemenes, M. 1991. Some morphological adaptations in bees (Hymenoptera, *Ludwigia elegans* (Onagraceae). *New Bot.* 35, 413–422.

Girling, R. D., Lusebrink, I., Farthing, E., Newman, T. A., and Poppy, G. M. 2013. Diesel exhaust rapidly degrades floral odours used by honeybees. *Sci. Rep.* 3, 2779.

Gonçalves, A., Goufo, P., Barros, A., Domínguez-Perles, R., Trindade, H., Rosa, E. A., and Rodrigues, M. 2016. Cowpea (*Vigna unguiculata* L. Walp), a renewed multipurpose crop for a more sustainable agri-food system: Nutritional advantages and constraints. *Journal of the Science of Food and Agriculture* 96(9), 2941–2951.

Graham, P. H. and Ranalli. P. 1997. Common bean (Phaseolus vulgaris L.). *Field Crop Research* 53, 131–146.

Gwathmey, C. O., Hall, A. E., and Madore, M. A. 1992. Adaptive attributes of cowpea genotypes with delayed monocarpic senescence. *Crop Science* 32(3), 765–772.

Hall, A. E. 2001. *Crop Responses to Environment.* CRC Press LLC, Boca Raton, FL.

Hall, A. E. 2004. Breeding for adaptation to drought and heat in cowpea. *European Journal of Agronomy* 21(4), 447–454.

Heuzé, V. et al. 2013. Cowpea (Vigna unguiculata) forage. Available at Feedipedia.org: A Programme by INRA, CIRA, AFZ, and FAO, www.feedipedia.org/node/233

Hogendoorn, K., Gross, C. L., Sedgley, M., and Keller, M. A. 2006. Increased tomato yield through pollination by native Australian *Amegilla chlorocyanea* (Hymenoptera: Anthophoridae). *J. Econ. Entomol.* 99, 828–833.

Hordzi, W. H. K. 2011. May. Insects observed on cowpea flowers in three districts in the central region of Ghana. *African Journal of Food, Agriculture, Nutrition and Development* 11(3), 4880+. Available at Gale Academic OneFile, link.gale.com/apps/doc/A258994306/AONE?u=anon~53113d2c&sid=googleScholar&xid=e8db4494 (accessed 21 April 2022)

Ige, O., Olotuah, O., and Akerele, V. 2011. Floral biology and pollination ecology of cowpea (*Vigna unguiculata* L. Walp). *Modern Applied Science* 5, 74–82.

IITA (International Institute of Tropical Agriculture). 2003. Crop and farming systems. Available at http://www.iita.org/crop/cowpea.htm

Kay, D. E. 1979. *Food Legumes.* Tropical Development and Research Institute, London.

Kennedy, C. M., Lonsdorf, E., Neel, M. C., Williams, N. M., Ricketts, T. H., Winfree, R., Bommarco, R., Brittain, C., Burley, A. L., Cariveau, D. et al. 2013. A global quantitative synthesis of local and landscape effects on wild bee pollinators in agroecosystems. *Ecol. Lett.* 16, 584–599.

Kevan, P. G., Clark, E. A., and Thomas, V.G. 1990. Insect pollinators and sustainable agriculture. *American Journal of Alternative Agriculture* 5, 13–22.

Klein, A. M., Vaissiere, B., Cane, J. H., Steffan-Dewenter, I., Cunningham, S. A., Kremen, C., and Tscharntke, T. 2007. Importance of pollinators in changing landscapes for world crops. *Proceedings of the Royal Society B* 274(1608), 303–313.

Kouam, E., Pasquet, R., Campagne, P., Tignegre, J.-B., Thoen, K., Gaudin, R., Ouedraogo, J., Salifu, A., Muluvi, G., and Gepts, P. 2012. Genetic structure and mating system of wild cowpea populations in West Africa. *BMC Plant Biology* 12, 113.

Kwapong, P. K., Danquah, P. O. A., and Asare, A. T. 2013. Insect floral visitors of cowpea (*Vigna unguiculata* L.). *Ann. Biol. Res.* 4(4), 12–18.

Ladeinde, T. A. O. and Bliss, F. A. 1977. Identification of the bud stage for pollinating without emasculation in cowpea (*Vigna unguiculata* (L.) Walp.). *Nigerian Journal of Science* 11, 183–194.

Le Conte, Y. and Navajas, M. 2008. Climate change: Impact on honey bee populations and diseases. *Revue Scientifique et Technique-Office International des Epizooties*, 27(2), 499–510.

Luo, Y. and Li, Z. 1999. Pollination ecology of Chloranches serratus (Thunb) Roemet Schult and Ch. Fortunei (A. Gray) SolmsLaub. (Chloranthaceae). *Ann. Bot.* 83, 489–499.

Lush, W. M. and Evans, L. T. 1980. The seed coats of cowpeas and other grain legumes structure in relation to function. *Field Crops Research* 3, 267–286.

Lush, W. M., Evans, L. T., and Wien, H. C. 1980. Environmental adaptation of wild and domesticated cowpeas (Vigna unguiculata (L.) Walp.). *Field Crops Res.* 3, 173–187.

Mackie, W. W. 1946. Blackeye beans in California. *Univ. California, Agr. Expt. Sta., Bull.* 696.

Mackie, W. W. and Smith, F. L. 1935. Evidence of field hybridization in beans. *Journal American Society Agronomy* 27, 903–908.

Maréchal, R., Mascherpa, J. M., and Stainier, F. 1978. Etude taxonomique d'un groupe complexe d'espèces des genres Phaseolus et Vigna (Papilionaceae) sur la base de données morphologiques et polliniques, traitées par l'analyse informatique. *Boissiera* 28, 1–273.

McGregor, S. E. 1976. *Insect Pollination of Cultivated Crop Plants*, Vol. 496. Agricultural Research Service, US Department of Agriculture, Washington, DC.

Milivojević, T., Glavan, G., Božič, J., Sepčić, K., Mesarič, T., and Drobne, D. 2015. Neurotoxic potential of ingested ZnO nanomaterials on bees. *Chemosphere* 120, 547–554.

Murdock, L. L., Seck, D., Ntoukam, G., Kitch, L., and Shade, R. E. 2003. Preservation of cowpea grain in sub-Saharan Africa – Bean/cowpea CRSP contributions. *Field Crops Research* 82, 169–178.

Musa, A. K., Liadi, M. T., and Adegbite, O. R. 2013. Impact of honeybees (*Apis mellifera adansonii*) (Hymenoptera: Apidae) pollination on pod and seed set of cowpea (*Vigna unguiculata* L. Walp) in Ilorin, Southern Guinea Savanna of Nigeria. *J. Agr. Res.* 1(7), 83–87.

Nepi, M. and Pacini, E. 1993. Pollination, Pollen viability and pistil receptivity in Cucurbita pepo L. *Ann. Bot.* 72, 527–536.

Nghia, N. T. and Srivastava, P. 2015. Biodiversity of beneficial insect associated with cowpea at Pantnagar—Uttarakhand—India. *Omonrice* 20, 73–79.

Ngowi, A. V. F., Mbise, T. J., Ijani, A. S. M., London, L., and Ajayi, O. C. 2007. Smallholder vegetable farmers in northern Tanzania: Pesticides use practices, perceptions, and cost and health effects. *Crop Prot.* 26, 1617–1624.

Noubissié, T. J. B., Bell, J. M., Nassourou, A. M., Njintang, Y. N., and Youmbi, E. 2011. Genetic analysis of seed proteins contents in cowpea (Vigna unguiculata L. Walp.). *Afr. J. Biotechnol.* 10, 3077–3086.

Ogunkanmi, L. A., Taiwo, A., Mogaji, O. L., Awobodede, A., Eziashi, E. E., and Ogundipe, O. T. 2005. Assessment of genetic diversity among cultivated cowpea (Vigna unguiculata L. Walp.) cultivars from a range of localities across West Africa using agronomic traits. *J. Sci. Res. Dev.* 10, 111–118.

Padulosi, S. and Ng, N. Q. 1997. Origin, taxonomy, and morphology of Vigna unguiculata (L.) Walp. In: Singh, B. B. et al. (eds.) *Advances in Cowpea Research, Japan International Research Center for Agricultural Sciences and IITA*, Ibadan, Nigeria, pp. 1–12.

Pandey, R. K. and Ngarm, A. T. 1985. Agronomic research advances in Asia. In: Singh, S. R. and Rachie, K. O. (eds.) *Cowpea Research, Production and Utilization*. Wiley, Chichester, UK, pp. 297–306.

Partap, U. and Partap, T. 2002. *Warning Signal from the Apple Valley of the HKH Region: Productivity Concerns and Pollination Problems*. Intern. Centre for Integrated Mountain Development (ICIMOD), Kathmandu, p. 106.

Philippe, J. M. 1991. *Pollination by Bees: Establishment of Colonies within Flowering Crops to Enhance Their Production*. Edisud, La calade, Aix-en-provence, France, p. 178.

Phillips, R. D., McWatters, K. H., Chinnan, M. S., Hung, Y. C., Beuchat, L. R., Sefa-Dedeh, S., and Komey, N. S. 2003. Utilization of cowpeas for human food. *Field Crops Research* 82(2–3), 193–213.

Potts, S. G., Biesmeijer, J. C., Kremen, C., Neumann, P., Schweiger, O., and Kunin, W. E. 2010. Global pollinator declines: Trends, impacts and drivers. *Trends Ecol. Evol.* 25, 345–353. (CrossRef) (PubMed)

Purseglove, J. W. 1968. *Tropical Crops Dicotyledons 1*. John Wiley and Sons, Inc., New York.

Quin, F. M. 1997. Introduction. In: Singh, B. B., Mohan Raj, D. R., Dashiel, K. E., and Jackai, L. E. N. (eds.) *Advances in Cowpea Research*. International Institute of Tropical Agriculture (IITA) and Japan International Research Center for Agricultural Sciences (JIRCAS), Ibadan, Nigeria.

Quinn, J. 1999. *Cowpea: A Versatile Legume for Hot, Dry Conditions*. Jefferson Institute, Columbia, MO.

Rachie, K. O., Rawal, K., Franckowiak, J. D., and Akinpelu, M. A. 1975. Two outcrossing mechanisms in cowpeas, Vigna unguiculata (L.) Walp. *Euphytica* 24, 159–163.

Rachie, K. O. and Silvester, P. 1977. Grain legumes. In: Lealey, C. L. A. and Wills, J. B. (eds.) *Food Crops of Lowland Tropics*. Oxford University Press, Oxford, UK, pp. 41–74.

Richards, A. J. 2001. Does low biodiversity resulting from modern agricultural practice affect crop pollination and yield? *Annals of Botany* 88(2), 165–172.

Robbins, W. W. 1931. *The Botany of Crop Plants*, 3rd edn. P. Blakiston's Sons & Co. Inc., Philadelphia, 639 pp.

Roubik, D. 1995. Pollination of cultivated plants in the tropics. Food and agriculture organization of the United Nations. *Bulletin* 118.

Rusike, J., Brand, G. Van Den, Boahen, S., Dashiell, K., Kantengwa, S., Ongoma, J., and Abaidoo, R. 2013. N2Africa putting nitrogen fixation to work for smallholder farmers in Africa. In: Vanlauwe, B. (ed.) *Agro-Ecological Intensification of Agricultural Systems in the African Highlands*. Routledge Piet van Asten, Guy Blomme, London.

Sebetha, E. T., Ayodele, V. I., Kutu, F. R., and Mariga, 2010. Yields and protein content of two cowpea varieties grown under different production practices in Limpopo province, South Africa. *African J. Biotech.* 5, 628–634.

Sheahan, C. M. 2012. *Plant Guide for Cowpea (*Vigna unguiculata*)*. USDA-Natural Resources Conservation Service, Cape May Plant Materials Center, Cape May, NJ.

Simão, F. A., Waterhouse, R. M., Ioannidis, P., Kriventseva, E. V., and Zdobnov, E. M. 2015. BUSCO: Assessing genome assembly and annotation completeness with single-copy orthologs. *Bioinformatics* 31, 3210–3212.

Singh, A., Baoule, A. L., Ahmed, H. G., Dikko, A. U., and Aliyu, U. 2011. Influence of phosphorus on the performance of cowpea (*Vigna unguiculata* L.) varieties in the Sudan savanna of Nigeria. *Agricultural Sci.* 2(3), 313–317.

Singh, B. 2003. Improving the production and utilization of cowpea as food and fodder. *Field Crops Res.* 84, 169–177.

Singh, B. B., Ehlers, J. D., Sharma, B., and Freire Filho, F. R. 2002. Recent progress in cowpea breeding. In: Fatokun, C. A., Tarawali, S. A., Singh, B. B., Kormawa, P. M., and Tamo, M. (eds.) *Proceedings, the World Cowpea Conference III, Challenges and opportunities for Enhancing Sustainable Cowpea Production*, International Institute of Tropical Agriculture (IITA), Ibadan, pp. 22–40.

Singh, B. B., Mohanraj, D. R., Dashiel, K. E., and Jackai, L. E. N. 1997. Advances in Cowpea Research Post-Harvest Storage of Cowpea in Sub-Saharan Africa. IITA/JIRCA Publication, Ibadan, Nigeria, pp. 302–312.

Singh, B. B. and Tarawali, S. A. 1997. Cowpea and Its Improvement: Key to Sustainable Mixed Crop/Livestock Farming Systems in West Africa. International Institute of Tropical Agriculture, IITA Kano Station, PMB 3112, Kano, ILRI International Livestock Research Institute, PMB 5320, Ibadan. Available at http://www.ilri.org/InfoServ/Webpub/Fulldocs/Cropresidues/chap%20

Singh, S. R. and Rachie, K. O. (eds.). 1985. *Cowpea Research, Production and Utilization*. Chichester, Wiley, p. 460.

Steele, W. M. 1972. Cowpeas in Nigeria. PhD Thesis, University of Reading, UK.

Stephanie, K. B., Albert, N., and Fernand-Nestor, T. F. 2015. Pollination and yield attributes of (cowpea) Vigna unguiculata L. Walp. (Fabaceae) as influenced by the foraging activity of Xylocopa olivacea Fabricius (Hymenoptera: Apidae) and inoculation with Rhizobium in Ngaoundere, Cameroon. *International Journal of Agronomy and Agricultural Research (IJAAR)* 6(2), 62–76.

Stone, G. N., Willmer, P., and Rowe, J. A. 1998. Partitioning of pollinators during flowering in an African acacia community. *Ecol.* 79(8), 2808–2827.

Suzo, M. J., Pierre, J., Moreno, M. T., Esnaut, R., and Le Guen, J. 2001. Variation in outcrossing levels in faba bean cultivars: Role of ecological factors. *J. Agric. Sci.* 136, 399–405.

Tarawali, S. A., Singh, B. B., and Gupta, S. C. 2002. Cowpea as a key factor for a new approach to integrate crop-livestock systems research in dry savannas of West Africa. In: *Challenges and Opportunities for Enhancing Sustainable Cowpea Production*. Proceedings of the world cowpea conference III held at the International Institute of Tropical Agriculture (IITA), Ibadan, Nigeria.

Tchuenguem Fohouo, F. N., Djongwangwe, D., and Bruckner, D. 2008. Foraging behaviour of the African honey bee (*Apis mellifera adansonii*) on *Annona senegalensis, Croton macrostachyus, Psorospermum febrifugum* and *Syzygium guineense var. guineense* flowers at Ngaoundere (Cameroon). *Pakistan J. Biol. Sci.* 11(5), 719–725.

Tindall, H. D. 1983. *Vegetables in the Tropics*. Macmillan Press, London.

Vas, C. G., De Oliveira, D., and Ohashi, O. S. 1998. Pollination contribution to the production of cowpea in the Amazon. *Hort. Sci.* 33(7), 1119–1135.

Warnock, S. J. and Hagedorn, D. J. 1954. Stigma receptivity in peas (Pisum sativum B.). *Agron. Jour.* 46, 274.

Watt, E. E., Kueneman, E. A., and de Araújo, J. P. P. 1985. Achievements in breeding cowpeas in Latin America. In: Singh, S. R. and Rachie, K. O. (eds.) *Cowpea Research, Production and Utilization*. Wiley, New York, pp. 125–128.

Westphall, E. 1974. Pulses in Ethiopia: Their taxonomy and agricultural significance. *Field Crop Abstracts* 24, 213–232.

Williams, C. B. and Chambliss, O. L. 1980. Outcrossing in southernpea. *Hort. Science* 15, 174.

Witthöft, C. M. and Hefni, M. E. 2016. Folic acid and folates: Physiology and health effects In: Caballero, B., Finglas, P., and Toldra, F. (eds.) *The Encyclopedia of Food and Health*, Elsevier, Amsterdam, the Netherlands, pp. 724–730.

Wousla, E. N., Andargie, M., Pasquet, R. S., Mondon, M., Menez, V., Cochin, C., Paul, L., et al. 2019. Is bigger better? Apidae (Xylocopinae), megachilidae and cowpea (*Vigna unguiculata*) pollination. *Plant Breeding* 139(1), 156–166.

Wray, M. K., Mattila, H. R., and Seeley, T. D. 2011. Collective personalities in Honeybee colonies are linked to colony fitness. *Animal Behaviour* 81(3), 559–568.

Xiong, H., Shi, A., Mou, B., Qin, J., Motes, D., Lu, W., and Wu, D. 2016. Genetic diversity and population structure of cowpea (*Vigna unguiculata* L. Walp). *Public Library of Sciemce One* 11(8).

20 Pigeon Pea

20.1 PIGEON PEA, *CAJANUS CAJAN* (L.) MILLSP.

20.1.1 Pigeon Pea

20.1.1.1 *Cajanus cajan* (L.) Millsp., Family Leguminosae

By 2030, India's population is predicted to reach 1.68 billion, and the estimated pulse requirement for that year is 32 million tons, with an anticipated annual growth rate of 4.2%. Pigeon pea, a significant legume crop, is cultivated in various countries across the tropics and subtropics, including India, Malaysia, Indonesia, the Philippines, the Caribbean, East Africa, and West Africa. Commonly referred to as "Arhar" in Northern India, this protein-rich pulse crop is in high demand in Asia. In India, pigeon pea is the second most important kharif grain legume, following chickpea, and is primarily grown under rainfed conditions. Pigeon pea, also known as red gram (*Cajanus cajan* (L.) Millspaugh), holds great importance as a food legume in the semi-arid tropics of Asia and Africa. It plays a crucial role in sustainable farming systems for smallholder rainfed farmers.

Pigeon pea holds a prominent position in rainfed agriculture in India. It plays a vital role in various agro-ecologies across the country, often grown alongside cereals, pulses, oilseeds, and millets. Pigeon pea is the second most important pulse crop after chickpea, encompassing approximately 4.42 million hectares of land (representing around 14.5% of the area dedicated to pulses). It contributes to 2.86 million metric tons of production (equivalent to 16% of total pulse production), with a productivity of approximately 707 kg per hectare. Pigeon pea is primarily consumed as dry split dhal throughout India, while various parts of the plant are utilized for other purposes. Increasing the productivity of pigeon pea is crucial in India, particularly to address protein malnutrition, as it serves as a primary protein source for the predominantly vegetarian population. Unfortunately, pigeon pea productivity has remained stagnant and low over the past few decades, prompting scientists to explore innovative methods for crop improvement.

The kernels are nutritionally rich, containing 20–22% protein. Its cultivation would be able to provide 40–60 kg N/ha to the subsequently grown crop. The green manure can utilize both leaves and young stems, while fallen leaves can serve as mulch, aiding in improving the soil's ability to retain water. Its beneficial effect as an intercrop has also been reported. Sometimes, ground dry peas are mixed with wheat flour just to improve the flour quality and its protein content. Furthermore, the pigeon pea plays a crucial role in both India's import and export of pulses.

Pigeon pea (*Cajanus cajan* (L.) Millsp.) is a member of the family Leguminosae. It is an important pulse crop in the semi-arid tropics. Globally, pigeon pea is commercially grown in over a dozen countries on about 5.25 million hectares, and India accounts for over 80% of the total area and production. It contains 23–28.60% protein which is rich in amino acids, like cystine, tryptophan, and methionine. It also contains 52.50 to 59.90% starch by weight (Singh and Jambunathan 1982). Pigeon pea forms a large many-branched shrub of about 1 to 3 m in different agro-ecological zones of India (Saxena 2008). Pigeon pea is a shrub with a pubescent, many-branched structure, reaching heights of 4 to 10 ft or more. Its vibrant yellow or orange papilionaceous flowers give way to long-beaked pods that are green or brown in color, covered in hair, and contain four to seven seeds. These pods measure approximately 2 to 3 in in length and are around 1/2 in thick (Figure 20.1). The 1 1/2 to 5 in terminal and/or axillary raceme bears several flowers that resemble the common sweet pea in shape, but the color is yellow, red, or yellow-purple, with a brownish back. The filaments elongate in the bud, and pollen is shed the day before the flower opens. Whether actual pollination of the stigma occurs before the flower opens is not clear. According to Purseglove (1968), most flowers tend to bloom from 0100 to 1500 h, typically staying open for approximately 6 hr.

DOI: 10.1201/9781032656724-22

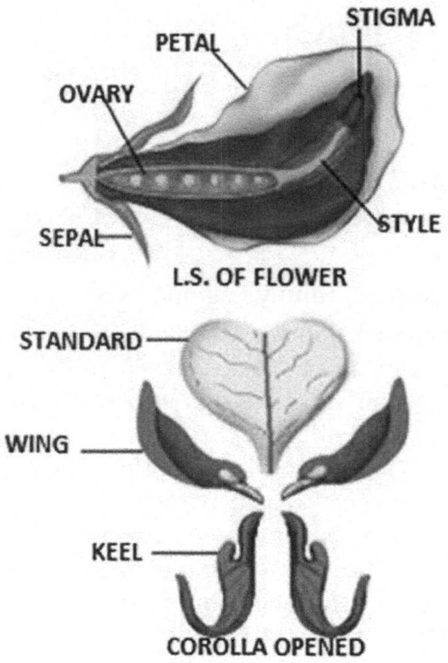

FIGURE 20.1 Flower of pigeon pea (*Cajanus cajan (L.) Millsp.*).

Various researchers have reported a wide range of natural cross-pollination rates, which can range from 0.1% to 48%. Factors such as location, cultivar, time of year, and pollinator populations contribute to the variable levels of cross-pollination (Abrol and Shankar 2015). The stigma remains receptive for up to 12 hr, while pollen germination occurs 24 hr after dehiscence, allowing foreign pollen to have an opportunity to germinate. Foreign pollen tubes tend to grow at a faster rate compared to self-pollen. Flowers play a crucial role in providing food resources such as pollen and nectar to sustain pollinators. Insects collect these substances and, in the process, transfer pollen from one flower to another. Cross-pollination facilitates gene recombination and the production of hybrid seeds, thereby capitalizing on the benefits of heterosis and hybrid vigor in crop seed production.

Krauss (1932) and Wilsie and Takahashi (1934) initially classified the pigeon pea as a predominantly self-pollinating species. However, subsequent observations by Wilsie and Takahashi (1934), Purseglove (1968), Matta and Dave (1931), Krauss (1927), Shaw (1932), and Abrams (1967) have revealed a significant level of cross-pollination, ranging from 5 to 40%. Although bees are known to visit the flowers in large numbers (Wilsie and Takahashi 1934), it remains unreported whether their purpose is to collect pollen, nectar, or both. Honeybees, megachilids (a type of solitary bee), and wasps are known for their involvement in cross-pollination, which contributes to improving the quantity and quality of pigeon pea yield. However, there are certain challenges associated with each of these pollinators.

Honeybees have a tendency to visit pigeon pea flowers more for nectar rather than pollen. This behavior leads to a lower rate of pollination because they learn to avoid the tripping apparatus, resulting in only around 1% of their visits, leading to successful pollination. However, when honeybees specifically search for pollen, they can be efficient pollinators.

For high pigeon pea yields, it is often necessary to have more bees in the field than required for honey production. This can be costly for beekeepers, since maintaining a larger number of bees requires additional resources and management. Wild bees, on the other hand, primarily visit pigeon pea flowers for pollen. While they can be faster in their movements and trip at a higher rate than pollen-collecting honeybees, they are usually not abundant enough on most fields to provide thorough pollination. Additionally, the reliability of wild bees is low because their populations fluctuate from season to season.

In order to address these challenges, it is desirable for farmers to cooperate with neighbors in their region, especially in acquiring beekeeping services. This collaboration can help overcome the limited availability of wild bees and ensure a more consistent and effective pollination process. However, since beekeeping services require effort and resources from beekeepers, remuneration is often necessary, especially when honey yields are low. It is worth noting that pigeon pea flowers are often cross-pollinated, with natural cross-pollination percentages ranging from 0.1% to 48.0%, as reported by various researchers (Howard et al. 1919).

Variable amounts of cross-pollination are associated with differences in locality, cultivar, and time of year and pollinator populations. Stigma is receptive up to 12 hr and pollen germination only 24 hr after dehiscence, thus creating a chance for foreign pollen to germinate. The foreign pollen tube grows much faster than self (Onim et al. 1979). Flowers provide subsistence for pollinators, as these produce pollen and nectar. Insects collect these food substances, and during the process, they transfer pollen from one flower to another. Cross-pollination provides chances of gene recombination and hybrid seed production, thus exploiting the heterosis and hybrid vigor in crop seed (Kevan and Baker 1983). Honeybees, megachilids, and wasps are known for their active involvement in cross-pollination and improving quantity and quality of pigeon pea yield.

Studies conducted at Hisar revealed that 24 insect species belonging to Hymenoptera (17), Lepidoptera (2), and Diptera (5) have been reported to visit pigeon pea blossoms. *Megachile* spp. was the most important insect visitor, accounting for 42.66% of the total visitors (Salunke 2003). Pollination studies in two cultivars, namely, Paras and Manak, under both caged and uncaged conditions, revealed that the open-pollinated crop yielded a significantly higher number of pods per plant. Specifically, in the Paras cultivar, 58.4 pods were produced with insect pollination compared to 42.1 without it. Similarly, in the Manak cultivar, 53.9 pods were produced with insect pollination compared to 38.7 without it. Thousand seed weight (g) was also significantly higher in open-pollinated crop (65.39 g and 65.06 g in Paras and Manak, respectively) than without insect-pollinated crop (63.46 g and 62.94 g in Paras and Manak, respectively. The vigor index showed a notable increase in open-pollinated crops, with values of 1,589.2 and 1,618.0 for Paras and Manak, respectively, compared to insect-pollinated crops, where the values were 1,118.6 and 1,009.6. However, no differences in percent germination and protein content of seed produced under open and without insect pollination were noticed (Table 20.1).

TABLE 20.1
Effect of Different Modes of Pollination on Yield Parameters in Two Cultivars of Pigeon Pea (*Cajanus cajan* (L.) Millsp.)

	Paras		**Manak**	
Yield Parameters	**Without Insect Pollination (Caged)**	**Open Pollination (Without cage)**	**Without Insect Pollination (Caged)**	**Open Pollination (Without cage)**
Number of pods per plant	42.1	58.4	38.7	53.9
Number of seeds per pod	3.9	4.2	3.7	4.0
1,000 seed weight (g)	63.46	65.39	62.94	65.06
Seed weight (g) per plant	11.4	15.3	10.1	13.1
Seed yield (kg) per plot (3 × 3m)	1.4	1.9, 1.3, 1.7		
Seed yield (q) per hectare	16.1	21.2	14.7	19.2
CD (P = 0.05)	**Cultivars**	**Type of pollination**	**Cultivars x Type of pollination**	
Number of pods per plant	NS	5.60	N.S.	
Number of seeds per pod	NS	N.S.	N.S.	
1,000 seed weight (g)	NS	0.75	N.S.	

An experiment was conducted in Jorhat (Assam) involving pigeon pea and *Apis cerana* bees. The experiment consisted of five treatments: open pollination (OP), without bee pollination (WBP), bee pollination 1 (BP1) with three hives per hectare, bee pollination 2 (BP2) with four hives per hectare, and bee pollination 3 (BP3) with five hives per hectare. The results of the experiment indicated that honeybee pollination had a significant impact on increasing the yield of pigeon pea. The plots that were pollinated by honeybees consistently yielded higher compared to those without bee pollination and open pollination. Notably, the use of five colonies of *A. cerana* was found to be sufficient for achieving the highest increase in pigeon pea yield compared to the other treatments (Table 20.2).

Rathi and Sihag (1993) found that number of pods per plant, number of seeds per pod, and weight of 100 seeds were similar for bee-pollinated and open-pollinated plots in *Cajanus cajan*. However, seed weight per plant, seed yield per plot, and seed yield per hectare were significantly different. All these parameters were significantly higher in plots caged with *Chalicodoma lanata* as compared to plots caged without bees (Table 20.3).

TABLE 20.2
Effect of Honeybee, *Apis cerana*, Pollination of Seed Set and Yields of Pigeon Pea

Treatments	Yield/Plot (kg)	Yield/Hectare (q)	Yield Increase over WBP (%)	Yield Increase over OP (%)
WBP	1.50±0.36	6.0±1.44	-	-
OP	2.0±0.44	8.0±1.62	33.3	-
BP1	2.5±0.37	10.0±2.04	66.6	25.0
BP2	3.0±0.19	12.0±0.74	100.0	50.0
BP3	3.5±0.29	15.0±1.18	150.0	87.5
SEd+	0.310	1.24	-	-
CD (*P* = 0.05)	**0.675**	**2.7**	**-**	**-**

TABLE 20.3
Yield Gain under Different Conditions of Pollination Treatments in *Cajanus cajan*

	Pollination Treatments			
Yield Parameters*	Plots Caged with *C. lanata*	Open-Pollinated Plots	Plots Caged without Bees	CD (*p* = 0.05)
No of pods per plant	82.43 + 1.33	77.73 + 0.80	67.13 + 0.93	5.02
No of seeds per pod[1]*	3.16 + 0.10	3.07 + 0.03	2.57 + 0.03	0.47
100 seed weight [1](g)*	6.40 + 0.06	5.97 + 0.06	5.47 + 0.03	0.47
Seed weight per plant[1] (g)*	16.69 + 0.57	14.20 + 0.16	9.42 + 0.23	2.46
Seed yield per plot [2](kg)	1.78 + 0.10	1.43 + 0.10	0.91 + 0.20	0.16
Seed yield per hectare [2] (kg × 100)*	19.76 + 0.70	15.90 + 0.70	10.13 + 1.90	1.91

[1] Mean + s.e. of three replicates.
[2] Mean + s.d. of three replicates
* $P < 0.05$: significant (ANOVA).
Source: Rathi and Sihag (1993).

20.2 PIGEON PEA (*CAJANUS CAJAN*)

20.2.1 Insect Visitors

Pigeon pea, a significant grain legume cultivated in tropical and subtropical regions, exhibits higher mineral content, ten times more fat, five times more vitamin A, and three times more vitamin C compared to regular peas (Madeley 1995). In Africa, the green pods and seeds are the most commonly utilized forms, although there is a growing trend in the consumption of dry seeds (Damaris 2007).

The flower of the pigeon pea plant exhibits protogynous hermaphroditism, as described by Choudhary et al. (2012). It has an early-morning blooming pattern and closes before noon, as noted by Songok et al. (2010). The flower produces both nectar and pollen, which serve as attractants for insects, as mentioned in studies conducted by Reddy et al. (2004), Martins in (2008), and Pando et al. (2011a).

X. olivacea (Mensah and Kudom 2011) has been found to have a positive impact on crop pollination due to its size. As stated by Pando et al. (2011b, 2013), this bee shakes flowers while collecting nectar and/or pollen, which aids in the release of pollen from anthers, leading to optimal stigma occupation. The bee can be effectively managed for pollination purposes (Roubik 1995). Moreover, Martins (2008) observed that *X. olivacea* carries abundant pollen loads, and its swift dispersal across large areas suggests efficient cross-pollination of pigeon pea flowers. William (1977) reported that among the 48 insect species visiting Cajanus flowers, Megachile species accounted for 31% of the total insect visitors, with other visitors including *A. dorsata*, *A. florea*, and *Xylocopa* spp.

In Punjab, India, Grewal et al. (1990) found that the pollination of *Cajanus cajan* flowers involved 11 bee species from four different families. Among these, the Megachilidae family consisted of 5 species, the Apidae had 3 species, the Xylocopidae had 2 species, and the Anthophoridae had 1 species. The most abundant species was Apis, accounting for 50.2% of the captured bees, followed by Megachilidae at 30.7%. At Hisar, India, Yogesh et al. (2009) observed a total of 21 insect species visiting *Cajanus cajan* flowers, which belonged to the Hymenoptera (14 species), Diptera (5 species), and Lepidoptera (2 species) orders. Out of these species, 12 collected both nectar and pollen, while the remaining 9 solely gathered nectar. The most frequent visitor to the flowers was *Megachile* sp., constituting 42.11% and 42.66% of the total visitors on respective cultivars. During a two-year study, Pando et al. (2011a) recorded 18 species of insects visiting *C. cajan* flowers. The most frequently encountered species were *Chalicodoma cincta cincta*, *Xylocopa calens*, *Apis mellifera adansonii*, *Megachile ituberculata*, *Camponotus flavomarginatus*, and *Xylocopa albiceps*. Additionally, apart from bees, wasps were also observed as potential predators.

In their 2012 study, Li et al. observed over 46 species of insects that visited pigeon pea (*Cajanus cajan* (L.) Millspaugh) flowers in open fields for pigeon pea production, excluding China. Among the various flower-visiting insects, *Megachile* spp. were found to be the primary pollinators. During the blooming stage, these flower-visiting insects visited each primary branch at a rate of 2.8 times per 10 min for the CGMS male sterile line, whereas they visited the CGMS restorer line at a rate of 5.2 times per 10 min. This indicates a preference for the flowers of the restorer line by the flower-visiting insects. Despite the significant difference in visitation frequency between the male sterile and restorer lines, this did not affect the dry seed yields significantly. Both the male sterile line (383.7 g per plant) and the restorer line (357.0 g per plant) yielded similar amounts of dry seeds, suggesting that sufficient pollen was transferred from the restorer line to the male sterile line by insect pollinators, even with fewer visits to the flowers of the male sterile line compared to the restorer line.

In Jorhat, Assam, India, Rajkhowa and Deka (2013) conducted a study on insect pollinators visiting pigeon pea flowers. The researchers identified several honeybee species, including *Apis cerana*, *A. mellifera*, *A. dorsata*, as well as other pollinators, such as *Xylocopa leucothorax*, *X. fenestrata*, *Polystis habrious*, *Vespa magnifica*, *Papilio demoleus*, *Musca domestica*, and *Monomorium indicum*. Among these, *A. cerana* was the most prevalent, accounting for 51.65% of the visitors, followed by *A. mellifera* (10.20%), *A. dorsata* (4.24%), *X. fenestrata* (5.32%), and *X. leucothorax* (4.28%). Additionally, *M. indicum* (9.72%) and *P. demoleus* (8.52%) were also frequent visitors. The researchers observed that various meteorological factors influenced the foraging activities of *A. cerana*, with temperature displaying a positive correlation with the honeybee's foraging behavior. Conversely, relative humidity exhibited a non-significant negative relationship. In another study by Tchuenguem et al. (2014) conducted in Nkolbisson, they investigated the impact of the carpenter bee *Xylocopa olivacea* on pod and seed yields of *Cajanus cajan* in Cameroon. The researchers found that the foraging activity of *X. olivacea* significantly increased the FR by 22.26%, the number of seeds per pod by 34.98%, and the percentage of normal seeds by 12.09% in both study years. As a recommendation, placing *X. olivacea* nests near *C. cajan* fields is suggested to enhance pod and seed production.

Tchuenguem Fohouo et al. (2014) reported that 19 species of insects were found visiting the flowers of *Cajanus cajan. Chalicodoma cincta cincta* was the most prominent one, followed by *X. olivacea*, *Apis mellifera adansonii*, *Megachile bituberculata*, and *Xylocopa albiceps*, and the bees were the most frequent floral visitors.

More et al. (2015) studied the effect of different modes of pollination, *viz*, open pollination, four-framed *Apis cerana*, six-framed *Apis mellifera*, and a colony of *A. florea* in seed production of pigeon pea and found that four-framed *Api indica* was superior for number of pods per plant and weight of 1,000 seeds, whereas six-framed *Apis mellifera* was superior for number of seeds per plant, yield in kilogram per hectare, and percent seed setting, and a colony of *Apis florea* was superior for percent crinkled seeds, while percent of germination was high under open pollination.

Vakaliya and Somani (2018) recorded 11 species of insect visiting Cajanus flowers. Eight species belonging to Apidae and three non-Apidae. The average count of Apidae insects was recorded as 1.18 ± 1.36 insects per 5 min per square meter, while for non-Apidae insects, it was 1.03 ± 1.04 insects per 5 min per square meter. Among the Apidae insects, the Trigona bee was observed in higher number, followed by Indian bee, Italian bee, and rock bee. The overall Apidae and non-Apidae visitors did not significantly fluctuate over time.

Pando et al. (2011a) reported that the foraging activity of *X. olivacea* resulted in a significant increase in the fruiting rate by 22.26%, as well as the number of seeds per pod by 34.98% and the percentage of normal seeds by 12.09%. Placing *X. olivacea* nests close to *C. cajan* fields is recommended to increase pod and seed production in Cameroon

Ichpal et al. (2017) recorded 15 species of insects belonging to five families, seven genera, and three orders visiting *Cajanus cajan* flowers. The insect belonging to order Hymenoptera were the most abundant. Of all the bees, megachilid bees were the most abundant and compared more than 50% of the total flower-visiting insects, followed by honeybees, 39.4%, and anthophorid bees, 6.6%. The insects belonging to order Lepidoptera and Diptera were reported in very small numbers at interrupted hours. The studies revealed that bee pollination was superior to all other treatments in improving qualitative and quantitative parameters of *C. cajan* crops in the order: bee pollination > open pollination > hand-pollination > self-pollination > Carbaryl 50 WP 0.1% + Open pollination > Imidacloprid 17.8 SL 0.05% + open pollination.

Singh (2016) reported that seven insect species, including *Megachile* spp., *X. tenuiscapa, A. zonata*, and *Nomia* sp., were the major pollinators of pigeon pea. Similarly, Chaudhary and Jain (1978) reported that *Megachile lamata* was the important pollinator of pigeon pea throughout India, Sri Lanka, and Burma. Williams (1977) reported that honeybees were the major pollinators of pigeon pea (*Cajanus cajanus*. L). Similar observations were also made by Brar et al. (1992) in pigeon pea.

Padhy et al. (2019) observed 11 species of pollinators visiting *Cajanus cajan* flowers, which included Apidae (45.5%), Megachilidae (36.4%), and Vespidae (18.2%), belonging to order Hymenoptera, and few insects belonging to Dipterans and Lepidopterans. Species-wise, diversity indicated that *Megachile lanata* Fab. was the most dominant one (29.1%), followed by *Megachile disjuncta* Fab. (20.5%), *Tetragonula iridipennis* Smith (14.2%), *Apis cerana indica* Fab. (10.95%) *Megachile bicolor* Fab. (10.7%), *Xylocopa latipes* Drury (8.5%), *Megachile hera* Bingham (2.1%), *Eumenes* spp. (1.7%), *Xylocopa aestuans* Lin. (1.2%), *Vespa tropica* Lin. (0.8%), and *Apis dorsata* Fab. (0.2%).

Julianna and Rufus (2010) reported 18 species of pollinators visit the flowers of *Cajanus cajan*; *C. c. cincta* is the most abundant species (28.99%), followed by *Xylocopa calens* (22.04%), *Apis mellifera adansonii* (11.41%), and *Megachile bituberculata* (09.55%); 6 species have a frequency of less then 2%, and 8 species have a frequency between 2 and 8%.

During the flowering period of pigeon pea, Rashmi et al. (2010) observed a total of 30 insect pollinator species. Out of these, 21 species were classified under the order Hymenoptera, 4 under Lepidoptera, 2 under Diptera, and 1 each under Coleoptera, Hemiptera, and Thysanoptera. Effect of the different modes of pollination in seed production of CMS line of pigeon pea, such as open pollination, *Apis cerana*, *A. florea*, and *Trigona iridipennis* pollination on yield parameters of pigeon pea revealed that the maximum pod set, pod length, pod volume, 100 pod weight, number of seeds per pod, seed size, seed weight, and germination percent were recorded in open pollination, followed by *A. florea*, *A. cerana*, and *T. iridipennis* pollination.

20.2.2 Fruit/Seed Setting

In Nkolbisson, Tchuenguem Fohouo et al. (2014) conducted a study examining the effects of the carpenter bee, *Xylocopa olivacea* Fabricius 1787, on the yield of *Cajanus cajan* (L.) Millsp. 1900, specifically on pod and seed production. Their findings indicated that *Xylocopa olivacea* primarily gathered nectar and pollen, with a foraging speed of 11.75 flowers per minute. The results showed that unprotected inflorescences had significantly higher fruiting rates (Fr), a greater number of seeds per pod, and a higher percentage of normal seeds compared to inflorescences protected from insects. The foraging activities of *X. olivacea* resulted in a notable increase in Fr by 22.26%, as well as a 34.98% increase in the number of seeds per pod, and a 12.09% increase in the percentage of normal seeds over the course of the two-year study. Consequently, it is recommended to position *X. olivacea* nests near C. cajan fields in Cameroon to enhance pod and seed production.

The size of *X. olivacea* (Mensah and Kudom 2011) has been reported to play a positive role in crop pollination. According to Pando et al. (2011b, 2013), when collecting nectar and/or pollen, *X. olivacea* shakes flowers, and this movement could facilitate the liberation of pollen by anthers for the optimal occupation of the stigma. This bee can be managed for pollination (Roubik 1995). Martins (2008) reported that *X. olivacea* bees carried a significant amount of pollen. The bee's ability to travel quickly over vast distances indicates its potential for regular and efficient cross-pollination of pigeon pea flowers.

Pando et al. (2011a) studied the impact of the Megachile bee, *Chalicodoma cincta cincta*, on pod and seed set of *Cajanus cajan* L. Mill sp. The treatments included unlimited floral access by all visitors, bagged flowers to deny all visits, and limited visits by *C. c. cincta* only. *Chalicodoma cincta cincta* mainly foraged for nectar and pollen resource. The mean foraging speed was 10.33 flowers per minute. Flowers visited by *C. c. cincta* had higher fruiting rate compared with the others, while those bagged had the lowest (Table 20.4). In addition, seed formation was higher in *C. c. cincta*–visited flowers compared with all others. The findings indicate that even in regular conditions, this particular crop faces a shortage of pollination. This is evident as the flowers visited by *C. c. cincta* displayed greater yields compared to those accessible to all visitors without limitations. Moreover, inflorescences that were not protected from insects exhibited significantly higher fruiting rates, greater numbers of seeds per pod, and a higher percentage of normal seeds compared to the protected ones. *Chalicodoma cincta cincta* foraging resulted to a significant increment of the fruiting rate by 19.65%, as well as the number of seeds per pod by 24.33%, and the percentage of normal seeds by 11.52%. Conservation of *C. c. cincta* nests close to *C. cajan* fields could be recommended to improve pod and seeds production in the region.

TABLE 20.4
Fruiting Rate, Mean Number of Seeds Yield per Pod, and Percentage of Normal Seeds According to the Treatments of *C. cajan*

Treatments	Years	Flowers	Pods	Fruiting/ Rate	Seeds/Pod Mean	Seeds/Pod SD	Total Seeds	Normal Seeds	% Normal Seeds
1. Unlimited visits	2008	3,188	2,801	87.86	5.93	0.72	15,271	14,503	94.97
2. Bagged flowers	2008	3,154	2,544	80.66	4.96	0.65	11,707	10,191	87.05
3. Unlimited visits	2009	3,313	3,021	91.19	5.99	0.59	16,818	15,971	94.96
4. Bagged flowers	2009	3,118	2,404	77.10	4.75	0.72	11,211	9,653	86.10
5. *C. c. cincta* flowers	2008	67	66	98.51	6.44	0.50	425	413	96.94
6. *C. c. cincta* flowers	2009	91	89	97.80	6.39	0.59	569	562	98.77

Source: Pando et al. (2011).

REFERENCES

Abrams, R. 1967. Studies on natural cross-pollination in pigeon peas (Cajanus cajan). *Puerto Rico Univ. Jour. Agr.* 51(1), 1–3.

Abrol, D. P. and Shankar, U. 2015. Role of pollination in pulses. *Advances in Pollen Spore Research* 33, 101–103.

Brar, H. S., Jhaji, H. S., and Gatoria, G. S. 1992. Abundance and activity of the bee visitors of Pigeonpea (*Cajanus cajan* (L.) Mill sp.) and role of *Apis mellifera* L. in its pollination. *Indian Bee Journal* 54(1–4), 76–80.

Chaudhary, J. P. and Jain, J. L. 1978. Nesting and foraging behaviour of a mason bee Megachile lanata Lepel (Megachilidae: Hymenoptera). *Indian Journal of Entomology* 40, 405–411.

Choudhary, A. K., Iquebal, M. A., and Nadarajan, N. 2012. Protogyny is an attractive option over emasculation for hybridization in pigeonpea. *Journal of Breeding and Genetics* 44, 138–148.

Damaris, A. O. 2007. The potential of pigeonpea (Cajanus cajan (L.) Millsp.) in Africa. *Natural Resources Forum* 31, 297–305.

Grewal, G. S., Singh, G., and Kandoria, J. L. 1990. Insect pollinators of pigeon pea around Ludhiana. *Indian J. Agric. Sci.* 60, 227–228.

Howard, A., Howard, G. C., and Khan, A. R. 1919. Studies in the pollination of Indian Crops. *I. Memoirs.* Department of Indian Crops, India (Botanical Series) 10, 195–200.

Ichpal, S., Shankar, U., Abrol, D. P., and Mondal, A. 2017. Diversity of insect pollinators associated with Pigeonpea, *Cajanus cajan* L. Mill sp. and their impact on crop production. *Int. J. Curr. Microbiol. App. Sci.* 6(9), 528–535. https://doi.org/10.20546/ijcmas.2017.609.063

Julianna, K. T. and Rufus, I. 2010. Weather during affects pollination and yield of highbush blueberry. *Journal of Economic Entomology* 103(3), 557–562.

Kevan, P. and Baker, H. G. 1983. Insects as flower visitors and pollinators. *Ann. Rev. Entomol.* 28, 407–453.

Krauss, F. G. 1927. Improvement of the pigeonpea, genetic analysis of cajanus indicus and the creation of new varieties through hybridization and selection. *Jour. Hered.* 18, 227–232.

Krauss, F. G. 1932. The pigeonpea (cajanus indicus) its improvement, culture and utilization in Hawaii. *Hawaii Agr. Expt. Sta. Bul.* 64, 46 pp.

Madeley, J. 1995. Pea with a differences. *Appropr. Technol.* 22(1), 5.

Martins, D. J. 2008. *Pigeon Pea (Cajanus cajan) in Mwanza, Western Tanzania.* Initial Survey of Good Pollination Practices. Food and Agriculture Organization of the United Nations, Rome.

Matta, D. N. and Dave, B. E. 1931. Studies in Cajanus indicus. *India Dept. Agr., Mem. Bot. Ser.* 19, 25 pp.

Mensah, B. and Kudom, A. 2011. Foraging dynamics and pollination efficiency of *Apismellifera* and *Xylocopa olivacea* on *Luffa aegyptiaca* Mill (Cucurbitaceae) in Southern Ghana. *Journal of Pollination Ecology,* 4, 34–38.

More, B. C., Latpate, C. B., and Jadhav, M. S. 2015. Pollinators diversity with special reference to role of honeybees in seed production of pigeon pea, Cajanus cajan L. *J. Ent. Res.* 39(2), 119–122.

Onim, J. F. M., Pathak, R. S., and Van-Eijnatten, C. L. M. 1979. Influence of insect pollinators on the degree of outcrossing in pigeon pea in Kenya. Proceedings of IV International Symposium on Pollination. 1978, Maryland Agricultural Experiment Station, Maryland, pp. 211–218.

Padhy, D., Satapathy, C. R., and Mohapatra, R. N. 2019. Insect fauna associated with pigeon pea (*Cajanus cajan* L.) in Odisha. *Journal of Pharmacognosy and Phytochemistry* 8(4), 766–768.

Pando, J. B., Tchuenguem Fohouo, F.-N., and Tamesse, J. L. 2011a. Pollination and yield responses of pigeon pea (Cajanus cajan L. Millsp.) to the foraging activity of *Chalicodoma cincta cincta* (Hymenoptera: Megachilidae) in Yaoundé´ (Cameroon). *Journal of Animal and Plant Science* 11, 1346–1357.

Pando, J. B., Tchuenguem Fohouo, F.-N., and Tamesse, J. L. 2011b. Foraging and pollination behaviour of Xylocopa calens Lepeletier (Hymenoptera: Apidae) on *Phaseolus coccineus* L. (Fabaceae) flowers at Yaoundé´ (Cameroon). *Entomological Research* 41, 185–193.

Pando, J. B., Tchuenguem Fohouo, F.-N., and Tamesse, J. L. 2013. Activité´ de butinage et de pollinisation de Xylocopa olivacea Fabricius 1787 (Hymenoptera: Apidae) sur les fleurs de *Vigna unguiculata* (L.) Walp. 1843 (Fabaceae) a' Y.

Purseglove, J. W. 1968. *Trop. Crops: Dicotyl.* Longman, London.

Rajkhowa D and Deka, M. K. 2013. Insect foragers and foraging behaviour of honey bee, *Apis cerana* on pigeon pea. *Indian Journal of Entomology* 75(3), 232–235.

Rashmi, T., Kuberappa, G. C., and Thirumalaraju, G. T. 2010. Pollinators diversity with special reference to role of honeybees in seed production of CMS line of pigeonpea, *Cajanuscajan* L. *Mysore Journal of Agricultural Sciences* 44(2), 295–299.

Rathi, A. and Sihag, R. C. 1993. Differential attractiveness of *Brassica campestris* L and *Cajanus cajan* L (Millsp) to two honey bee species. In: *Proceedings of an International Symposium on Pollination in Tropics, Bangalore,* Oxford & IBH Publishing Co., New Delhi, India, pp. 113–119.

Reddy, L. J., Chandra, S., Pooni, H., and Bramel, P. J. 2004. Rate of outcrossing in pigeonpea under intercropped conditions. In: Bramel, P. J. (ed.) *Assessing the Risk of Losses in Biodiversity in Traditional Cropping Systems: A Case Study of Pigeonpea in Andhra Pradesh.* International Crops Research Institute for the Semi-Arid Tropics, Andhra Pradesh, India, pp. 133–141.

Roubik, D. W. 1995. Pollination of cultivated plants in the tropics. *FAO Agricultural Services Bulletin* 118, 1–198.

Salunke, S. 2003. Studies on insect pollination in Pigeonpea (Cajanus Cajan (L.) Millsp). MSc thesis Entomology, CCSHAU, Hisar, 77 pp.

Saxena, K. B. 2008. Genetic improvement of pigeonpea: A review. *Tropical Plant Biology* 1, 159–178.

Shaw, F. J. F. 1932. The breeding of Cicer and Cajanus. *Imp. Burl. Plant Genet., Herbage Plants Bul.* 7, 14–16.

Singh, A. K. 2016. Pollinating efficiency of native bee pollinators of Pigeonpea (*Cajanus cajan*) in Nagaland. *Russian Journal of Ecology* 47(3), 310–314.

Singh, U. and Jambunathan, R. 1982. Distribution of seed protein fractions and amino acids in different anatomical parts of chickpea (*Cicer arietinum* L.) and pigeonpea (*Cajanus cajan* L.). *Qual. Plant Plant Foods Hum. Nutr.* 31, 347–354.

Songok, S., Ferguson, M., Muigai, A. W., and Silim, S. 2010. Genetic diversity in pigeon pea [Cajanus cajan (L.) Millsp.] landraces as revealed by simple sequence repeat markers. *African Journal of Biotechnology* 9, 3231–3241.

Tchuenguem Fohouo, F.-N., Pando, J. B., and Tamesse, J. L. 2014. Pollination efficiency of Xylocopa olivacea (Hymenoptera: Apidae) on Cajanus cajan (Fabaceae) flowers at Yaoundé, Cameroon. *International Journal of Tropical Insect Science* 34, 138–148.

Vakaliya, M. A. and Somani, M. D. 2018. Insect pollinators of pigeonpea, [*Cajanas cajan* L.] crops in middle Gujarat condition. *Trends in Biosciences* 11(8), 1838–1843.

Williams, I. H. 1977. Behaviour of insects foraging on Pigeonpea (*Cajanus cajan* L. Millsp.) in India. *Tropical Agriculture* 54, 353–363.

Wilsie, C. P. and Takahashi, M. 1934. Natural crossing in the pigeonpea. *Jour. Agr. Res.* 49, 923–927.

Yogesh, K., Salunke, S., and Sharma, S. K. 2009. Abundance of insect visitors and nectar sugar production in two cultivars of pigeonpea [*Cajanus cajan* (L.) millsp.] flowers at Hisar (Haryana). *Annals of Biology* 25(1), 53–58.

21 Chickpea

21.1 INTRODUCTION

One of the ways to achieve food security is through focused and practical work by reviving and expanding research strategies and activities beyond traditional stars of food grains, such as wheat, rice, barley, etc., to crops like pulses. The production of pulse crops will play an important role in overcoming food insecurity and malnutrition problems. Chickpea, *Cicer arietinum* L. (Fabaceae), is a self-pollinated annual grain legume crop. Chickpea serves as a major source of protein in the diet. Chickpea helps maintain soil fertility through biological nitrogen fixation. It obtains 80% of its nitrogen (N) needs through symbiotic nitrogen fixation, and it has the capacity to fix up to 140 kg N ha/1 from the atmosphere. It leaves a substantial amount of residual nitrogen for subsequent crops and adds plenty of organic matter to maintain and improve soil health. Taproot system in chickpea can help overcome drought periods by extracting water from deeper layers of the soil. Dal (split chickpea without seed coat) and flour are used extensively in India as a thick soup for making breads. Sprouted seeds are also used in salads. Chickpea seeds may be milled or directly used for animal feed. Leaves yield an indigo-like dye. The acidic secretions released by the leaves can serve both medicinal purposes and be utilized as vinegar. India holds the leading position as the primary producer, responsible for 64% of the overall chickpea production.

21.2 GENERAL ACCOUNT OF PULSES AND CHICKPEA

Pulse crops contribute around 25–28% of the total global production (Goyal et al. 2018). While developing nations like India have successfully achieved self-sufficiency in food grain production, they have unfortunately faced significant challenges in attaining the same level of self-sufficiency in pulses. This problem could be solved if attempts are made for improvement in indigenous production levels. While efforts have been made by increasing the area under pulses, more important is to increase production by exploiting yield potential of existing varieties through genetic manipulations.

Chickpea requires cool climate and high temperature for growth and maturity. It performs well when grown on sandy/loam soils associated to good drainage system. Being a self-pollinating crop, mutation breeding could be rewarding for broadening the genetic base of total plant yield.

Chickpea, *Cicer arietinum* L., is an important legume pulse crop grown in dry and semi-dry areas of the world. This crop has been domesticated since 8500 BC (Tanno and Willcox 2006). It ranks third in world production after dry beans (*Phaseolus vulgaris* L.) and dry peas (*Pisum sativum* L.) (FAOSTAT 2013). This crop is grown in more than 50 countries of the world. The major countries include Afghanistan, Australia, Chile, Ethiopia, India, Iran, Mexico, Pakistan, and Spain (Upadhyaya et al. 2008; Kassie et al. 2009; FAO 2010). Chickpeas serve as a significant source of vegetable protein, phosphorous, and calcium for individuals with limited financial means who are unable to purchase meat (FAO 2005–2006; ICARDA 2017). They are consumed by humans in both dry and green seed forms and are also utilized as a forage crop for livestock (Dana et al. 2000). During the last few decades, its yield has not increased substantially. There could be many biotic and abiotic factors that affect yield adversely (Singh et al. 1994). This may include low yield capability of the varieties grown, as reported by Paolini et al. (2006). Chickpea has one of the highest nutritional compositions of any dry edible legume (Mallikarjuna 2001). Seeds of chickpea are considered antibilious and also yield 21% starch (Duke 1981).

DOI: 10.1201/9781032656724-23

21.3 BIOLOGY OF *CICER ARIETINUM L.*

Chickpea is an important source of protein for millions of people in developing countries, particularly in South Asia, who are largely vegetarian in food habits. In India, the major chickpea production areas are in Madhya Pradesh, Rajasthan, Uttar Pradesh, Haryana, Maharashtra, Telangana, Andhra Pradesh, Karnataka, and Punjab. It is a hardy crop and can be grown on marginal lands, where the high-input crops fail to give economic returns

21.4 GENERAL PLANT CHARACTERISTICS

Chickpea is an herbaceous annual, and the plant height ranges from 30 to 70 cm. It has a taproot system which is deep and strong. The lateral roots develop symbiotic nodules with the *Rhizobium* bacteria, which are capable of fixing atmospheric nitrogen in plant-usable form. The nodules (slightly flattened fan-like lobes) are visible about one month after seedling emergence and are confined to the top 15 cm of the soil surface. The leaves are imparipinnate with serrated leaflets and arise alternatively from the third node, varying from 5 to 17. However, some varieties have simple leaves. The entire surface of the plant, except the corolla, is densely covered with fine hairs, known as trichomes, which are glandular and secrete a highly acidic exudate containing malic, oxalic, and citric acids. These acids play an important role in protecting the plant against insect pests (Sharma et al. 2008; Narayanamma et al. 2008). Five growth habits, based on angle of branches from the vertical stem, are: erect, semi-erect, semi-spreading, spreading, and prostrate. The erect and semi-erect varieties are suitable for mechanical harvesting.

21.5 FLORAL BIOLOGY

Chickpea flowers are complete, are bisexual, and have a papilionaceous corolla. They are white, pink, purple, or blue in color. In colored flowers, the peduncle may be of different color, the floral part purplish and the racemal green. The calyx consists of five sepals, and the corolla is comprised of standard two wings and two lower petals that lie inside the wings, which are united at the lower margins to form a keel. There are ten stamens, which surround the pistil. The anthers open lengthwise and shed the pollen directly onto the stigma (Figure 21.1).

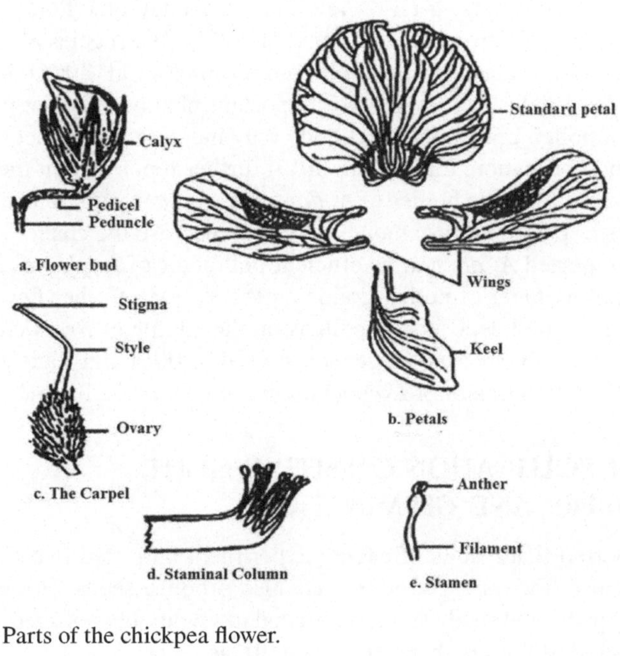

FIGURE 21.1 (A–E) Parts of the chickpea flower.

21.6 POLLINATION AND FERTILIZATION

Chickpea is a diploid and predominantly self-pollinated legume, but cross-pollination by insects sometimes occurs (Purseglove 1968). Pollination occurs one to two days before flower buds open (van der Maesen 1972). Insects are reported to visit chickpea flowers but rarely mediate cross-pollination. Honeybees (*Apis* spp.), bumblebees (*Bombus* spp.), and wild bees are attracted to open chickpea flowers (van der Maesen 1972; Tayyar et al. 1996). There is no evidence that insect pollination increases seed production in chickpea (Klein et al. 2007). The natural rate of pod setting has been reported between 18 and 59%. Singh and Auckland (1975) reported 24% pod setting when artificial pollination was carried out on the same day of emasculation, and 15% pod setting when it was done one day after emasculation.

Chickpea is a self-pollinated crop, but cross-pollination may occur to some extent (Arain 2012). Pollinators can play a significant role to increase crop yield by ensuring proper pollination. There are several studies indicating increased crop yield by insect pollinators in self-compatible crops. Effect of cross-pollination has also been studied in pulse crops (Aouar-sadli et al. 2008; Etcheverry et al. 2008; Fohouo et al. 2009), but such studies are lacking. There is also variation in plant–pollinator interaction among populations of different regions (Thompson 1994, 1997; Ollerton and Louise 2002). For example, Fausto et al. (2001) reported variations in flower density, and pollinators' abundance and visitation rates among populations of *Clarkia xantiana* A. Gray (Onagraceae). In another study, variations in the pollinator species of *Phaseolus coccineus* L. were reported by Búrquez and Sarukhán (1980), where hummingbirds were found at higher elevation, and hummingbirds and bumblebees at lower elevation. Therefore, plant–pollinator interaction in a specific location should be studied to find out exact mechanism. In the current study, we hypothesized that bees play a significant role in the pollination biology of *C. arietinum*.

Latif et al. (2019) reported *A. florea, A. dorsata, Amigella* sp., and *E. aeneus* as the most efficient pollinators of chickpea based on their abundance, visitation frequency, visitation rate, and pollen load. The open-pollinated flowers that were mostly visited by insects have the highest capsule weight, seed weight, seed numbers per pod, and seed germination (%). They further found that *A. dorsta* and *A. florea* were the most abundant species among all floral visitors.

Two floral visitors, *A. florea* and *E. aeneus*, visited the maximum number of flowers, and their visitation frequencies were significantly the highest among the 14 insect species captured on chickpea during both years. In general, visitation rate and frequency of a floral visitor are important predictors of its pollinating effectiveness (Zameer et al. 2017). The species with high visitation rates and frequencies are considered as effective pollinators (Singh et al. 2006), but this is not always true. Pollen load on the floral visitor is another important factor to predict pollination efficiency. Maximum numbers of pollen grains on insect body may increase the chances of pollen deposition on the other flower and pollination. Latif et al. (2019) further reported that the maximum numbers of pollen grains recovered from the bodies of *A. florea*, *A. dorsata*, and *Amegilla* sp. indicated their greater ability to transfer pollen grains among flowers to increase the chances of cross-pollination. Zameer et al. (2017) reported *A. dorsata* as efficient pollinator of *Raphanus sativus* L. because it deposited the maximum number of pollen grains on the stigma of other flowers. Ali et al. (2011) found that *E. aeneus* is a good depositor of pollens on the stigma of *Brassica napus* L. flowers as compared to other syrphid fly species. Hogendoorn et al. (2006) and Latif et al. (2016) reported *Amegilla* sp. as an efficient pollinator of *Lycopersicon esculentum* Mill. And *C. aphylla*.

21.7 EFFECTS OF POLLINATION ON SEED WEIGHT, SEED NUMBERS, AND GERMINATION

Latif et al (2019) reported that results of caging experiment indicated that *C. arietinum* is a self-compatible plant because flowers caged with bags also produce seeds. However, open-pollinated flowers have higher capsule and seed weight, produced more numbers of seed, and have better seed germination as compared to flowers that were caged (Table 21.1).

TABLE 21.1
Mean (± SE) Capsule Weight, Seed Weight, Numbers of Seed per Pod and Germination (%) in Caged and Open-Pollinated Flowers of Chickpea

Pollination	Capsule Weight (g)	Seed Weight (g)	No. of Seeds/Pod	Germination (%)
Caged	0.17 ± 0.01	0.11 ± 0.01	1.42 ± 0.07	65.50 ± 2.77
Open	0.27 ± 0.01	0.18 ± 0.01	1.67 ± 0.07	83.0 ± 1.96

Source: Latif et al. (2019).

This shows the effect of cross-pollination done by floral visitors. Open flowers were frequently visited by pollinators, mostly by *A. florea* and *E. aeneus*, which led to better cross-pollination of chickpea flowers. In previous studies, cross-pollination by insects resulted in higher crop yield. For example, Free (1966) reported that the number of seeds per plant were less (15.1) in the absence of pollinated insects as compared to those in visited by insects. In conclusion, chickpea flowers are visited by a variety of insects, including bees, wasps, flies, moth, and butterflies, but all were not effective pollinators. Bees were the most abundant floral visitors as compared to all other groups. Based on the highest values of visitation rate, visitation frequency, and pollen load, *A. florea* and *E. aeneus* were considered as the efficient pollinators of chickpea. Insect visitation on the open-pollinated flowers caused cross-pollination, ultimately leading to higher crop yield.

21.8 CHICKPEA CULTIVATION IN INDIA

As chickpea has a deep taproot, which enhances its capacity to withstand drought conditions, it is usually suited to areas having relatively cooler climatic conditions and low rainfall. It yields best when grown on sandy loam soils having an appropriate drainage system, as it is very sensitive to excess water. Chickpea production is also affected by excessive cold and heat. In India, chickpea is a winter season crop, but severe cold and frost are injurious. Frost at the time of flowering results in failure of flowers to develop into seeds or kills the seeds inside the pod. It is generally grown under rainfed conditions but gives good returns under irrigated conditions. Excessive rains soon after sowing or at flowering and fruiting or hailstorms at ripening cause heavy crop loss. It is best suited to areas having moderate rainfall (60–90 cm per annum). Chickpea is sown between September and November in India. The maturity of Desi-type chickpea is 95–105 days, and of the Kabuli types, 100–110 days. In India, it is harvested in February, March, and April, depending on the region where it is grown. Chickpea is often cultivated as a sole crop, but at times, it is also grown in rotation with other crops, such as sorghum, pearl millet, wheat, and coriander.

REFERENCES

Ali, M., Saeed, S., Sajjad, A., and Whittington, A. 2011. In search of the best pollinators for canola (*Brassica napus* L.) production in Pakistan. *J. Appl. Ent. Zool.* 46, 353–361.

Aouar-sadli, M., Louadi, K., and Doumandji, S. E. 2008. Pollination of the broad bean (*Vicia faba* L.var. major) (Fabaceae) by wild bees and honey bees (Hymenoptera: Apoidea) and its impact on the seed production in the Tizi-Ouzou area (Algeria). *Afr. J. Agric. Res.* 3, 266–272.

Arain GN. 2012. Chick peas cultivation in Pakistan. Available at www.valleyirrigationpakistan.com/wp-content/uploads/2012/09/CHICK-

Búrquez, J. A., Sarukhán, J. 1980. Biologiá floral de poblaciones silvestres y cultivadas de Phaseolus coccineus L. I. Relaciones planta-polinizador. *Boletín de la Sociedad Botanica de México* 39, 5–27.

Dana, N., Shenkoru, T., and Tegegne, A. 2000. Growth rates and testicular characteristics of Ethiopian highland sheep offered chickpea haulm supplemented with incremental level of Leucaena leucocephala leaf hay. *Livest. Prod. Sci.* 65, 209–217.

Duke, J. A. 1981. *Handbook of Legumes of World Economic Importance*. Plenum Press, New York, pp. 52–57.

Etcheverry, A. V., Aleman, M. M., and Fleming, T. F. 2008. Flower morphology, pollination biology and mating system of the complex flower of *Vigna caracalla* (Fabaceae: Papilionoideae). *Ann. Bot.* 102, 305–316.

FAO. 2005–2006. FAO statistical yearbook. Available at www.faostat.fao.org/ (accessed 17 February 2008)

FAO. 2010. Production year book, 2009. Food and Agriculture Organization of the United Nations (FAO), Rome, Italy. Available at http://faostat.fao.org/

FAOSTAT. 2013. Available at http://faostat3.fao.org/faostatgateway/go/to/download/Q/QC/E

Fausto, J. A., Eckhart, V., and Gebber, M. A. 2001. Reproductive assurance and the evolutionary ecology of self-pollination in *Clarkia xantiana* (Onagraceae). *Am. J. Bot.* 88, 1794–1800.

Fohouo, F. N. T., Ngakou, A., and Kengni, B. S. 2009. Pollination and yield responses of cowpea (*Vigna unguiculata* L. Walp.) to the foraging activity of *Apis mellifera* adansonii (Hymenoptera: Apidae) at Ngaoundéré (Cameroon). *Afr. J. Biotechnol.* 8, 1988–1996.

Free, J. B. 1966. The pollination requirements of broad beans and field beans (*Vicia faba*). *Jl. Agric. Sci.* 66, 395–398.

Goyal, M., Singh, J., Kumar, P., and Sirohi, A. 2018. *Pulse Improvement Physiological, Molecular and Genetic Perspectives*. Wani, S. H. and Jain, M (eds.). Springer Nature, Switzerland AG, pp. 1–11.

Hogendoorn, K., Gross, C. L., Sedgley, M., and Keller, M. A. 2006. Increased tomato yield through pollination by native Australian Amegilla chlorocyanea (Hymenoptera: Anthophoridae). *J. Econ. Entomol.* 99, 828–833.

ICARDA. 2017. Chickpea. Available at www.icarda.org/crop/chickpeas

Kassie, M., Shiferaw, B., Asfaw, S., Abate, T., Muricho, G., Ferede, S., Eshete, M., and Assefa, K. 2009. Current situation and future outlooks of the chickpea sub-sector in Ethiopia. ICRISAT and EIAR. Available at www.icrisat.org/tropicallegumesII/pdfs/Current_Situation.pdf

Klein, A.-M., Vaissière, B. E., Cane, J. H., Steffan-Dewenter, I., Cunningham, S. A., Kremen, C., and Tscharntke, T. 2007. Importance of pollinators in changing landscapes for world crops. *Proceedings of the Royal Society B: Biological Sciences* 274, 303–313.

Latif, A., Malik, S. A., Alvi, A. M., Saeed, Q., Saeed, S., Shuja, M. A., Ejaz, M., and Iqbal, N. 2016. Pollinators' community of Capparis aphylla at Dera Ghazi Khan, Punjab, Pakistan. *J. Entomol. Zool. Studies* 5, 967–969.

Latif, A., Malik, S. A., Saeed, S., Iqbal, N., Saeed, Q., Khan, K. A., Ting, C., Hamed, A., and Ghramh, H. A. 2019. Diversity of pollinators and their role in the pollination biology of chickpea, Cicer arietinum L. (Fabaceae). *Journal of Asia-Pacific Entomology* 22, 597–601.

Mallikarjuna, N. 2001. Prospects of using *Cicer canariense* for chickpea improvement. *International Chickpea and Pigeonpea Newsletter* 8, 23–24.

Narayanamma, V. L., Sharma, H. C., Gowda, C. L. L., and Sriramulu, M. 2008. Incorporation of lyophilized leaves and pods into artificial diets to assess the antibiosis component of resistance to pod borer, *Helicoverpa armigera* (Hubner) (Lepidoptera: Noctuidae) in chickpea. *Int. J. Trop. Insect Sci.* 27(3/4), 191–198.

Ollerton, J. and Louise, C. 2002. Latitudinal trends in plant-pollinator interaction: Are tropical plantsmore specialized? *Oikos.* 98, 340–350. https://doi.org/10.1034/j.1600–0706.2002.980215.x

Paolini, R., Faustini, F., Saccardo, F., and Crino, P. 2006. Competitive interactions between chickpea genotypes and weeds. *Weed Res.* 46, 335–344.

Purseglove, J. W. 1968. *Cicer arietinum* L. In: *Tropical Crops. Dicotyledons.* Longman Group Limited, London, pp. 246–250.

Sharma, H. C., Varshney, R. K., Gaur, P. M., and Gowda, C. L. L. 2008. Potential for using morphological, biochemical, and molecular markers for resistance to insect pests in grain legumes. *Journal of Food Legumes* 21(4), 211–217.

Singh, J., Agrawal, O. P., and Mishra, R. C. 2006. Foraging rates of different Apis species visiting parental lines of Brassica napus L. *Zoo's Print J.* 21, 2226–2227.

Singh, K. B. and Auckland, A. K. 1975. Chickpea breeding at ICRISAT. Proceedings of the International Workshop on Grain Legumes. January 13–16, 1975, ICRISAT Center, Hyderabad, India, pp. 3–18.

Singh, K. B., Malhotra, R. S., Halila, M. H., Knights, E. J., and Verma, M. M. 1994. Current status and future strategy in breeding chickpea for resistance to biotic and abiotic stresses. *Euphytica* 73, 137–149.

Tanno, K. and Willcox, G. 2006. The origins of cultivation of *Cicer arietinum* L. and *Vicia faba* L.: Early finds from Tell el-Kerkh, north-west Syria, late 10th millennium B.P. *Veg. Hist. Archaeobot.* 15, 197–204.

Tayyar, R., Federici, C. V., and Waines, G. J. 1996. Natural outcrossing in chickpea (Cicer arietinum L.). *Crop Science* 36, 203–205.
Thompson, J. N. 1994. *The Coevolutionary Process*. University of Chicago Press, Chicago, IL.
Thompson, J. N. 1997. Evaluating the dynamics of coevolution among geographically structured population. *Ecology* 78, 1619–1623.
Upadhyaya, H. D., Dwivedi, S. L., Baum, M., Varshney, R. K., Udupa, S. M., Gowda, C. L. L., Hoisington, D., and Singh, S. 2008. Genetic structure, diversity, and allelic richness in composite collection and reference set in chickpea (*Cicer arietinum* L.). *BMC Plant Biol.* 8, 106. doi: 10.1186/1471-2229-8-106
van der Maesen, L. J. G. 1972. Cicer L., a monograph of the genus, with special reference to the chickpea (Cicer arietinum L.), its ecology and cultivation. PhD thesis. Wageningen University, the Netherlands.
Zameer, S. U., Bilal, M., Fazal, M. I., and Sajjad, A. 2017. Foraging behavior of pollinators leads to effective pollination in radish *Raphanus sativus* L. *Asian J. Agri. Biol.* 5, 221.

22 Pulses Production Way Forward

The problem of inadequate access to good sources of nutrition has significant implications for humanity and overall economic growth. The relationship between poverty and food and nutritional insecurity is an important factor. In India, the country has a relatively high rate of poverty and, indeed, the largest number of malnourished people in the world. Because pulses not only have beneficial nutritional properties but can also enhance agricultural productivity through improved technology, they have a significant role to play in agriculture for nutrition and health (Singh and Saxena 2016).

During the past half century, India has enhanced food availability. The Green Revolution has brought a significant change in the production of food grains. But there has been little progress in nutritionally important crops, such as pulses. Therefore, improvement in the production of pulses remains an important issue in India (Pande et al. 2012; Chakraborty et al. 2016).

In India, pulses continue to be quite important in relation to other noncereal protein sources. Moreover, compared with those other sources, pulses remain cheap. That demand for pulses owes to several factors, such as increases in population, in urbanization, and in the demand for processed products that use pulses as an ingredient (Joshi and Kumar 2011). Despite this rising demand, pulses have suffered marginalization as a crop, pushed out of traditional growing areas into non-irrigated, rainfed areas.

The supply of pulses and technological advancements have been greatly impacted by this movement (Niranjan et al. 1998; Asare-Marfo et al. 2010). In the past decade, the pulse processing industry has been characterized by rapid growth in the number of mills and the output of the organized sector, which now accounts for three-quarters of processed pulses in the market. Although the organized sector operates large plants, their technology may not be as advanced as that of international producers.

The processing costs and processors' margins are high, accounting for about two-thirds of the difference between farmgate prices and consumer prices. Therefore, the promotion of processing/cost-cutting policies has the potential to bring about a lower consumer price. Lack of processing is yet another drawback that has resulted in fewer differentiated products. Product differentiation has the potential to increase consumer demand for pulses and pulse-based products.

Despite India's being the largest producer and consumer of pulses, its domestic production consistently falls short of demand. Consequently, the imports of pulses have been increasingly significant. Important issues include (1) improving technology for enhanced production in marginalized areas, (2) adjusting the trade in pulses with the needs of poor consumers in mind, (3) developing the value chain to better link consumers with producers, and (4) that the pollination requirements of most of the pulse crops are unknown, whether they are self-fertile or self-sterile. Pollination has been reported to benefit production of pulse crops both qualitatively and quantitatively. All pulse crops are open-pollinated to varying degrees, which means in order to achieve seed set, pollen must be transferred between flowers. Mechanisms like wind can achieve pollen transfer or cross-pollination, but the most effective method is utilizing insects to carry the pollen from one flower to the next

DOI: 10.1201/9781032656724-24

22.1 IMPORTANT SUGGESTIONS FOR BOOSTING PRODUCTION OF PULSES

22.1.1 Reorientation of Pulses Production Systems

Due to various policy measures for the Green Revolution, pulses have been pushed into marginal environments and into non-traditional areas. Realistic options for improving outcomes for the pulses sector need to take into account the centrality of the cereals sector in the production systems. For example, pulse-growing technologies should provide for the option of using rice fallow systems, which means adjusting to the need for short-duration and very-short-duration pulse crops (Asare-Marfo et al. 2010). In addition, pulses are needed for planting as intercrops along with certain noncereal crops. The pulse sector requires technology development, production systems, and extension services to address its needs effectively. Moreover, there is still a requirement to expand the cultivation of pulses by bringing more land under their cultivation.

22.2 TAKING PULSES TO NEW AREAS WITH WATERSHED DEVELOPMENT

Notwithstanding the concentration of pulses in rainfed areas, in order to reduce inter-temporal variability in yields and also make pulses a more attractive crop, pulses growing in non-irrigated areas should also be grown in irrigated areas. Efforts should be made to encourage the cultivation of pulses in districts where successful watershed development has been achieved in recent years.

22.3 TECHNOLOGY INTERVENTIONS

Though much technological progress has been made in pulses, it has not achieved the desired level. There is still ample opportunity to advance pulse varieties by further utilizing the genetic diversity and tailoring technologies to align with the specific needs of the pulse industry within the agricultural system (Roy et al. 2016).

22.4 THE WAY FORWARD

The main responsibility of the government in the pulse sector should be to focus on investing in the development of infrastructure and research on technology, as well as spreading awareness, all of which would contribute to the growth of pulse production and expansion of markets. The pricing policy of pulses needs to be reassessed (Vyas 2003; Birthal et al. 2015).

22.5 INCREASING NUTRITIONAL INTAKE FROM PULSES

Steps are needed to be taken to help pulses meet their full nutrition potential. The decreasing preference for pulses can be increased by developing pulse-based products. The Pulse Innovation Partnership (PIP) is an example of an initiative that brings together public and private organizations, civil society agencies, and academia on a global scale. Its goal is to collaborate with food companies of all sizes across various regions, offering tailored knowledge services for innovation and marketing. This partnership aims to create a significant avenue for the development and promotion of pulse-based products (Kumar et al. 2010; Feder et al. 2011; Birthal et al. 2015).

REFERENCES

Asare-Marfo, D., Birol, E., Karandikar, B., Roy, D., and Singh, S. 2010. Varietal Adoption of Pearl Millet (Bajra) in Maharashtra and Rajasthan, India: A Summary Paper. HarvestPlus Project Report. International Food Policy Research Institute (IFPRI), Washington, DC.

Birthal, P. S., Kumar, S., Negi, D. S., and Roy, D. 2015. The impacts of information on returns from farming: Evidence from a nationally representative farm survey in India. *Agricultural Economics* 46(4), 549–561.

Chakraborty, S., Kishore, A., and Roy, D. 2016. Effectiveness of Food Subsidies in Raising Healthy Food Consumption: Public Distribution of Pulses in India. IFPRI Discussion Paper 1523. International Food Policy Research Institute (IFPRI), Washington, DC.

Feder, G., Birner, R., and Anderson, J. R. 2011. The private sectors role in agricultural extension systems: Potential and limitations. *Journal of Agribusiness in Developing and Emerging Economies* 1, 31–54.

Joshi, P. K. and Kumar, P. 2011. Food demand and supply projections for India: 2010–2030." In: *Trade, Agricultural Policies and Structural Changes in India's Agrifood System: Implications for National and Global Markets.* Report for TAPSIM Project KBBE212617. International Food Policy Research Institute, New Delhi.

Kumar, A. G., Roy, D., and Gulati, A. 2010. *Liberalizing Food Grains Markets: Experiences, Impacts, and Lessons from South Asia.* Oxford University Press, New Delhi.

Niranjan, S. K. D. F. F., Bantilan, M. C. S., Joshi, P. K., and Saxena, K. B. 1998. Evaluating hybrid technology for pigeon pea. Proceedings of the International Workshop on Joint Impact Assessment of NARS/ICRISAT Technologies for the Semi-Arid Tropics, December 2–4, 1996. ICRISAT Center, Patancheru, India, pp. 231–240.

Pande, S., Sharma, M., and Ghosh, N. 2012. *Role of Pulses in Sustaining Agricultural Productivity in the Rainfed Rice-Fallow Lands of India in Changing Climatic Scenario.* International Crops Research Institute for the Semi-Arid Tropics (ICRISAT), Patancheru, India.

Roy, D., Kishore, A., Joshi, P. K., and Mishra, B. (2016). Pulses for food and nutritional security: The technology perspective. *Indian Journal of genetics and plant breeding*, 76(04), 375–387.

Singh, N. P. and Saxena, M. C. 2016. Towards Self-Sufficiency of Pulses in India. Base Paper for National Academy of Agricultural Sciences (NAAS) Brainstorming Workshop, April 8, 2016. New Delhi.

Vyas, V. S. 2003. Market reforms in Indian agriculture. Paper presented at Jawahar Nehru University NU-IFPRI Workshop on Economic and Agricultural Reforms in China and India. March 25–26, New Delhi.

Index

A

B

C

D

E

F

V

W

X

Y

Z